South Africa's Economic Crisis

This book belongs to
Donald F Power

AVAILABLE AT:
FRONTIER PUBLICATIONS
INCWADI PUBLIKASIES
ROOM 201 CAPITOL (JET STORES) BLDG.
NORTH END P.E. 6001
TEL: (041) 546329

South Africa's Economic Crisis

AVAILABLE AT:
FRONTIER PUBLICATIONS
INGWADI PUBLIKASIES
ROOM 201 CAPITOL (JET STOKES) BLDG.
NORTH END, P.E. 6001
TEL: (041) 546328

South Africa's Economic Crisis

EDITED BY
STEPHEN GELB

David Philip • Cape Town
Zed Books Ltd • London and New Jersey

First published 1991 in southern Africa by David Philip Publishers (Pty) Ltd, 208 Werdmuller Centre, Claremont 7700, South Africa

Published 1991 in the United Kingdom and North America by Zed Books Ltd, 57 Caledonian Road, London N1 9BU, UK, and 163 Atlantic Highlands, New Jersey 07716, USA

© 1991 David Philip Publishers

ISBN 0-86486-149-4 (paper, David Philip)
ISBN 1-85649-022-X (cased, Zed Books)
ISBN 1-85649-023-8 (paper, Zed Books)

Cover design by Alison Hartmann

Printed by Clyson Printers (Pty) Ltd, 11th Avenue, Maitland, Cape

British Library Cataloguing in Publication Data:

South Africa's economic crisis.
  1. South Africa. Economic conditions
  I. Gelb, Stephen
  330.968

  ISBN 1-85649-022-X
  ISBN 1-85649-023-8 pbk

Library of Congress Cataloging in Publication Data is available from the Library of Congress

Contents

List of Contributors	vii
List of Abbreviations	ix
Preface	xi

1. South Africa's economic crisis: an overview   STEPHEN GELB   1
2. State, capital and growth: the political economy of the national question   MIKE MORRIS   33
3. The crisis and South Africa's balance of payments   BRIAN KAHN   59
4. The politics of South Africa's international financial relations, 1970–1990   VISHNU PADAYACHEE   88
5. South African gold mining in transformation   BILL FREUND   110
6. Coal mining: past profits, current crisis?   JEAN LEGER   129
7. Manufacturing development and the economic crisis: a reversion to primary production?   ANTHONY BLACK   156
8. The South African capital goods sector and the economic crisis   DAVID KAPLAN   175
9. The accumulation crisis in agriculture   MIKE DE KLERK   198
10. The restructuring of labour markets in South Africa, 1970s and 1980s   DOUG HINDSON   228
11. Unemployment and the current crisis   DAVID LEWIS   244

Notes and References	267
Index	286

Contributors

ANTHONY BLACK	School of Economics, University of Cape Town
MIKE DE KLERK	School of Economics, University of Cape Town
BILL FREUND	Department of Economic History, University of Natal
STEPHEN GELB	Institute for Social and Economic Research, University of Durban-Westville
DOUG HINDSON	Institute for Social and Economic Research, University of Durban-Westville
BRIAN KAHN	School of Economics, University of Cape Town
DAVID KAPLAN	Department of Economic History, University of Cape Town
JEAN LEGER	Department of Sociology, University of the Witwatersrand
DAVID LEWIS	Department of Economic History, University of Cape Town
MIKE MORRIS	Centre for Social and Development Studies, University of Natal
VISHNU PADAYACHEE	Institute for Social and Economic Research, University of Durban-Westville

Abbreviations

AMM	Association of Mine Managers of South Africa
ANC	African National Congress
ARMSCOR	Armaments Development and Production Corporation
ASSOCOM	Associated Chambers of Commerce
BD	*Business Day*
CGB	*Coal, Gold and Base Minerals of South Africa*
COSATU	Congress of South African Trade Unions
CPS	Current Population Census
DCDP	Department of Constitutional Development and Planning
ECLA	Economic Commission for Latin America
EEC	European Economic Community
ESKOM	Electricity Supply Commission
FC	Federated Chambers of Industry
FM	*Financial Mail*
FT	*Financial Times*
GDP	gross domestic product
GENCOR	General Mining and Finance Corporation
GNP	gross national product
IC	integrated circuit
ICOR	incremental capital–ouput ratio
IMF	International Monetary Fund
ISCOR	Iron and Steel Corporation of South Africa
ISI	import–substitution industrialisation
JMC	Joint Management Centre
LDC	less developed country
LIBOR	London inter-bank rate
MAR	*Mining Annual Review*
NAMPO	National Maize Producers' Organisation
NIC	newly industrialising country
NSMS	National Security Management System
NUM	National Union of Mineworkers

ABBREVIATIONS

NUMSA	National Union of Metalworkers of South Africa
NYT	*New York Times*
OPEC	Organisation of Petroleum–Exporting Countries
RDM	*Rand Daily Mail*
SAAU	South African Agricultural Union
SACP	South African Communist Party
SAM	*South African Mining, Coal, Gold and Base Minerals*
SAMES	South African Microelectronic Systems
SANLAM	South African National Life Assurance Company
SAPO	South African Post Office
SASOL	South African Coal, Oil and Gas Corporation
SATS	South African Transport Services
SBDC	Small Business Development Corporation
SCC	Standstill Co-ordinating Committee
TCOA	Transvaal Coal Owners' Association
TNC	transnational corporation
UDF	United Democratic Front
UIF	Unemployment Insurance Fund
UWUSA	United Workers' Union of South Africa

Preface

This book is the product of the Economic Trends Research Group, a group of progressive economists, economic historians and social scientists based in Johannesburg, Durban and Cape Town. The group, which meets several times each year, started in early 1987 with 8 members, and has grown to comprise 21 in late 1990.

The initial stimulus for the project arose in late 1986 when new economic sanctions were being imposed on South Africa by the United States, the Commonwealth and the European Economic Community countries. While progressive organisations in South Africa supported these actions by foreign governments, there was also a need, felt especially by the Congress of South African Trade Unions (Cosatu), to assess the likely impact on the economy in general and on their members' economic situation in particular.

The request from Cosatu for in-depth analysis of the effects of sanctions led in late 1986 to the formation of the Economic Trends Group, co-ordinated by Stephen Gelb. The Economic Trends Group was initiated under the auspices of the Labour and Economic Research Centre (LERC), a trade-union service organisation in Johannesburg, established at that time by Stephen Gelb and two colleagues involved in other union-linked research projects. In January 1989, the project moved, with its co-ordinator, to the Institute for Social and Economic Research at the University of Durban-Westville. Stephen Gelb resigned as co-ordinator in April 1990, and was succeeded by David Lewis, of the Economic History Department at the University of Cape Town, from where the project is now administered.

It soon became evident that there was a need to broaden the scope of the Economic Trends Group's inquiry beyond sanctions as such, since their effects were fundamentally influenced by ongoing economic developments. A more general analysis of the South African economy would also contribute to deeper understanding within the trade-union movement, and in this way to the unions' capacity to intervene more effectively in the wider debate on the economy's current problems and future directions.

Close links between the project and the trade unions have been essential to shaping and reshaping the project throughout its four-year history. These links have included the regular participation of trade unionists in the group's meetings, as well as the contribution by project members to union education on economic issues. Although it is invidious to single out individuals, I feel it is necessary to emphasise at this point the contribution of Alec Erwin, Cosatu Education Secretary, at the time the project was begun.

A report entitled 'The Economic Crisis: Recent Economic Trends in South Africa' was presented to Cosatu in November 1988. In addition to an overview section, the report examined South Africa in the world economy; the mining, manufacturing and agricultural sectors of production; the labour market; and state policy on unemployment. This volume is the edited and revised version of that report.

The report was presented to workshops with the federation's leadership in November 1988 and again in February 1989. Workshops have subsequently been held with the leadership of some of Cosatu's affiliates.

The Economic Trends project was considered a significant success, in terms of the content of its report, and as an important exercise in co-operation between progressive academics and the trade-union movement. As Cosatu and its affiliates have taken up more actively the issues of the current state and future policies of the South African economy, substantial use has been made of the completed research of the Economic Trends Group.

As a result, the group was requested by Cosatu to begin a second phase, focussed on 'Economic Prospects and Future Policies for South Africa'. Broadly speaking, the starting-point for the second phase is the question of whether the South African economy is likely, or able, to emerge from its crisis during the next decade, and if so, what its underlying structure is likely to be. This inquiry will then provide a basis for developing a coherent framework for national economic policy. The report of the second phase of the project is expected to be completed in late 1991.

It is appropriate to acknowledge here the contributions and assistance of a number of individuals and organisations to the Economic Trends Group project, and to this volume in particular.

By its very nature, the project has depended on the contributions of its members, in terms of their own research; in assisting each other, and me, with comments and suggestions; and their participation in, and assistance with the organisation of, the group's meetings. I would like to thank all the other contributors for this.

Thanks are due also to my former colleagues at LERC, especially Judy Maller and Cindy Cupido; and to my present colleagues at ISER, especially Vishnu Padayachee and Doug Hindson, fellow members of Economic Trends, and John Butler-Adam, the Director. Many officials of Cosatu and its affiliates have assisted and supported the work in a variety of ways. The role of Alec Erwin, Numsa Education Secretary, has been crucial, through his consistent participation in the work of the project, and his always constructive approach to smoothing the relationship between researchers and mass-based organisations. His successor as Cosatu Education Secretary, Khetsi Lehoko, and others in that department have also been very supportive.

Financial assistance for the project has been received from Oxfam (United Kingdom), Oxfam-Canada, Friedrich Ebert Stiftung (West Germany), the South African Council of Churches, the Canadian Labour Congress, and War on Want (United Kingdom).

Finally, I wish to thank Russell Martin, of David Philip Publisher, for his assistance, and even more for his patience and forbearance.

STEPHEN GELB

1
South Africa's economic crisis: an overview

STEPHEN GELB

South Africa enters the 1990s with a legacy of economic problems that makes a long and depressing list. The 1980s were marked by stagnation in output growth; inflation entrenched at over 13 per cent per annum; a weak rand; a permanent decline in foreign exchange reserves; and historically low personal savings ratios. There has been massive and growing unemployment, with no net creation of new jobs in the manufacturing sector through the 1980s. Other serious difficulties have emerged in the labour market: wage increases, measured by employers (and sometimes the state) relative to productivity growth, were perceived as being 'too high'. Workers, on the other hand, see their wage gains as 'too low', when measured by their purchasing power. Growing poverty has been expressed as well in the severe shortages of essential consumer items, most prominently housing.

A range of views has been put forward to explain these problems. Implicit in most of these views is the assumption that the market economy itself is essentially stable. In other words, it is essentially self-correcting in response to disturbances, if left to itself. The basic cause of the difficulties, in this view, is inappropriate intervention in the economy, in the form of government policy or, more broadly, politics. One group sees apartheid as the major culprit, either directly because of the limits it has placed on the operation of the 'free market', or else indirectly because of the impact of sanctions imposed by other countries. A second view rests on similar assumptions about the stability of the market economy, but argues that recent fiscal and monetary policy has been too lax in responding either to shocks emanating from the world economy or to pressures to raise fiscal spending.

In contrast, the approach taken in this volume rests on the idea that capitalist economies are inherently subject to extended phases of decline and disruption, alternating with periods of stability and more sustained growth. In other words, the economic system endogenously generates instability, and intervention is consequently required to restore stability.

The problems identified above are, it is argued, best understood in the context of a long-run perspective on capitalist growth and development, as the manifestations of an accumulation crisis in South African capitalism that dates back to 1974. This crisis has been expressed in part in the form of structural problems which have developed over the past fifteen to twenty years in the productive sectors – industry, agriculture and mining – of the South African economy; in the labour and financial markets; and in South Africa's international trade and financial relations. This introductory chapter examines growth and decline at the aggregate macroeconomic level.

The term 'crisis' popularly connotes an idea of collapse or breakdown. But the original, and more useful, meaning of the term is 'turning-point'. In this sense a crisis implies that the capitalist economy cannot continue to develop in the same form and along the same path as before. The existing growth model – the combination of patterns of production, distribution and consumption – has begun to decay, and a resumption of sustained accumulation requires the emergence of a new growth model.

South African capitalism reached such a turning-point in the mid-1970s, reflected at the macroeconomic level both as a decline in the long-run growth rate and as a change in the cyclical pattern of aggregate economic growth, towards greater fluctuation and instability, as compared with the period after 1945. These trends are examined in section 1 below. The changes in the pattern of GNP growth have been linked to the failure of the post-war 'growth model' – the combination of patterns of production, distribution and consumption, in other words, the form of capitalist growth. This and other concepts are elaborated in section 2, which develops the conceptual framework of the chapter.

The growth model which emerged in South Africa in the post-war period is discussed in section 3. Defined there as 'racial Fordism', it focused on extending industrialisation by means of the production of (previously imported) sophisticated consumer goods primarily for the white South African market. Racial domination comprised the pre-eminent factor shaping economic institutions. Whites occupied a similar position to that of the working classes in the advanced industrial countries, their living standards steadily rising, while blacks (especially Africans) remained relatively impoverished, though their incomes did rise slowly. What provided the foundation of the model was the expansion of exports of gold and other precious metals, and their stable prices on world markets. This exerted a stabilising influence on accumulation, limiting its fluctuations.

This model fitted well with growth patterns in the major industrialised economies, and like these countries, South Africa grew rapidly, with an average GDP growth rate of 4.9 per cent per annum between 1945 and 1974. By the start of the 1970s, however, accumulation began to falter, as the contradictions in the growth model – the failure of the economy to absorb surplus labour, and the import-dependence of manufacturing industry, for example – came to obstruct growth. The emergence of the crisis is spelt out in section 4, as a process of interaction between developments in the international capitalist economy and changes in the domestic social structure arising from the long-wave accumulation boom itself.

The onset of crisis disrupted the stable relations which had characterised the growth model – between export revenues and investment, and between income distribution and productivity growth. The result, as indicated in section 5, has been stagnation, rising production costs and declining investment, though with marked fluctuations and instability in these trends. 'Racial Fordism' proved rigid and inflexible in responding to the difficulties it faced. The effects of accumulation had themselves sharpened some of the model's contradictions, for example its tendency to produce rising unemployment, and the increasing weight in imports of the manufacturing sector's needs for capital equipment and intermediate input. Overall, the inability of the old growth model to adapt to the difficulties has brought about its decline and collapse.

Much of the 'reform' process in South African politics since the late 1970s can be read as an effort to adapt and shore up the old growth model. But 'reform' involved structural changes which were only partial in their extent: the efforts to preserve the racial definition of much of the institutional structure placed narrow limits on policy options. Furthermore, in many instances the 'reform' policies themselves exacerbated the difficulties, by making the economy more vulnerable to 'shocks' from the international economy, or by deepening the economic material hardships faced by the poor and so stimulating political conflict.

The collapse of 'racial Fordism' has been accompanied by the appearance of elements of a possible successor. A new growth model will have to emerge in the context of a fundamentally transformed social structure. Much greater class differentiation has developed within racial groups in South Africa in the course of the crisis. Substantial middle classes and large armies of unemployed have formed amongst urban blacks, while employed workers (at least in the higher-skill grades) now occupy, relative to the mass of urbanised unemployed, well-paid and secure employment.

At the other end of the spectrum, there has been a significant increase in the concentration of ownership, and the major conglomerates have in recent decades greatly extended their economic power. The reason for this is that the mining and financial sectors have been relatively favoured by macro-economic stabilisation policies, in comparison with the manufacturing and commercial sectors.

As noted above, the crisis is not yet resolved: we are still in the midst of the transition. The ending of apartheid – the last important element of the old growth model – through the introduction of a non-racial democratic state may finally open the way for a future growth model to emerge, though there is no guarantee this will occur. Section 6 of this chapter considers some of the preconditions for a new growth model. It also spells out in broad terms the features of two alternative approaches to future accumulation in South Africa, one being advocated by 'big' business and the present government, and the second increasingly becoming the perspective of the African National Congress and related organisations.

GROWTH TRENDS AND CYCLES

The perspective adopted in this chapter focuses on the long run. The objects of analysis are the long-run swings in capitalist growth and development, each wave of which persists over a period of several decades. The analysis does recognise that short-run cyclical fluctuations (business cycles) occur through the upward and downward phases of the longer cycles. Though the short-run movements are not my primary concern, they are explicitly integrated into the conceptual framework.

This section identifies the period since 1974 as a distinct phase in South Africa's economic growth, the downswing of a long-run wave. This conclusion is based on an examination of overall growth rates and other indicators of economic performance. My starting point is the country's poor record of economic growth during the period. In comparison, during the preceding period growth was both more rapid and more stable. Between 1946 and 1974, South Africa's real GDP (gross domestic product, adjusted for inflation) grew at an average rate of 4.9 per cent per annum. In the subsequent decade to 1984, the rate dropped to only 1.9 per cent, while during the 1980s as a whole, annual GDP growth averaged a mere 1.5 per cent. The decline is evident in Table 1 which shows average growth rates calculated from one peak of the business cycle to the next.[1]

Table 1. Average growth rates, real GNP (1980=100)

Business cycle peaks	Growth per annum per cent
1946–8	5.6
1949–51	6.8
1952–5	5.1
1955–7	4.5
1958–60	4.1
1960–5	5.8
1965–7	4.2
1967–70	4.9
1971–4	7.8
1974–81	2.6
1981–4	0.0
1984–8	1.8

Source: Calculated from South African Reserve Bank, *Quarterly Bulletin*, various issues and supplements; International Monetary Fund, *International Financial Statistics*, various issues.

Figure 1 shows the growth rate of real GNP (gross national product) in the long-run, that is, averaged over three-year periods. What is especially noteworthy is the remarkable stability of the rate of long-run growth (reflected by the slope of the line) up until 1974. The figure furthermore indicates clearly that 1974 marked the end of this persistent and steep rise during the post-war period.

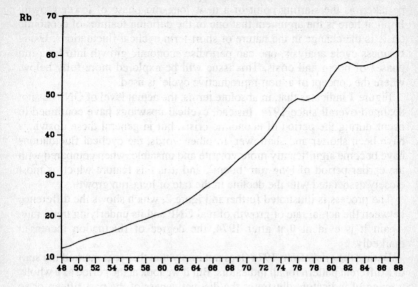

Fig 1. Real GNP (1980 prices): 3-year moving average

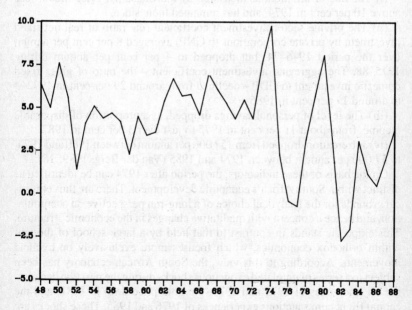

Fig. 2. Real GNP (1980 = 100): deviation from 3-year trend

This suggests that the business-cycle peak which occurred in 1974 should be taken as the starting-point of a new, long-run phase of lower growth. Implicit here is the argument that one of the defining features of economic crisis is the change in the nature of short-term cyclical fluctuations. Using business cycle analysis, one can periodise economic growth into long-run phases of boom and crisis. This issue will be explored more fully below, where the concept of a 'non-reproductive cycle' is used.

Figure 1 indicates that, in absolute terms, the actual level of GNP has not declined overall since 1974. Instead, cyclical upswings have continued to occur during the period of economic crisis, but in general these upswings have been shorter and shallower. In other words, the cyclical fluctuations have become significantly more volatile and unstable, when compared with the earlier period of long-run 'boom', and it is this feature which is most closely associated with the decline in the rate of long-run growth.

The process is illustrated further in Figure 2, which shows the difference between the actual rate of growth of real GNP and its underlying trend rate. Again it is evident that after 1974, the degree of fluctuation increased markedly.

Dating the crisis from 1974 does not imply that this year marked the start of a different pattern for all indicators. But if we look at the period as a whole, a range of indicators illustrates the distinctiveness of the past fifteen or so years:[2]

(i) The rate of inflation, as measured by the Consumer Price Index, rose above 10 per cent in 1974, and has remained there since.

(ii) The private-sector investment coefficient (the ratio of real net fixed investment by private corporations to GNP) averaged 8 per cent per annum over the period 1946–74, but dropped to 4 per cent per annum during 1975–88. The aggregate investment coefficient – the ratio of gross fixed domestic investment to GDP – declined from around 29 per cent in 1972–4 to around 19 per cent in 1986–7.

(iii) The level of personal savings dropped, as a proportion of disposable income, from about 11 per cent in 1975 to just over 3 per cent in 1987.

(iv) Job creation dropped from 157 000 per annum between 1960 and 1974, to 57 000 per annum between 1974 and 1985 (Van der Berg, 1989: 16).

On the basis of these indicators, the period after 1974 can be identified as distinct within South Africa's economic development. There are thus empirical grounds for the analytical choice of a long-run perspective on accumulation, and hence a concern with qualitative changes in the economic structure. This emphasis stands in contrast to that held by a large school of thought within orthodox economics, which focuses more exclusively on cyclical movements. According to this view, the South African economy has been subject to a series of unrelated exogenous shocks during the past two decades: the oil price rises of 1973 and 1979, the gold price drop in 1981, and the capital flight cum sanctions experiences of 1976 and 1985. These shocks are held to account for cyclical fluctuations over the period (see, for example,

Cloete, 1986; De Kock, 1986a).

Our immediate concern, however, is with those orthodox economists who themselves focus on what they argue are longer-run structural issues, and are critical of a preoccupation with short-run cycles (Lombard, 1988; Lombard and Van den Heever, 1989; RSA Department of Finance, 1990). Here we need to return to the debate about alternative explanations for South Africa's economic decline, briefly referred to above. Two broad schools of thought on this issue were identified there, distinguished by their understanding of the capitalist (market) economy as inherently stable or unstable.

Orthodox (neo-classical) economic analysis starts from the premise that the market economy is an optimal mechanism for the allocation of resources, and is also endogenously stable or self-correcting when disturbed from a position of equilibrium. Structural problems, in this view, are essentially the consequence of market imperfections, introduced by inappropriate interventions by external agents, in particular the government. These are expressed as price (or cost) 'distortions', which obstruct allocative efficiency, and thus reduce growth rates. The 'distortions' are seen to arise in the first instance from excessive or inappropriate policy interventions (or both) in markets, that are frequently intended to shift the distribution of incomes in favour of groups politically supportive of the government. This in turn gives rise to prices for factors of production (labour and capital) or for foreign exchange that do not reflect actual factor endowments.

Restoring growth requires structural reforms to 'get prices right'. Such reforms, it is argued, will facilitate a spontaneous correction of the growth problems, as economic agents respond to the more appropriate price signals, and adjust their use of productive factors in line with the economy's comparative advantage. In the South African case, as in others, the most common policy prescription is the argument in favour of lower real wages of labour, in order to reduce unemployment, and a higher rate of interest, so as to economise on the use of capital (which is seen to be in short supply).

This chapter, and indeed this volume as a whole, present an alternative perspective on economic growth, rather than a thoroughgoing critique of the neo-classical view. It is, however, worth identifying two of the fundamental differences between the neo-classical approach and that adopted here. The neo-classical argument contrasts markets, which are 'free', with institutions, which impose constraints. In this volume, markets are viewed as social institutions, which organise the process of exchange, and thereby impose constraints (as well as being enabling in their effects). In addition, different kinds of markets are recognised, imposing different kinds of constraints (Hodgson 1988). By implication, one cannot assume that all markets will clear, that is, reach an equilibrium between demand and supply, even if left to themselves. This has important macroeconomic consequences.

What is more, the emphasis of the analysis in this volume does not fall on the efficient allocation of scarce resources (this being implicitly identified with the optimality of the growth process), but on the process of growth and development over time (Walsh and Gram, 1980). In the former case, the focus of the analysis is the process of the determination of relative prices. 'Inappropriate' relative factor prices (wages being 'too high' relative to the cost of capital) lead, it is argued, to the excessive use of the scarce resource (capital), as substitute for the abundant factor of labour. In the latter approach, by contrast, what is of interest is the impact, on the size and distribution of the aggregate surplus, of changes in absolute cost levels, which has implications for economic growth. These changes in costs, we argue, result from the operation of markets themselves, and to counteract them thus requires changes in the way markets operate.

We turn now to a discussion of the 'regulation' school, which has developed the idea of the market as a social institution, and which is concerned, like marxist economic theory more generally, with analysis of the economic surplus in the aggregate.

CONCEPTUALISING ACCUMULATION AND CRISIS: THE REGULATION APPROACH

The 'regulation' approach emerged in France during the 1970s, as a theory of capitalist accumulation which aimed to explain how and why capitalist economies come to be transformed in the course of their development. The immediate context was the world economic crisis – the ending of the long post-war boom, first signalled in the late 1960s. This had produced a revival of both monetarist economic theory and marxist economics, as the prevailing orthodoxy – the neoclassical–Keynesian synthesis – proved increasingly unable to explain the decline in growth and accompanying stagflation.

Debates amongst marxists provided a range of explanations for the economic crisis.[3] Each of these was partial and incomplete, in the sense of focusing on a specific feature of the capital accumulation process – either production or distribution – rather than the overall process.

They had two further defects in common. The first and most critical was that the various theories were concerned primarily with the possible causes of fluctuations in the rate and path of capital accumulation, and failed to examine adequately the different types of fluctuations that actually occurred in capitalism. In other words, they were able to explain only that cyclical fluctuations would occur, but were unable to differentiate, except on an *ad hoc* basis, between the secular (long-run) tendencies of capital accumulation, and the shorter-run fluctuations.[4]

There was a second common problem related to this: many analyses implied that a crisis reflected insurmountable difficulties for the capitalist system, a 'terminal disease'. Even though capitalism's survival through several crises in the past was acknowledged, this could not be adequately accounted for. In consequence, these approaches were unable to consider the central implication of capitalism's repeated survival of major crises – its mutability. Changes in the form of capitalism – that is, in the institutional context of accumulation – were generally ignored; and as a result, capitalist social relations were presented as existing within an ahistorical vacuum.

In attempting to overcome these problems, the regulation approach marked a fundamental departure in understanding of the theory of capitalist development within the marxist tradition.[5] Its starting-point is the idea of 'regulation', which in its French usage, as in this context, has a more complex meaning than in English. 'Regulation' expresses the view that the behaviour of individual economic agents is socially determined, though the meaning of 'determination' here owes more to 'influence' than to 'control'. 'Regulation' involves, according to Lipietz, 'internalised rules and social procedures which incorporate social elements into individual behaviour' (Lipietz, 1987: 15).[6] These rules and procedures may be partially formalised through legislation or other means, but this is not necessarily the case.

What is being suggested is that the actions of individuals in society are neither random nor arbitrary; rather, they are routinised (or 'canalised') by means of institutions, norms and networks, and thus reflect, and simultaneously constitute, social relations between larger aggregates within society.

Much (though not all) of the activity of economic agents takes place in the context of markets. What is implicit here, as noted above, is a view of markets as institutions – socially produced processes – as opposed to the somewhat 'naturalistic' understanding of markets in neo-classical theory. The focus of analysis is not on the issue of whether markets 'clear' (with non-clearing implying some form of market imperfection), but rather how the market, as an institutional process, changes over time.

In this view, the forms taken by capitalist social relations – 'class struggles' – are 'historically specific and contingent': 'existing structures reflect and institutionalise, in forms that make capitalism workable, past conflicts between collective actors, between the classes as they are organised' (Noel, 1987: 323). Capitalist 'regulation' thus resolves the system's contradictions, making it 'workable', and enabling accumulation to proceed. The 'regulation of a mode of production is ... the way in which the determinant structure of a society is reproduced' (Aglietta, 1980: 13). 'Regulation' implies that class conflict takes particular forms of expression, which shape (constrain) the process, making some outcomes more likely than others. But it is also vital to emphasise that the resolution of these contradictions is only provisional and temporary: 'regulation' does not suspend or eliminate class conflict, or dissolve the contradictions of capitalism. Indeed, 'regulation' theory focuses above all on the process of exploitation in class societies, that is, the appropriation by one class of the surplus produced by another. The various processes through which this surplus is expanded or contracted, comprise the major driving force in the accumulation process.

A high rate of accumulation may endure over an extended period. But capitalism remains an inherently unstable system, as accumulation itself transforms over time the nature and composition of the various classes, and as a result class conflict begins to express itself in new ways, beyond the capacities of existing institutions. Thus, any resolution of the contradictions embodied in existing institutions will ultimately break down, producing a crisis.

A crisis, then, implies that the form of capitalist accumulation – that is, the institutions which stabilise accumulation – is no longer adequate to sustain accumulation. In other words, 'regulation' no longer occurs in a smooth manner, and needs to be transformed. For this reason, crises represent periods of heightened conflict in society, as social groups struggle to dominate this process of change: 'crises are moments of collective action, of social creation. No outcome is predetermined. It is in fact precisely because capitalism has so many problems building adequate institutional frameworks for itself that crises last so long' (Noel, 1987: 319).

These 'new' conflicts can, however, be resolved (and generally have been) by transforming existing structures and institutions, without changing the fundamentally capitalist nature of the social relations themselves. An economic crisis should thus be seen more appropriately as a turning-point in the form of capitalism, rather than as the inevitable end of the capitalist system, as in the more traditional marxist approach. It is possible that the resolution of a crisis could be the replacement of capitalism by a different economic system. But in principle it is as likely (and historically more so) that the resolution will involve a transformation in the form of capitalist accumulation.

Capitalist development, therefore, in both the world economy as a whole and in individual national economies, is understood by the 'regulation' school as undergoing a series of transformations over time. The establishment of a specific form of accumulation makes possible a phase of relatively stable[7] accumulation in the long run. The transitional periods, during which one form of capitalism gives way to another, are characterised as crises. Crises are long-run periods of far-reaching institutional change, during which accumulation is uncertain and has a lower average rate.

The successive phases of capitalist development are specified more concretely by three concepts: regime of accumulation; growth model; and mode of regulation. The central focus is the process of exploitation, that is, the division of the (net) aggregate product into two parts: necessary product and surplus product, or, expressed in value terms, variable capital and surplus value. The proportions of this division are defined as the rate of surplus value. Changes in this rate occur through either distributional shifts or changes in the production process.

The rate of surplus value places a ceiling on the rate of accumulation. The concept 'regime of accumulation' summarises the stabilisation of this rate over time. More precisely, it 'describes the fairly long-term stabilisation of the allocation of social production between consumption and accumulation' (Lipietz, 1987: 14).[8] Implicit in a 'regime of accumulation' is an interaction – 'correspondence' – between transformations in the conditions of production and transformations in the conditions of realisation (sale) of the resulting output. The regime of accumulation 'connects the individual decisions of producers and the socially determined effective demand they must confront' (Noel, 1987: 311).

Production transformations refer to changes in technology and the labour process; changing quantities of capital invested; and the latter's distribution between sectors of production. These changes result from decisions by capitalist firms, seeking surplus profits deriving from lower production costs, or from new products. Changing production conditions affect the rate of surplus value, through their impact on the cost of reproduction of labour-power (the value of wage goods).

The latter variable also depends on the conditions of realisation, which comprise the consumption 'norms' of the working class, collective social spending, exports, and so on, and which are independently determined from changes in production. Hence, the rate and direction of change in reproduction costs, and thus in the rate of surplus value, may be inappropriate for sustaining accumulation (Driver, 1981: 155).

The regime of accumulation abstracts from specific national economies. The 'growth model' of a particular country is the specific expression of a regime of accumulation within its economy, and in turn reflects the incorporation of the economy into this regime, understood as a global phenomenon. In other words, the growth model describes the form of capitalist accumulation within that particular economy, as well as the nature of its insertion into the world economy as a whole (the 'international division of labour'). The concept of a 'growth model' is thus less abstract than that of 'regime of accumulation', and helps to clarify the ambiguity between the international and national dimensions of accumulation.[9]

The third concept, the 'mode of regulation', provides the essential link between the regime of accumulation and a nationally specific growth model. The mode of regulation comprises the complex of social institutions, structures and implicit norms which act to 'regulate' the behaviour of economic agents, in the sense defined above. In other words, the mode of regulation comprises the wide range of processes which organise and influence the multi-faceted conflicts amongst classes and amongst other social groups.[10] Elements of the mode of regulation relate to all aspects of the accumulation process: the wage relation (in the labour process and the labour market), the structure of demand (including the 'norms' of working-class consumption, foreign trade and the public sector), competitive interrelations between capitals, and the financial system.[11] The mode of regulation is clearly a nationally distinct phenomenon, even though some of its elements may be international. The nature of the state, and state policy in relation to each of these aspects of accumulation, are self-evidently important in shaping the mode of regulation (Lipietz, 1987: 17ff; Jessop, 1989: *passim*).

A growth model, then, comprises a regime of accumulation (or, more precisely, *the* regime of accumulation), as well as a mode of regulation. A long wave of sustained accumulation in a national economy is characterised by the relative stability and persistence of the growth model. During such a period, a 'match' – dynamic harmony – is maintained within the regime of accumulation.

This 'match' does not imply a static or constant situation. The equilibrium persists over the long run, with repeated short-run departures in the form of cyclical fluctuations. But the nature of the short-run fluctuations varies between long-run booms and crises. During the former, economic agents (including policy-makers) respond to the downswing in a fairly predictable manner, consistent with the coherent mode of regulation. Hence, conditions for the resumption of capital accumulation are quickly brought about. The various institutions which comprise the mode of regulation constrain and influence agents' behaviour so as to maintain long-run dynamic equilibrium. The cyclical downswing during the boom is a stabilising process. Hence, the business cycle during this phase of accumulation is labelled 'reproductive', or well-behaved: the downturn 'endogenously restores conditions for rapid accumulation without requiring fundamental changes in the structure of the accumulation process' (Gordon *et al.*, 1983: 152).

Crises, in contrast, reflect a lack of harmony – disequilibrium – within the regime of accumulation, linked to growing incoherence in the mode of regulation, and a 'malfunctioning' of some or all of its constituent institutions. In other words, the growth model breaks down, slowing down accumulation, and further exacerbating the disintegration process.

The crisis phase is distinguished from the boom by a change in the nature of the business cycle, from reproductive to non-reproductive, or perverse. The latter implies that conditions for rapid accumulation are no longer 'automatically' restored – the institutions of the mode of regulation 'determine' agents' behaviour in ways that are not consistent with this outcome. Variables move in the reverse direction, as compared with their movement during the reproductive downswing. Upswings which do occur are likely to be hesitant and short-lived. A return to a sustained high rate of accumulation – the resolution of the crisis – requires the successful structuring of a new mode of regulation.

The shift in the nature of business cycles makes it possible to identify crises with some precision. The actual shift from long-run boom to crisis is a complex process, however. The regulation approach explicitly excludes a general theory of the causes of economic crisis, that is, a theory which is seen to apply to all economies and through all historical periods. What can be said is that crises occur when a regime of accumulation becomes exhausted, and reaches its inherent limits. The precise nature of these limits depends on the particular regime being analysed.

The reaching of limits arises from the accumulation process itself, which is uneven, and produces a variety of structural changes which eventually make impossible the maintenance of the dynamic equilibrium reflected in the regime of accumulation. These structural changes are reflected both in growing pressures emanating from the international economy, as well as in greater domestic pressures resulting from changes in the relative size and strength of social forces. Thus class conflict begins to express itself in new ways, undermining the viability of existing institutions.

When a downswing occurs in these circumstances of growing pressure as a regime of accumulation approaches its limits, policy-makers and other economic agents will adopt their 'usual' course of action in response to a downswing – the actions which made previous cycles reproductive in form. Now, however, these actions produce perverse results and a non-reproductive downswing, or crisis. 'Because of the interaction between external and internal pressures, established routines and rules of thumb no longer give the expected results' (Mjoset, 1987: 412).

However, in the first phase of the crisis, the problems are regarded as temporary, and not fundamentally different in kind from those experienced in the past. Economic agents, including policy-makers, therefore tend to stick to their old strategies. Two further phases of crisis have been identified (Boyer, 1988; Mjoset, 1987; Lipietz, 1989). In the second, there is a realisation that the problems are structural, rather than conjunctural. Policy-makers 'grope' for more far-reaching solutions, which generally aim to reorganise elements of the mode of regulation, but within the context of the old growth model. The attempts are often partial and independent of each other, and simply accelerate the growth model's disintegration. Once the model has disintegrated, however, a third phase becomes possible, in which a new growth model can emerge, a process involving the development of new institutions of regulation.

The framework described in this section has been used by the 'regulation' school to analyse the rise and decline of successive regimes of accumulation in the development of capitalism, most particularly the 'Fordist' regime in the post-1945 period. The next section examines the Fordist growth model which emerged in South Africa.

RACIAL FORDISM: POST-WAR ACCUMULATION IN SOUTH AFRICA

From the perspective of regulation theory, the post-war combination of apartheid and import-substitution industrialisation in South Africa can be seen as the defining characteristics of a 'racial Fordist' growth model. Like Fordism in the advanced countries, accumulation in South Africa during this period involved linking the extension of mass production with the extension of mass consumption; in South Africa, however, both production and consumption were racially structured.

In the Fordist regime of accumulation, transformations in the labour process involved the widespread introduction of production-line methods, yielding substantial improvements in productivity and increases in output. This pattern of production had already become common during the 1920s, but at that time had not been matched by real wage improvements. This incompatibility between changes in production and changes in consumption patterns led, it is argued, to the underconsumption-induced Depression of the 1930s.

From the late 1930s in some countries, but especially after the Second World War, a new mode of regulation was 'constructed', institutionalising mass consumption. Elements of this included the spread of collective bargaining, state provision of social security and a range of social services, and easier access to various types of credit. Mass production and mass consumption were linked in a virtuous circle: productivity improvements offset rises in both capital intensity and real wages, so that the rate of exploitation was stabilised (in the long run) at a high level. From being solely a cost of production, wages came to be regarded also as a component of demand. Installation of new technologies and expansion of capacity by industrial firms were increasingly based on internal finance supplemented by easy credit, which helped to stabilise aggregate investment.

This regime of accumulation was combined with a diverse range of nationally specific modes of regulation, giving rise to different growth models, though a series of general features can be identified for the advanced capitalist economies (Jessop, 1989: 263-4). At the international level, some elements of cohesion are worth mentioning. The first is the multinational enterprise, which obviously pre-dates Fordism, but became much more common in the post-war period. This was an essential vehicle for the diffusion of technological change across economies, as well as the spread of the 'American way of life', the norms of mass consumption.

The international monetary system similarly embodied American domination. Codified in a set of rules adopted at Bretton Woods in 1944, it was based on a gold-exchange standard, which fixed the price of gold in US dollars, and effectively made the dollar the international reserve currency. The fixed price for its main export was of critical importance to South Africa, as we shall spell out below. Bretton Woods also adopted a fixed exchange-rate rule, as well as a set of procedures for balance of payments adjustment which prioritised external balance over internal, at least in the medium term (Williamson, 1985; De Vroey, 1984).

It should be self-evident, but is perhaps worth making explicit, that labelling the growth model 'Fordism' should not be interpreted as an exclusive focus on the assembly-line labour process. Fordism refers both to the relation between the expansion of productivity, and thus of output, which the production-line reorganisation of the labour process made possible, and to the expansion of consumption 'norms', which became necessary to absorb the additional output. In addition, use of the label is not meant to underplay other processes of production, aside from the assembly line. Indeed, the original analyses of Fordism by the regulation school emphasised the rapid growth and technological development of the investment goods sector (equipment and inputs) as the major pressure disturbing the dynamic equilibrium, from the late 1960s (Aglietta, 1979: 100ff).[12]

In South Africa, it was evident at the start of the post-war period that the initial phase of import-substitution industrialisation was levelling off. Economic growth and industrial expansion had accelerated from 1933 with South

Africa's departure from the gold standard, and concentrated on the production of non-durable consumer goods (food, clothing and shoes), which was relatively labour-intensive.

From the late 1940s, the emergence of Fordism worldwide as a regime of accumulation combined with specific features of South African society to favour the continuation of the import-substitution process. At the same time, the local economy moved more actively into expanding the production of consumer durables, as well as intermediate products. The same path was followed by several other primary commodity exporters which had reached the end of primary import substitution, particularly Latin American economies like Brazil, Argentina and Chile.

The alternative was the option taken during the 1950s by the south-east Asian NICs (newly industrialising countries), such as South Korea and Taiwan. They moved from primary import substitution to the primary export substitution stage, concentrating on developing foreign markets for labour-intensive non-durable consumer goods. By following this route, they were able to absorb into industrial employment the 'labour surplus' moving out of the agricultural sector (Ranis and Orrock, 1985; Evans and Alizadeh, 1984).

Two crucial domestic factors pushed South Africa towards its 'choice' of accumulation strategy. The first was domestic politics – in broad terms, white minority domination exerting pressure on policy-makers to adopt strategies that would enhance white living standards. The other was the insertion of the economy into the 'international division of labour' as a mineral exporter, which made possible the importation of the capital equipment necessary to expand manufacturing.

The fixed international price of gold minimised the fluctuations of South Africa's export earnings. This distinguished South Africa's international situation during the post-war era from that of other primary-commodity exporters. In the latter, growth was repeatedly and severely destabilised by export earnings fluctuations that occurred in line with the business cycle in the advanced countries. As elaborated below, South Africa's stable export earnings were crucial in maintaining long-run growth. While this situation persisted, the failure of South African manufacturing to become internationally competitive and develop significant export capacity did not present a problem. During the 1950s, mineral output and productivity expanded far more rapidly than manufacturing, which in turn made possible an inversion of this position during the 1960s.

The growth model which emerged in South Africa was of the type that Lipietz (1987: 62) has called 'sub-Fordism', the qualification translating naturally into a racial one for the South African case: 'import-substitution policies ... did result in a real social transformation, and the emergence of a modern working class, modern middle strata and modern industrial capitalism. [The results] might be described as a "sub-Fordism", as a caricature of Fordism, or as an attempt to industrialise by using Fordist technology and its model of consumption, but without either its social labour processes or its mass consumption norms.' South African policy-makers followed the common international approach to import substitution at this time, inasmuch as their tariff policy was intended to promote rapid output growth and industrial expansion, by concentrating in the first instance on consumer goods production. This meant that rates of effective protection (measuring not just tariffs, but the overall impact on prices of trade and exchange-rate policy) on capital equipment and intermediate inputs were low (and often negative). The importation of these categories of goods was encouraged, and local production actively discouraged. This facilitated rapid industrial expansion, and therefore aggregate growth, but ultimately increased 'dependence' on the world economy, as we shall see below (Bell, 1975: 499–500; Helleiner, 1972: 101–6, 120ff).

As a result of the limited local production of capital equipment, more advanced technologies were generally incorporated as additions to capacity, and existing equipment and technologies remained in use. This link between capital-deepening and capital-widening (extending production capacity) caused the capital–labour ratio and productivity to grow at a lower rate (in absolute terms) than in the advanced capitalist economies, where implementation of new technologies generally involved scrapping existing equipment. At the same time, on account of the inverse relation between productivity growth and employment growth, the latter rate was higher in South Africa than in the advanced economies.

This pattern of changing conditions of production fitted well with the development of the wage relation on the basis of rigid apartheid differentiation in the labour market. In this respect, too, the South African economy represented a mutation of Fordism as compared with the advanced economies. The situation of the white working class was institutionalised along very similar lines to that of the working classes of the Western economies: an increased proportion of this group moved into skilled and supervisory positions in the labour process, with steady rises in their real wages making possible the spread of mass consumption of housing and locally produced consumer durables. Structures of collective bargaining, a social welfare system and very favourable subsidy and consumer credit arrangements all underpinned the process. In this fashion, underconsumption was made impossible and at the same time whites captured the lion's share of overall productivity gains.

The African working class did obtain some portion of these gains, however. As is well known, Africans occupied a subordinate position in the labour market, with restricted mobility, and without legal collective-bargaining power. Strict control at the point of production was complemented by equally severe limits in the consumption sphere. Urban Africans (including those from the middle classes) were strictly excluded from the mass consumption 'norms' which applied to whites and, at a later stage, to the Indian and coloured groups. Consumption levels in the urban townships were not significantly different from rural standards.

Nevertheless, as industrialisation proceeded, it produced a stratum of semi-skilled, permanently urbanised workers, and in time urban African workers became increasingly differentiated (Hindson, 1987). Firms seeking surplus profits tended to lower their unit labour costs by 'floating the colour bar' – substituting low-wage black labour for high-wage whites – when they adopted new technologies. On this basis, the real income of black workers did grow during the post-war period, though at a far slower rate than that of whites.

Because the growth model was biased towards capital-intensive industry, black labour extruded from the agricultural sector was not absorbed in the process of economic growth, as was the case in the south-east Asian economies. The resulting high levels of unemployment were instead handled politically, by means of containment within the bantustan system.

Although state intervention to manage the wage relation and the labour market was universal in the Fordist era, apartheid policies used to achieve this in South Africa were clearly unusual. The role of the South African state extended also to direct and heavy involvement in extending the manufacturing sector, especially the growth of industries producing intermediate goods. Their growth was in turn closely tied to the mining industry, as the major source of demand and often as the prototype for the wage relation (Clark, n.d.).

The racial Fordist growth model embodied within itself a 'match' in two important relationships in the accumulation process in South Africa during this period. The first was the harmonious interaction between the growth of the mining and intermediate goods industries ('Department I'), on one hand, and the extension of consumer goods industries ('Department II'), on the other. This made possible the stability of the investment coefficient – the ratio between (fixed) investment and aggregate production.

The second was the linkage between transformations of production and the extension of consumption patterns, or the change in the rate of surplus value. Changes in this variable can be estimated from the growth rates of real wages, labour productivity and the capital–output ratio. It is striking that the long-term trends of these variables exhibited a similar interaction in South Africa to that in the advanced countries.[13] The increase in the capital–labour ratio (the 'organic composition of capital') was approximately equivalent to the rate of productivity growth.

Changes in the capital–output ratio in South Africa are also affected by the price level for imported capital equipment and intermediate goods. This rose slowly during the 1960s (see Table 2). Given these trends, the capital-output ratio proved stable over the long run.

In manufacturing, the wage share remained stable over the long run, at around 46 per cent, while in mining and in commercial agriculture, the wage share actually fell during the 1960s (Nattrass, 1981: Table 8.1; Knight, 1979).

The long-run dynamic equilibrium was reflected also in reproductive cyclical fluctuations with short downswings. Their character is suggested by examining a stylised cycle (see De Kock, 1975; Smit and van der Walt, 1970, 1973, and 1982). As the pace of investment increased during the upswing, the growth rate of the capital–labour ratio increased, so that eventually the capital–output ratio also began to rise. In other words, the cyclical upswing itself, starting with a rising profit rate, would induce downward pressure on profitability.

Table 2. Trend growth rates of capital composition variables

% change per annum	Labour productivity	Capital–labour ratio	Output–capital ratio
1961–70	2.70	3.15	0
1970–3	0.30	4.95	−4.35
1973–7	−2.37	3.48	−5.74

Source: Report of the Study Group on Industrial Development Strategy (Kleu), Tables 6.5 & 6.6.

This was reinforced by a rise in unit labour costs, as the growth in capital utilisation and investment also tightened the labour market for white, coloured and Indian workers, raising their real wages. 'Registered' unemployment – covering these groups – represented a sensitive index of the business cycle. The nature of the mode of regulation was such, however, that downward pressure on profitability gave rise to 'bottlenecks', which determined the upper turning-point of the business cycle.

Firstly, the acceleration of fixed investment tightened the foreign-exchange constraint, notwithstanding the inflow of net capital which often occurred. This in turn led to tighter liquidity conditions in the monetary system, pushing up interest rates. During the 1960s fixed investment in manufacturing was financed by bank credit (in so far as external finance was required), so that the rise in the cost of borrowing slowed the rate of investment (Dickman, 1973).

It is worth pointing out that the impact of the foreign-exchange constraint during upswings was conditioned not only by the import of investment goods, but also by the relative stability of export earnings, owing to the fixed world price of gold. Without the fixed price, cyclical upswings would have been accompanied by balance of payments surpluses, and at most a delayed decline in liquidity (Krogh, 1985).

Similarly, 'skilled labour shortages' would develop near the top of the cycle, the upward floating of the colour bar making productivity improvements possible only with a lag. This process reinforced the growing disincentives confronting further fixed investment and capacity expansion.

By thrusting the business cycle into the downswing and limiting the long-run rise in the capital–labour and capital–output ratios, the foreign-exchange and skilled-labour bottlenecks served to stabilise and maintain the long-run growth trend. This is the converse of the conventional view that these 'bottlenecks' reflected structural contradictions in the accumulation process and limited economic growth over the long run. During the downswing, the 'bottlenecks' would be relaxed. Foreign exchange reserves rose as imports declined and the balance of payments deficit dropped. Easier money conditions were restored. As production levels dropped, registered unemployment began to rise, so that the growth of real unit labour costs slowed. Eventually, a revival of accumulation occurred, affirming the reproductive character of the downswing.

THE ORIGINS OF THE ACCUMULATION CRISIS

The crisis in the South African economy was triggered by the first 'oil shock' of November 1973, when the OPEC-led rise in the oil price produced a generalised recession in the world economy. But the oil price rise acted merely as the catalyst: the real causes of the crisis lay much deeper in the structure of the economy and had been developing for some time.

It was argued above that an economic crisis developed when a regime of accumulation reached its inherent limits. In the case of Fordism, which was centred on productivity growth through increasing returns from mechanisation, these limits began to become binding from the late 1960s, when productivity growth at the initial rate no longer proved possible. The roots of this decline lay in the nature of the Fordist labour process, in which 'the majority of workers have no control over their own work. ... The only way [to] increase overall productivity is to invent ever more complex machines. We can thus see why the downturn in productivity [growth] goes hand in hand with a rising coefficient of per capita fixed capital' (Lipietz, 1987: 44; also Lipietz, 1986; Boyer, 1988: 194ff). The second cause identified was the effect of the growing scale of production. As the need for expanding into world markets became more pressing, firms were forced into international competition outside the context of national growth models (Boyer, 1988: 196).

By the start of the 1970s, racial Fordism was also reaching its limits. At this time, the long-run trend variables – the growth rate of labour productivity, the capital–labour ratio and the capital–output ratio – began to diverge from their previous time-paths. The growth of the capital–labour ratio accelerated, while productivity growth began to level off. The growing gap between these two rates of change was reflected in an increasingly rapid rise in the capital–output ratio. As was the case with the interaction of these variables during the boom, the new pattern in South Africa echoed that of the advanced economies.

The limits to racial Fordism were, in part, the consequence of the apartheid-based mode of regulation. Already by the early 1970s, a link between skill shortages and declining productivity growth was recognised. In his 1971 chairman's address, Harry Oppenheimer of the Anglo American Corporation argued (in a statement remarkably reminiscent of the approach taken in this chapter): 'we are approaching the stage where the full potential of the economy, as it is at present organised, will have been realised, so that if structural changes are not made, we will have to content ourselves with a much lower rate of growth. ... Prospects for economic growth will not be attained so long as a large majority of the population is prevented by lack of formal education and technical training or by positive prohibition from playing the full part of which it is capable in the national development' (cited in Nieuwenhuysen, 1972: 160).

A more pressing problem in the early 1970s was the failure of the manufacturing sector in South Africa (as in most 'sub-Fordist' economies) to have developed a substantial export capacity. This prompted the establishment of the Reynders Commission on export trade. Its report diagnosed the problem not as excess capacity in manufacturing due to a lack of foreign markets for industrial output, but rather as cyclical constraints posed by current account deficits on the balance of payments.

The underlying problem was that imports had become dominated by production goods: import-substitution policies had changed the composition of South Africa's imports, but not their weight in overall supply, to any significant degree (Black, 1990). But sustained economic growth had meant that the quantity of investment, and hence of imports, had grown much faster than the overall export earnings, which were now beginning to impose a ceiling on further growth.

The mounting internal pressures upon racial Fordism were exacerbated by external pressures from the developing international crisis of Fordism. The productivity slowdown in the advanced economies raised the unit value of their output, and these effects were transmitted to the South African economy, in particular the manufacturing sector, through a rise in the price level of imported machinery, from a low point in 1968 (see Table 3). This process accelerated from 1970: the compound growth rate of the price index being 10.8 per cent per annum between 1970 and 1974, and then 19.7 per cent per annum up to 1980.

Table 3.   Import price index (1964=100)

	All imports	Machinery		All imports	Machinery
1966	109	118	1974	168	184
1968	103	111	1976	280	304
1970	109	122	1978	362	445
1972	130	156	1980	520	541

Source: *SA Statistics*, various issues (SITC classifications).

This rise in machinery costs was one of the causes of the South African economic crisis. Although the rising cost of capital equipment did not lead to a fall-off in demand (the price elasticity of machinery imports into South Africa being very low), it had two significant effects. The first was financial: more funds had to be committed for investment projects. This had an immediate impact on the growth of depreciation figures (see Table 4). Pari passu, the rate of inflation rose, from an average of 2.5 per cent per annum during the 1960s to 8.1 per cent during the first half of the 1970s (most of this before the oil 'shock'). This then had the further effect of forcing down the real interest rate – the 'real cost' of finance – which conventional theory presents as the cause of rising capital intensity during the 1970s.

Table 4. Provision for depreciation (current prices, replacement value)

% change	1946–60	1960–70	1970–5	1975–80
Total economy	10.90	9.27	19.93	18.31
Private sector	10.83	9.16	19.54	17.50

Source: *South African Reserve Bank Quarterly Bulletin*, Supplement, September 1981

Another result of the higher cost of capital equipment imports was a rise in the cost, in terms of capital invested, per rand of productivity improvement. In other words, the 'marginal efficiency of investment' declined in value terms, undermining the underlying profitability of investment projects, as the new pattern of relations between the trend variables showed.

The effect of this on the capital–labour and capital–output ratios would have been reinforced by the substantial increase in state infrastructural investment from the early 1970s: investment in public corporations grew by 130 per cent between 1970 and 1974, while their fixed capital stock rose at an annual rate of about 14 per cent between 1970 and 1977, compared with 9 per cent during the previous decade. For the private sector as a whole, capital stock growth remained constant at 4.5 per cent per annum through the period from 1960, but for private manufacturing industry, the growth rate of the capital stock halved, from 9 per cent to 4.5 per cent, as investment fell off during the downswing after 1974 (Swanepoel and Van Dyk, 1978).

A second cause of the crisis in the South African economy was the development of instability in export earnings. Growing problems in the international financial system, linked in part to the productivity slowdown, finally led, in March 1973, to the collapse of the Bretton Woods system of fixed exchange rates (Parboni, 1981; Block, 1977). Already in August 1971 the US Administration had suspended the arrangement whereby the dollar was convertible into gold at a fixed price. As an immediate consequence the price of South Africa's major export, gold, began to fluctuate freely on the world market. Notwithstanding, the rising trade deficit and declining foreign-exchange reserves in late 1971 forced a 12.28 per cent devaluation of the rand in December 1971. On one hand, this indicated that the reproductive characteristics of the business cycle were not operating as smoothly as before. On the other hand, the devaluation prefigured the much higher variability of the rand, which has continued to the present (Holden, 1985). With the move to floating international exchange rates, the world prices of other primary commodity exports began to fluctuate more widely than before.

The effect was a shift in the pattern of South Africa's export earnings. No longer would the balance of payments and the foreign-exchange constraint exercise a stabilising long-run effect through their role in cyclical fluctuations; instead they became destabilising and unpredictable factors. In such circumstances, underlying long-run decline was more likely to be reinforced than counteracted, even when a rising gold price provided a stimulus to growth. In other words, given the fluctuations of the prices of gold and other mineral exports, and of overall export earnings, cyclical movements in the overall economy, both upswings and downswings, have been exaggerated. In this respect, the behaviour of the South African economy has become more like that of other primary exporters.

The transmission of crisis tendencies from the international economy created stresses within racial Fordism which could not be absorbed. A major reason for this 'inflexibility' was the growth and increasing concentration in large-scale production of the industrial working class during the previous two decades. The eruption in 1973 of wage strikes in Durban and elsewhere signalled that the 'racially despotic' labour relations system within production had reached its end. The strikes took place explicitly in response to rising inflation, itself a consequence of the tendencies discussed above. In the mining industry, African workers were granted wage increases in the wake of the rise in the gold price from late 1971. It seems that these were connected to both the labour unrest in industry, and to the actual or threatened withdrawal of foreign mine labour from South Africa. The mines had to offer wages which were competitive with other sectors (Yudelman, 1984). The growth of black real wages, notwithstanding the recessionary conditions (see Table 5), pushed up real unit labour costs, and imparted a non-reproductive character to the cyclical downswing of the mid-1970s.

An examination of the cyclical downswing that began in South Africa in mid-1974 reveals that a shift occurred in its character. The key variables are those which had indicated the reproductive nature of the downswing during the earlier phase: the fall in real unit labour costs, and the rise in foreign-exchange reserves. If the cycle had become non-reproductive, these should have moved in the opposite direction, as was indeed the case, during both the major recessions in the course of the crisis, from September 1974 until the end of 1977, and again from the end of 1981 until early 1986.

Evidence of the developing crisis was already present in the recession lasting through 1971 and most of 1972. In these years, the real unit labour cost showed only a very small decline, while a devaluation in late 1971 was necessary to protect the foreign reserves, suggesting that this mechanism was becoming increasingly unreliable.

The next recession, triggered by the 'oil shock' of late 1973, marked the start of the crisis proper. Real unit labour costs rose by an average of nearly 2,7 per cent per annum between 1973 and 1977 (Table 5), even though 'registered' unemployment more than tripled during the period. This was obviously the consequence of rising African real wage levels, reflecting the breakdown of the mode of regulation. Foreign-exchange reserves also followed a 'perverse' pattern, continuing to decline (in US dollar terms) despite two devaluations of the rand during 1975. Even in terms of the domestic currency, their rise in late 1975 was both limited and temporary.

Table 5. Trend growth rates of wage share variables

% change per annum	Labour productivity	Real wages labour cost	Real unit
1966–70	2.97	2.18	−0.65
1970–3	2.65		
1973–7	−0.14	2.54	2.68
1977–82	4.11	2.67	−1.44
1981–4	−0.06	2.04	2.10

Source: National Productivity Institute, *Productivity Focus 1986*.

Similar developments can be traced in these variables in the period after 1981 (Tables 3 and 5). During both periods, it should be noted, there were temporary and limited revivals of economic activity, linked in each instance to some recovery in the international gold price and thus some improvement in the foreign-exchange reserves. The fact that in neither case could the initial revival be consolidated into a more developed upswing through an improvement in investment, is highly suggestive of the non-reproductive nature of the preceding cyclical decline.

In both cases these temporary revivals only compounded the depth of the subsequent continued decline. A further indication of the non-reproductive nature of the downswings was the very slow pace with which the upswing (when it finally occurred) developed any endogenous momentum. Both 1978, and the year from mid-1986 to mid-1987, were 'officially' periods of cyclical upswing. But the improvement in the levels of business cycle indicators was minimal, and expectations of the overall rate of growth were repeatedly revised downwards. The poor state of business confidence, in both instances, can be taken as an index of the disintegration of the mode of regulation: one of the central consequences of a stable and inherent mode of regulation is its invigorating effect on capitalists' 'animal spirits'.

THE DEVELOPMENT OF THE CRISIS

I have suggested above that the crisis can be periodised on the basis of the changing diagnosis of the problems faced by the growth model, from a view that they were temporary and conjunctural to one that understood them to be more fundamental. This changing perception has been influenced by the short-term movements of the business cycle, although there is no direct causation between business cycle movements and shifts in policy approach.

On this basis, three phases of the crisis can be distinguished. The first two will be discussed in this section, and the third – contrasting approaches towards structuring a new growth model – in the concluding section of the chapter.

The first phase of the crisis, which can be identified as lasting from mid-1974 until 1978, was characterised by recession and the emergence of stagflation internationally, in the wake of the oil shock. In the advanced capitalist economies, traditional Keynesian deficit-spending policies were maintained, boosting aggregate demand, but accelerating inflation at the same time. The maintenance of demand contributed to the high growth rates achieved by the east Asian NICs, creating both a market for their manufactured exports, and a pool of international finance available for lending. The NICs had already developed internationally competitive manufacturing sectors, and so made significant 'advances' within the international division of labour during this period (Frieden, 1981; Lipietz, 1987).

South Africa was unable to do the same, and in its experience of stagflation continued to follow the pattern of the advanced capitalist economies. The weight of the 'racial Fordist' growth model – in relation to both the nature of manufacturing, and state involvement in shaping economic activity – ruled out any dramatic growth of manufacturing exports. The substantial rise in South Africa's public foreign debt during this period (though less marked than for some NICs) was used rather to finance major infrastructural projects, such as Sasol. Such 'strategic' considerations – relating to efforts to preserve racial domination – came to weigh more and more heavily during this period, and account for the rapid growth in public sector fixed investment spending. At the same time, this spending was the counterpart, in the racial Fordist context, of deficit spending in the advanced economies, helping to raise the inflation rate, but also, it should be said, preventing an absolute decline in overall growth.

The recessionary conditions did not produce declines in nominal profitability in industry, despite the fall in underlying profitability conditions. Instead, the structure of monetary regulation in the import-substitution growth model, oriented as it was towards credit expansion, made for rising nominal aggregate demand, and accelerating inflation. Real interest rates became negative. Attempts to deal with this through traditional methods – the limiting of bank lending by means of direct controls and high liquid asset requirements – did not succeed. Unsatisfied borrowers, if they were large enough, moved into the higher-cost 'grey' market for money, outside the officially regulated banking system.

But while this process could offset in financial terms the recessionary effects of trade deficits for major corporations, it could not restore real growth. The decline in real growth was reinforced and prolonged by the capital account shock following the Soweto uprising of 1976. Massive outflows of capital, combined with already depleted foreign exchange reserves, helped enforce a deflationary policy.

Soweto 1976, and the response to it, signalled clearly that the problems facing the economy were not simply temporary, and that greater efforts were needed to reshape the growth model. In moving into a second phase of the crisis, South Africa was ahead of the advanced countries, where this occurred only from 1979, in response to the second oil shock, which was linked to the Iranian revolution. In response the state appointed a series of commissions of inquiry during 1977, with most reporting by early 1979. In general, the state's approach was adaptive and partial – 'groping' for solutions – in the sense that policy initiatives undertaken to reorganise specific aspects of the mode of regulation were carried out independently of each other. Moreover, the overall context remained the racial Fordist model, even while it was acknowledged that the problems faced were structural.

These early restructuring policies were intended to overcome two particular sets of problems. On one hand, they aimed to overcome the limits imposed by apartheid social relations on the labour process and labour markets, at least for a stratum of the black population. By linking productivity improvements to wage increases for a well-trained, fully urbanised black industrial workforce, the intention was to boost the size of the overall domestic consumer market (especially for durable and semi-durable goods), while simultaneously domesticating the emergent black trade unions.

In other words, the logic of the 'sub-Fordist' linking of mass production with mass consumption was to be continued, but its limits were to be extended. At the same time, the manner in which the distinction drawn within the black population between 'insiders' and 'outsiders', and the spatial dimensions of apartheid were maintained, reflected continuity in policy towards unemployment.

The second set of issues related to the economy's links with the transformed international financial system, and concerned both exports and the implications for the domestic financial system. The thrust of the approach adopted here was price liberalisation in financial markets (De Kock, 1980) in line with the orthodoxy of the time, at least in English-speaking countries. Liberalisation formed a prerequisite for the use of interest and exchange rates as instruments of monetary policy, which was in turn seen to be necessary for obtaining control over the domestic monetary system and restabilising the price level and the external balance.

The approach to the new accumulation strategy which evolved during the late 1970s fitted well with the interests of 'finance capital', the small and interlocking group of mining finance houses and long-term insurance institutions which had come to dominate all sectors of the economy in the course of post-war accumulation. Their outlook was based on the 'small advanced economy' model, with mineral exports of rising value providing the basis for continuing industrial expansion along earlier lines, while increasing integration into the international financial system eliminated 'anomalies' such as negative real interest rates and 'overvalued' exchange rates. These rates had of course been critical for industrial development in the previous period, but were now counter-productive for the dominant group. This orientation was stated clearly by Harry Oppenheimer, in 1976: 'The increase in black wages reflects the beginning of a process, still actively continuing, of a changeover from a labour-intensive, low wage, low productivity economic system – typical of industrial development in its earliest stage – to the capital-intensive, high wage, high productivity system which characterises the advanced industrial countries.... The migrant labour system becomes less and less appropriate from an economic point of view ... this does not mean, however, that it is possible to foresee a time when migrant labour, particularly in the gold mining industry, can be completely phased out' (Seidman, 1980).

These policy measures were implemented from 1979, as the South African economy captured much larger 'international rents', following the leap in the gold price after the second oil shock. The optimism which this engendered formed an essential element of the policy context. The 1970s had been characterised by extreme indecision on the part of the South African monetary authorities concerning their response to the new international exchange-rate regime, and the attendant gold price situation. There had also been a capital account shock linked to political developments. Nevertheless, Dr Gerhard de Kock felt able to argue, in 1980, that 'the South African economy has, on balance, been strengthened by [international economic developments] ... [so that] the long-term secular trend of economic activity will probably be strongly upwards' (1980: 351).

During the next two years, this analysis appeared to be correct, as real GNP growth rates reached record levels. But what it ignored were effects that appear to have deepened the subsequent recession, illustrating the destabilising impact on economic growth of movements – in both directions – of the volatile gold price. A myriad of new investment projects was initiated by both the private and public sectors, in the confident atmosphere. The acceleration of the capital–labour ratio during this upswing reflected a significantly higher rate of capital-deepening investment, lowering the age profile of the capital stock. In effect, the massive rise in the gold price removed the financial constraints which had inhibited this type of investment at the start of the 1970s.

This may have produced a much larger decline in investment levels consequent on the falling aggregate demand and capacity utilisation from late 1981. A related indicator was the weaker financial position of manufacturing and commercial companies in 1979–80, compared with 1972–3, despite the improved nominal profitability during the 'gold boom'. Since the negative real interest rates favoured borrowers, the expansion of capacity was financed substantially by short-term credit; debt-equity levels rose, even though earnings records were extremely good. All in all, the increased financial demands imposed on companies by the upswing, after the earlier decline in underlying profitability, made them more vulnerable to the subsequent recession, as the massive rise in overall debt levels which occurred after 1981 indicates (Dickman, 1982).

The rise in the gold price had, until late 1981, cushioned the impact on

South Africa of the second oil shock. There was no such relief, however, from the recessionary effects on the international economy of the monetarist deflationary policies adopted by the United States Federal Reserve Bank in 1981, in an attempt to halt accelerating inflation. These policies produced an international debt crisis. For South Africa, the immediate consequence was the collapse of the gold price, and severe balance of payments instability which persisted until 1986.

The emphasis of stabilisation policy fell, inevitably, on restoring external balance, at the expense of domestic output and employment. Given the liberalisation of the financial markets from 1979 on, adjustment was achieved first by using interest rates as the policy instrument, from the beginning of 1981 to late 1982. Thereafter the exchange rate was used directly as a policy instrument, when the gold price began to drop again in September 1983 (Cloete, 1986).

In this context, the two aspects of the 'experimental' accumulation strategy of the late 1970s came to contradict each other. While their mutual reinforcement may have been possible under some circumstances – a high and stable gold price, for example – it proved impossible when the policy emphasised balance of payments stabilisation, with a very unstable gold price. Essentially the policy choice was to favour mining, as the export sector, over manufacturing, and in consequence the intention to achieve expansion and productivity growth in the latter sector was doomed to failure. Instead the rising cost of imports, as the rand fell, and of working capital, as interest rates rose, eliminated the 'safety net' strung below industrial profitability in the earlier phase of the crisis, and profits fell even in nominal terms. As circumstances for industrial and commercial companies deteriorated, financial fragility increased: levels of bankruptcies rose, and the foreign debt burden of the private sector grew.

Although a similar process occurred throughout the world economy, what was specific in South Africa was the uneven impact on manufacturing compared with mining, which boomed, at least in rand terms, and especially after mid-1983. While the mining and the financial sectors became very profitable in the short term, the associated rise in inflation made South African mining, even gold mining, relatively costly in international terms.

The uneven sectoral impact of policy did prevent a classic debt-deflation process, which 'should' have resulted from the financial crunch of 1984–5, when interest rates rose to record levels, as deflationary pressure was applied to cut imports of consumer goods. Instead of a massive wave of bankruptcies decimating the manufacturing and commercial sectors, the finance houses used their large liquid asset base to support illiquid companies in these sectors, which substituted equity financing for loans. This was a central feature of the wave of financial market activity, itself part of an international boom in these markets stimulated from 1982 by the monetarist shock, and continuing even beyond the stock market crash in October 1987.

One result of this 'casino-type' activity was a significant increase in concentration of asset ownership in South Africa. In 1985–6, there were 61 new listings on the Johannesburg Stock Exchange, and in 1987, 200 (including 90 in the last three months of the year). Moreover, in 1985–6, there were 60 rights offers, as companies raised new capital from existing shareholders to pay off debt, and 133 mergers and acquisitions. Many of these were the product of disinvestment by foreign-owned corporations, a process which had political dimensions, but in a more fundamental sense was linked to the crisis of Fordism.

The recession from 1981 was a major factor in the re-emergence of political conflict during the early 1980s. This in turn was important in undermining the restructuring of the labour market, at least in so far as the intentions of policy-makers were not met. The trade-union movement could not be easily bought by wage-productivity trade-offs, this at a time of recession and growing unemployment. However, the trade unions became more militant, and strikes for wage hikes were frequent. The emergence of a powerful trade-union movement, based upon skilled and semi-skilled African workers, was part of a process of increasing class differentiation within the black population in the course of the crisis.

Slower growth and retrenchments, together with township resistance, placed unemployment and urbanisation on the agenda as essential issues to be addressed in a new growth model. As is well known, political conflict also increased foreign political pressure for the removal of apartheid, contributing to the debt crisis in 1985.

Enforced repayment of outstanding debt was financed in part by the trade surplus (based on a rise in the gold price during most of 1985 through 1987). However, the primary source of funds has been corporate savings, which have replaced personal savings in a relatively stable overall savings level. Nevertheless, the rise in corporate savings has been a product of the dramatic slump (63 per cent between 1981 and 1986) in private-sector investment in new productive capacity. Public sector investment was similarly limited in the face of external financial constraints.

What this means is that the corporate sector has been slowly liquidating its real capital stock, lowering productivity levels over the long term, in order to finance the capital outflow, as well as current government spending, which was larger than current revenue for most of the 1980s.

In the second half of the 1980s stagnation returned, and it was acknowledged that the initial attempts to restructure racial Fordism had been unsuccessful. The crisis was not yet resolved, although the second phase of the crisis saw the disappearance of most elements of racial Fordism. The third phase of the crisis was ushered in by the dramatic political developments of late 1989 and early 1990, which opened the way to the ending of apartheid – the last important element of the old growth model. This would finally make it possible for a new growth model to emerge in the future.

TOWARDS RESOLUTION OF THE CRISIS[14]

What are the accumulation options on the table in South Africa? As in the 1940s, South Africa appears to be facing two alternative paths. Both are articulated in terms of a 'growth plus redistribution' framework, but they reflect the interests of different combinations of classes and groups. Their separate growth paths have strongly contrasting implications for the nature, extent and time-scale of redistribution.

Two general points can be made about preconditions for a new growth model. One is that since the crisis originated on the 'supply side', its resolution must be concerned with restructuring the process of production. Inadequate demand became a factor only during the course of the crisis. As a result, a new growth model cannot be based simply on an expansion of demand, for example through the redistribution of incomes.

The second point is that the high degree of class differentiation in South Africa, which increased during the economic crisis, defines the political terrain on which accumulation strategies are put forward. An accumulation strategy will necessarily have to bring together, in a positive-sum context, a widely divergent array of group interests, but it is unlikely that all groups in society could be incorporated. Indeed, important and powerful groups are likely to be excluded by any specific accumulation strategy.

The achievement of a constitutional settlement, and its terms, will naturally be a crucial factor in determining which of the two alternatives is ultimately pursued. It would reorganise (some of) the interests of different groups in society, in the process of transforming the balance of forces amongst them. It is also possible that a constitutional settlement could shift the balance between social forces in such a way as to prevent the emergence of any viable coalition linked to one or other accumulation strategy, and thereby prevent a new growth model from emerging.

The present government and big business are increasingly implementing an accumulation strategy which we call neo-liberal export-oriented growth. The De Klerk administration has introduced into government economic policy a much higher degree of internal coherence, as well as greater consonance with the expressed concerns of much of South African big business. While many of the policies incorporated into this strategy have been long advocated in the course of the crisis, they are now being implemented, while others that do not fit have been abandoned.

The strategy focuses on restructuring and regenerating the manufacturing sector in particular, by using 'neo-liberal' (market-based) policies to alter cost structures and restore profitability, and to expand markets for manufactures, above all through exports. The emphasis is on the export of beneficiated minerals (currently exported in a semi-processed form) and intermediate manufactures.

Neo-liberal policies involve the state limiting its own economic activity in relation to the provision of goods and services, for example by a process of privatisation. Secondly, state intervention in the activities of other economic agents is limited to defining the broad parameters of market processes – that is, the general cost levels of productive factors (labour and capital especially) and other incentives (such as tax allowances and subsidies). The desired transformation of the profitability and the structure of production will then emerge, it is claimed, from the autonomous responses of economic agents to the changes in price 'signals' in the market.

Those employed in the major sectors of this economy – mining, large-scale manufacturing, and finance – would clearly benefit substantially from their location. This would include both the black middle class and organised workers, who could achieve wage gains in line with productivity growth (which would be expected to be fairly rapid). At the same time, income inequalities amongst blacks, and indeed overall, would probably widen.

'Redistribution' within this strategy would take place on a differential basis, and over a relatively long time-scale. The intention would be to channel a fraction of the additional resources, available from the expected boost in growth, to the 'disadvantaged' sectors of the population. This is evidently the intention of mechanisms such as the recently announced 'Jan Steyn Fund'. Money of this kind would provide the initial capital, on the basis of which the 'development areas' (as the official jargon puts it) would be expected to 'develop themselves', more or less independently of the capital-intensive, large-scale manufacturing sector. In essence, the urban poor would have to meet their own needs for employment and basic subsistence.

The neo-liberal export-oriented strategy would, in sum, reinforce and extend the dualistic structure of South African society. While this strategy attempts to address the problems behind the crisis of economic growth over the past fifteen years – the poor profitability of production and the instability of export earnings – an alternative strategy would seek to address as well some of the contradictions – of unemployment and extreme poverty – which arose out of the 'choice' of racial Fordism in the 1940s, and which have been dramatically magnified by the crisis of that growth path.

The path abandoned in the 1940s would have aimed to absorb the labour surplus emerging from the agricultural sector, through the expansion of labour-intensive, basic consumer goods industries, for the domestic market in the first instance. The simultaneous creation of employment and the expanding production of basic consumer goods are precisely the objective, in broad terms, of the alternative strategy open to South Africa today, although the strategy itself must necessarily be based upon current circumstances.

Rather than separating redistribution and economic growth, as in the government–big business approach, the aim is to achieve growth through the more extensive and more rapid redistribution of incomes and wealth. This strategy would be based on a synthesis of interests, bringing together not only the employed working class and the mass of urban unemployed, but also the middle classes (black and white).

In the context of recent policy debates within government and business, these objectives have been articulated in the form of the 'inward industrialisation' strategy (Lombard, 1988; Dreyer and Brand, 1986). The nub of their argument is that black urbanisation has led to a shift in the distribution of incomes and in the composition of consumer demand, thereby providing both the scope and the need for expanding and transforming production.

While the inward industrialisation position appears to have lost considerable ground within business and government circles, an accumulation strategy with similar objectives is in the process of being elaborated by the ANC and its allies. This policy package differs substantially, however, from the neo-liberal perspective implicit in 'inward industrialisation'.

In the alternative strategy, a central role is accorded to the state, whose capacities are seen to provide an essential counterweight to the inevitable reluctance of extremely powerful private economic agents, especially the conglomerates, to bring about a fundamental shift in economic development. State intervention would, however, be targeted and selective, based upon sectoral planning, rather than overarching and based upon general principles, as in a central planning system. At the same time, where it was undertaken, intervention would be pervasive and far-reaching in shaping the activities of economic agents, in contrast to the neo-liberal reliance on autonomous responses.

The critical question facing this growth model would be relations between business firms and government planning institutions. Co-operation, rather than conflict, requires that firms be involved in the sectoral planning process (as indeed should labour). But it requires also that firms accept a subordinate role in this process, on the ground that the planning process will create the conditions for profitable production. This is a particular problem in South Africa, because of the highly concentrated corporate structure, which dominates the provision of external finance to industrial firms. The subordinate position of the banks to non-bank financial intermediaries within the conglomerate structure reinforces the non-bank intermediaries' relative immunity from state control. Restructuring private sector investment would probably require the use of an anti-trust policy to restructure the financial networks of the conglomerates.

This alternative accumulation strategy would possibly produce, at least in the medium term, a somewhat lower rate of economic growth than might be achieved by a strategy less concerned with improving equity. At the most abstract level, this is inevitable if consumption levels are raised within a relatively short time-scale, as the pressure of popular expectations will dictate. The challenge will be to ensure that some limits are placed upon aggregate consumption expenditure, to enable the projected expansion and shift in production emphases.

Employment creation through labour-intensive industrial expansion will also involve some trade-off with potential productivity growth, which could be obtained from increased applications of technology, at least until the labour surplus is absorbed. On the other hand, significant productivity enhancements and dynamic efficiency gains can be derived from the sectoral planning approach; it also allows for the more rapid diffusion of new forms of organisation, and methods of production, as well as the avoidance of wasteful competition. Comparative experience suggests that this approach presents a much more promising route to cost savings and greater profitability than neo-liberal attempts to 'get the prices right'.

2
State, capital and growth: the political economy of the national question

MIKE MORRIS

INTRODUCTION[1]

This volume examines the economic crisis which we argue emerged in South Africa in the early 1970s. At around the same time, the first significant political challenge to apartheid in nearly two decades emerged. Marxist analysis, using Gramsci's concept of organic crisis,[2] has viewed these two processes as inextricably linked. As the crisis has developed the links between the economic and political crisis have been drawn with greater frequency, so that by now almost the entire political spectrum accepts that fundamental political change is a prerequisite for further sustained economic growth. However, a close examination of the literature, including the frequently cited piece by Saul and Gelb (1986), reveals that the connections between these two crises have not been well established.

Politically the focus has been on the exclusion of blacks from the rights and obligations associated with citizenship of the sovereign state. The sovereign state is the institutionalisation of the political power of the 'South African nation'. This is why all conceptions of the struggle for national liberation, however they have been formulated by the various contending forces within the black disenfranchised majority, have focused on the transformation of the political character of the state. Consequently challenges to particular expressions of state power, even when they pertain to more economic or social issues (e.g. labour mobility, job classification, education, etc.), have taken the form of an oppressed nationalism challenging the very definition of the state and whomever it represents. In other words both the expressions of state power and the challenges to them have taken the form of competing and conflicting resolutions of the national question.

This chapter attempts to examine these expressions of state power, not only from the perspective of national political struggle, but also in terms of their forming a part of the accumulation process. By sketching out the links between different growth paths and resolutions of the national question it attempts to pose the connections between economic crisis on the one hand and political crisis on the other, as well as the efforts to resolve these crises.

THEORETICAL CONCEPTS

In order to analyse the relationship between the economic and political crisis in South Africa, we need to set out some basic concepts which allow us to establish the link between accumulation and the state, between the advancement of a new growth path and a possible resolution of the national question.[3]

In the process of accumulation, capital takes on the different forms of money capital, production capital and commercial capital in turn. The circuit of capital cannot however exist in isolation. It requires, for example, the ongoing provision of labour-power in appropriate quantities and qualities, and capital cannot necessarily provide this for itself. The intervention of an agency external to capital, like the state, is necessary for this.

The state intervenes in accumulation in a given society by regulating the operations of different markets – e.g. labour, finance, foreign trade. This occurs through state institutions setting the parameters and norms – i.e. rules and social procedures – governing the way that social classes, groups and individuals relate to one other. Since capitalist societies are characterised by specific forms of exploitation and class relations, institutional regulation attempts to produce stable economic and political reproduction under conditions where conflict and contradictions predominate. The stable operation of such norms requires institutional enforcement and political consent. The political process of ensuring legitimacy, compulsion and consensus in society is essentially a process of welding what is deemed to be the 'national interest' into a particular form in order to ensure social stability.

Since the state is inherently a part of accumulation, a crisis in accumulation is also a crisis of the institutional unity of the state, including the manner in which those who comprise the nation are institutionally unified within the state.

Struggles between various classes and political forces over different national growth models and political forms of creating national consent are therefore an integral aspect of resolving a crisis. Accumulation, the form of state and what is perceived to be the national interest are interconnected issues which any concrete analysis of a crisis has to deal with.

The modern nation cannot be separated from the capitalist state since it is given form by the concrete operations of the state. The nation is concretely institutionalised through the state defining national territory and inscribing national tradition in its apparatuses. The nation therefore tends to coincide with the state since it is concretely institutionalised within the state's attempt to structurally define and reproduce the national interest. One cannot discuss the nation without discussing the state's role in defining and materially constituting the 'nation-people'. That is why Poulantzas says that 'the capitalist State is functional to the nation' (1978: 99).

However, the nation extends beyond the state's attempts to encapsulate and define its constituent elements. For the composition of the 'nation-people' and of the 'national interest' is precisely a site of struggle between the various contending classes and social forces in any historical situation.

An organic crisis tends to call into question the institutionalised (i.e. consented or begrudgingly accepted) conception of the 'nation-people'. Depending on the particular historical situation and the severity with which the previously stable social relationships are undermined, it may well pose the need for a new resolution of the national question.

It is important to avoid positing deterministic relationships between economic crisis, political restructuring and redefinitions of the national question. As Jessop makes clear: 'If economic crises do sometimes act as a steering mechanism, they are not an automatic pilot. The state responds to the political repercussions of an economic crisis rather than to the crisis as such; and specific interests will no doubt try to interpret its causes and remedies to suit themselves.' (1983a:149) We have now entered a new theoretical terrain – one of struggle as opposed to structure; of transformation as opposed to stabilisation; of process as opposed to maintenance. If organic crisis throws into disarray existing economic conceptions (the orthodoxy of the current growth model) and political stabilisation (the implicit acceptance of the national interest) then an organic solution requires new strategies to regenerate accumulation and restructure a political basis of hegemonic consent. In the literature these two sets of strategies are referred to as accumulation strategies and hegemonic projects.[4]

Accumulation strategies refer to redefinitions of the basic aspects comprising the existing growth model or the positing of radical alternatives to this model with the strategic aim of resolving an economic crisis. Accumulation strategies are directly concerned with economic expansion on an international or national scale. They are oriented towards alterations in the relations of production, towards the balance of class forces.

Hegemonic projects refer to the construction and maintenance of the social basis of support for a particular form of state. This involves taking account of the balance among all social forces and the mobilisation of support behind a concrete, national-popular programme of action. Hegemonic projects are integrally concerned with unifying the nation around broad issues concerned, primarily but not exclusively, with non-economic objectives such as political stability, social reform, and national and military expansion. They are also integrally concerned with who constitutes the nation, and who, in practice or in theory, is excluded from the composition of the 'nation-people'.

Oriented towards different areas of social reality, these two sets of strategy are not necessarily functional to each other. They may be dissociated or even contradictory, in which case the societies are doomed to the economic and political downward spiral of many Third World countries. Alternatively, they may be mutually reinforcing and lead to the reconstitution of the basis of national support and a new growth path.

Organic resolution can take two broad forms – 'one nation' or 'two nation' strategies. The former, as in typically social democratic policies, 'aim at expansive hegemony in which the support of the entire population is mobilised through material concessions and symbolic rewards'. On the other hand, a two nation strategy, as in Thatcherism, aims at 'a more limited hegemony concerned to mobilise the support of strategically significant sectors of the population and to pass the costs of the project to other sectors'. In a situation of economic crisis, or where the possibility of making serious economic concessions is restricted, then 'two nation strategies are more likely to be pursued' (Jessop, 1983b: 103–4).

In South Africa, the crisis of the 1970s posed the national question in a particularly acute form. This gave rise to a variety of economic and political strategies from capital and the state as they, unsuccessfully and often contradictorily, attempted to deal with both the blockage in accumulation and the formulation of a new resolution of the national question. The task of this chapter will be to chart the links between restructuring accumulation and reorganising the form of state, and to show how the political crisis of the hegemonic system of Verwoerdian apartheid required a new resolution of the national question.

THE CHARACTERISTICS OF VERWOERDIAN APARTHEID

The National Party came to power in 1948 on an apartheid programme. Verwoerdian apartheid inaugurated a series of state interventions throughout the decade which resulted in a stable form of racial domination and economic growth. This was based on a restricted Fordist growth model that structured the extension of mass production and consumption along racial lines. It was combined with a hegemonic project to resolve the national question which maintained white supremacy by defining and institutionally restructuring the South African 'nation-people' as consisting of many ethnic nations with commensurate claims to national statehood in a variety of alternative dependent states. Apartheid or 'separate development' thus became the mechanism to unite whites around a policy of extreme racial domination and economic privilege.

In its concrete implementation, as opposed to its ideological formulations, Verwoerdian apartheid was historically founded on an accumulation strategy which had two critical aspects to it – import-substitution industrialisation (ISI) and the reproduction and exploitation of cheap black labour.[5] This gave rise to differential state interventions in the reproduction of capital and labour.

Local industrialisation policy was based on protective tariffs, exchange and import controls governing finance and trade, parastatal corporations in a number of industries (e.g. Iscor, Sasol), production of consumer durables for the mainly white home market, a state-regulated wage bargaining system which excluded Africans and ensured increasing standards of living mainly for the white population (but also to a more limited extent for coloureds and Indians), and the export of primary commodities. The state structured the emergence of a particular labour market through institutionalisation of a segregated land policy that squeezed the homelands into producing a plentiful supply of cheap, unskilled black migrant labour. Within the urban areas a variety of state regulations governed the daily existence of the fully proletarianised and migrant black labour force in both the spheres of production and reproduction.

State intervention in the economy was parametric rather than pervasive. It fostered local industrialisation by erecting defensive barriers (e.g. tariff protection, import controls, local content regulations) around local secondary industrial enterprises. However, the state did not intervene radically within private capitalist enterprises and sectors with the aim of restructuring the very process of capitalist production itself. There was not an interlocking of manufacturing capital and the state in the process of rapidly fostering accumulation. Behind the barrier of limited protection the state allowed capitalists to make all the major choices about products, research and development, technological choice and transfer, company integration and investment directions. The state did not act pervasively, in a coercive and coordinating manner, as in the case of the south-east Asian NICs (e.g. directing investment into specific channels), to bring about a radical restructuring of manufacturing capital and ensure its international competitiveness.

The labour aspect (supply, reproduction and control) of apartheid's accumulation strategy was, however, markedly different in tone and content. The state intervened within the relations of production, with the clear aim of ensuring the dominance of capitalist relations of production within the society. It was based, on the one hand, on maintaining a migrant unskilled black labour force with some subsistence roots in the homelands, whose social relations were partially reproduced via precapitalist social relations and the local bantustan state networks. On the other hand, the smaller, fully proletarianised black labour force, including the migrant labourers for the time they were allowed in the urban areas, was reproduced within tightly controlled urban townships. The state intervened radically through an intricate system of influx control measures to control the flow of African workers to the urban industrialised areas. Within these areas it ensured both the provision and tight control over the collective consumption requirements of the migrant and settled black working class residing there. Furthermore, it ensured a compliant and docile work force through the effective suppression of attempts to unionise the African working class, as well as ethnic segregation within existing unions of coloured and Indian workers from their white counterparts. The absorption of the latter categories of workers into craft unions and artisanal places in the division of labour provided skill and flexibility to complement the cheapness of African unskilled and semi-skilled labour.

This particular form of industrialisation had a major effect on the class character of capital's relations to the state. It produced a state–capital relationship which was simultaneously distanced as well as being dependent. The state maintained a distance from the capitalist enterprises, whilst capital simultaneously exhibited a dependency on state coercive labour practices and, to a lesser extent, on protective defensive barriers against foreign competition.⁶

These measures to a large extent ensured the successful operation of capitalist enterprises and were welcomed as such. However, they were not only the result of the state implementing a particular accumulation strategy. The apartheid policies concerning black urban rights, influx control, homelands, collective bargaining, etc. were also tied into the hegemonic project of racial domination. In this respect, the manner in which they were implemented often reflected the interests of classes other than industrial capital and produced a tension between state and organised industry.

Most white, coloured and Asian workers were fully proletarianised, dependent on wages or salaries for their reproduction, and mainly located in the metropolitan areas. These workers were integrated into a statutory industrial relations system (from which African workers were legally excluded) that regularised industrial wages and conditions of service. The industrial council system produced a uniformity of wages and conditions of service on an industry-wide basis: this allowed firms to forecast wage costs and consumer demand, and facilitated a regular rise in the standard of living of these workers and the limited but profitable expansion of the home market. This section of the labour force formed an integral part of a system of mass production and consumption, particularly in the consumer goods sector, based on the predominant use of machine-paced, semi-skilled labour, which by the late 1960s displayed many of the characteristics of Fordism in advanced capitalist countries.⁷

Within the terms of Verwoerdian apartheid, the South African 'nation' was narrowly defined in racial terms (as white) and incorporated into a racially defined social welfare state. State-provided social services were differentially allocated on racial lines. A social welfare state existed for whites, providing protection and subsidisation in the areas of education, health, housing, employment placement and unemployment benefits. Although differentially applied, many of the benefits of this racially defined social welfarism were also available to those classified as coloured and Asian. The extent to which they were applicable to Africans was so minuscule as not to be relevant.

The hegemonic project of the National Party was racial exclusion of all those not classified as 'white'. It operated to obstruct upward class mobility of blacks (both within the residential terrain and the sphere of production) and to ensure the economic integration of whites through the social welfare benefits of the apartheid state. This was the ideological cement that unified most classes amongst the enfranchised white population behind the ruling National Party's political programme. It projected and structured the national popular interest as a racially defined territorial and social segregation, summed up in the terms 'grand apartheid' and 'petty apartheid'. 'Grand apartheid' captured the system of bantustan-based territorial segregation, while 'petty apartheid' covered the racial exclusion of blacks from the social welfare reproductive institutions and amenities reserved for whites.

Under the policy of separate development the state attempted to resolve the national question by creating a 'multi-national unity' through newly created state apparatuses in the various bantustans. The South African national tradition and territory was statutorily compartmentalised into fixed ethnic nations and national states, each with its own 'traditional territory and territorial tradition'. Whites (with the Afrikaner *volk* elevated to an even more select status) were defined as the 'South African nation', and their privileged access to the political and social welfare apparatuses was formally inscribed into the state.

The black population was divided into a multiplicity of ethnically defined 'nations', each with its own newly created national ethnic cultures, traditions, territories and political rights inscribed into the structures of the 'national states' of the bantustans. These traditions (some real, some conveniently created) were supposed to give the individual citizens of these other 'nations' a new national historical path stretching back through the mists of a tribal past and forward into the new dawn of an ethnically promised national liberation. To some extent this project derived support from classes that directly benefited from the opportunities for limited accumulation, corruption and bureaucratic employment which opened up in the bantustans. However, the narrow definition of the South African 'nation' depended fundamentally on state coercion of blacks, particularly in the urban areas. There was never any serious possibility of Verwoerdian apartheid's resolution of the national question acquiring the consent of the mass of the black population.

For those sections of the people encompassed within the ambit of the 'nation' (i.e. whites), the South African state manifested many of the characteristics of capitalist democracy, whilst those excluded from the privileged 'nation' (i.e. blacks) experienced the state primarily as a repressive apparatus denying them political rights. Consequently the state most closely resembled an amalgam of a racially exclusive form of democracy dependent upon a highly developed battery of repressive powers directed at the excluded majority.[8] Although it was a limited form of democracy it still displayed many of the characteristics of a bourgeois democratic state. Power remained structurally located and rotated within typically flexible capitalist institutional structures – e.g. parliamentary forms, election of political parties, separation of the executive from the legislature – from which all the black popular classes were legislatively excluded.

Hence struggles within the white population were essentially about the circulation of power and were deflected into the institutional form of party politics. In this respect, reflecting its bourgeois democratic characteristics, the state displayed a marked degree of flexibility as power circulated between parties and state apparatuses. However, the struggles of the excluded black majority (whatever their content) became immediately politicised as they bounced up against the barriers of political exclusion and repression. Struggles lost their specificity, becoming generalised and potentially explosive since there was no way for them to become institutionally channelled. The national question seemed to permeate all forms of struggle for social, economic and political advancement, resulting in an oscillation towards selective cooptation on the one hand and demands for national liberation on the other.

CAPITAL AND VERWOERDIAN APARTHEID

The tone of the state–capital relation was set by the successful match between the accumulation strategy and the hegemonic project of Verwoerdian apartheid. The National Party's hegemonic project, which elevated Afrikanerdom to a special status within the white 'nation-people', defined a differential relationship of 'Afrikaner' and 'English' capital to the state. The state supported the economic interests of 'English' capital but it gave special status to 'Afrikaner' business enterprises which were defined as an integral part of the 'Afrikaner nation'. It consciously built up an Afrikaner bourgeoisie existing partly in state parastatals and partly in private businesses.[9] State corporations such as Iscor, Eskom, and Sasol concentrated a large segment of productive capital in the hands of the state. They provided an avenue for advancement for the Afrikaner managerial stratum, establishing this segment of the capitalist class in a dependent relationship to the National Party. 'Afrikaner' capital, fledgling and ideologically close to the governing National Party, from which it derived much needed political support, was reliant on the maintenance of a close and dependent relationship to the state.

'English' capital, on the other hand, had a different historical relationship to the state. Economically secure and historically dominant in mining and manufacturing, it operated on the basis of a fairly stark separation between economic and political arenas, assuming an attitude that capital's business was 'business' while the state's was 'politics'. These differential relations to the apartheid state manifested themselves in a substantially divided capitalist class; divisions that took the form of, but which ran much deeper than, differences of language and culture.

Language and culture played a major organisational role in the process of capitalist class formation. The separate linguistically based business associations organised and represented individual firms, making it possible for them to operate as a class.[10] However, in the process they were also responsible for reproducing organisational division and different trajectories of class formation between Afrikaner and English capital. These linguistic and organisational divisions had a profoundly limiting effect on the political role that capital played, and substantially contributed to the marked degree of political incoherence of capital in South Africa.

Capital's political incoherence was also exacerbated by some of the characteristics of the accumulation strategy of Verwoerdian apartheid and the consequent relation of distance–dependency between capital and the state. The dependency on state protection for their economic well-being meant that businessmen had an alternative option to competing in the marketplace – they in effect became special pleaders. Protection and the importance of state contracts in the domestic market allowed for a group of industrialists, retailers, manufacturers, bankers and insurance companies to increase their profit by cultivating a special relationship with the relevant section within the state. When faced with increased competition from domestic and foreign competitors, a manufacturer had the appealing alternative of going to Pretoria to plead for a higher control price, greater protection or increased government contracts.

The internationalisation of capital and the integration of South Africa into the world economy in the post-war period produced new divisions within capital. Most medium-range industrialised countries outside of Europe and the USA developed new relationships of dependency and conflict between foreign and local capital. In these dependent countries the internationalisation of capital increased the linkages, and hence dependency, of local capitals on the main industrial economies. This process radically undermined the former independence of national capitalist groupings and brought about the emergence of a new, more dependent fraction of local capital – the domestic bourgeoisie. This bourgeoisie is characterised by a greater dependence on foreign capital while still attempting to foster independent local industrialisation, increased mass consumption at home, and state-aided exports. The varied and differing levels of integration into, and dependency on, the world economy create, however, divisions within this capital and increase its incoherence. Hence, despite its local political orientation, this fraction of the bourgeoisie displays a major difficulty in constructing a political bloc which can represent the 'national interest' and cohere with other social forces.[11]

South Africa's linkages with the world economy increased markedly in the post-war period. ISI encouraged the production of sophisticated products for the more affluent (white) consumption sector. Foreign capital was welcomed on its own terms and the dependent linkages increased rather than decreased. The increasing dependency of South African capitalism on the world economy over the past few decades has resulted in a substantial degree of integration of local Afrikaner and smaller English capital into the world economy. Furthermore, older, large English capitals like Anglo American and Barlow Rand have become internationalised in their own right, with substantial foreign industrial and mining holdings as well as major operations on the world capitalist financial markets.

This has resulted in a further division, overlaying but also cutting across the historical divide between English and Afrikaner capital, between a more internationally oriented bourgeoisie and one which is more domestically oriented. The internationally oriented bourgeoisie has been deeply and conservatively embedded in a tradition of liberalism and free marketeerism, and displayed a hostility to social democratic traditions. It has been the powerhouse within organisations such as the Urban Foundation, has formulated its political role as improving the quality of life of black urban areas and has shied away from a radical reorientation of capital's role in the political arena. The domestically oriented bourgeoisie was represented prominently within Afrikaner capital by the previous head of Sanlam, the late Dr Fred du Plessis, who espoused a clear policy of 'neo-protectionism' with regard to foreign trade, foreign exchange and monetary policy.[12] The internationalisation of capital has thus served to deepen South African capital's political incoherence.

Finally, the representational form of the South African state has always acted to constrain and shape capital's political options, particularly in regard to formulating alternative resolutions of the national question. Capital does not exercise its influence over the state in a direct, unmediated form. In launching a hegemonic project it has to operate at the level of the 'national-popular'. This essentially means operating within the confines of political parties, ideological slogans, national-popular programmes, etc. Essentially the problem that faced any fraction of capital in South Africa was that the historical absence of blacks from the parliamentary apparatus imposed clear limits to a radical change in the course of state policy, since in order to do so it would have to mobilise within a political party a social basis of support (e.g. from amongst the black middle class) from those who have been constitutionally excluded from the 'nation-people'. Any fraction of capital seeking to formulate an alternative hegemonic project to resolve the national question was hence trapped by a contradiction located in the very form of the state.[13]

THE POLITICAL CRISIS OF VERWOERDIAN APARTHEID

Verwoerdian apartheid attempted to compress class differentiation amongst blacks outside of the bantustans by placing ceilings upon upward class mobility, confining all Africans within homogenised townships, and using the pass laws to channel and block migration to the cities. This policy of compression cut substantially across class and other differences and reinforced a black unity against apartheid. The Soweto uprising of 1976 brought this clearly home to the state and capital. The initial response was straight from the heart of Verwoerdian apartheid. Relying on the tried and trusted policy of repression, the government cracked down and banned a large number of organisations and people.

However, coinciding as it did with the beginnings of a structural crisis in the economy, various forces within capital and the state were also pushed, in a very tentative form, to seek an alternative accumulation strategy and resolution of the national question from that embodied in Verwoerdian apartheid. In short, the crisis, and the attendant popular struggles of the 1980s, brought to the fore the need for the dominant classes to confront the problems of both growth and a new resolution of the national question. The discourse of the dominant classes became permeated with attempted solutions to these problems. Phrases like 'new national unity', 'redistribution for political stability', 'addressing social, economic and political grievances', 'reconciliation', and 'building a new nation' abounded.

If the crisis posed the need for a restructuring of South African society it did not, and indeed could not, guarantee that a coherent accumulation strategy and new hegemonic project would emerge. On the contrary, the attempts by capital and the state to move along these lines were characterised by hesitancy and confusion as to what was required. Rather than facing the need for both a new accumulation strategy and hegemonic project, capital and the state veered to reducing the issue to a socio-economic problem. The Urban Foundation was set up in the immediate aftermath of Soweto 1976 with the initial aim of coopting the black middle class. This it hoped to achieve through improving the quality of life of the black urban environment, principally through township upgrading schemes, housing programmes and site and service experiments. Responding to the problems of economic crisis and ignoring the need to formulate an alternative resolution of the national question, the initial responses from these quarters were dominated by manoeuvres primarily aimed at formulating a new accumulation strategy.

Monopoly capitalism had come to dominate industrial capital with concomitantly more sophisticated requirements from the state-controlled system of reproducing labour-power. Manufacturing industry required more skilled and semi-skilled black labour. The state experienced increased pressure from the advanced sectors of capital calling for the reform of the pass laws in order to expand the settled urban population and thereby raise the rate of productivity of this labour force. Furthermore, as a result of the dissolution of precapitalist subsistence relations in the bantustans, the material basis of migrant labour had been significantly undermined. Bantustans increasingly became simply repositories of fully proletarianised surplus populations in the rural areas. Hence the state could no longer attempt to secure the social reproduction of this section of the working class by displacement of the social welfare costs of reproduction onto rurally based precapitalist subsistence societies. With increasing unionisation and militancy of black workers, this impacted onto employers, who faced growing demands for higher wages to cover the full costs of reproduction of migrant labour.

Thus the first reformulations of an accumulation strategy posed by manufacturing industry focused on reforming the labour aspects of Verwoerdian apartheid. Big capital, especially that owned by, or linked to, foreign capital began pushing for the liberalisation of the urban labour markets to increase mobility within the metropolitan urban areas and a restructuring of the statutory industrial relations system to enable African workers to become members of trade unions. This, plus the shake-up in the governing circles attendant upon P. W. Botha's coming to power at the end of the 1970s, produced the beginnings of a new policy. A new discourse, of the 'reform' of apartheid, came to dominate the decade of the 1980s.[14]

This process of state-initiated 'reform' was basically composed of three elements which, depending on the phase of 'reform', were differentially emphasised within state policy.[15] These elements were: initiating a limited 'democratisation' (opening up) of ideological and political life; implementing the 'deracialisation–reracialisation' of social and political life; and instituting a partial, and selective, 'redistribution' of social resources towards the black majority.

These elements were not uniformly or equally implemented. They acquired positions of dominance in state policy at different times in the process, depending on the phase through which the reform process was progressing. One can therefore trace the process of 'reform' of apartheid through its different phases both in terms of which elements were dominant as well as in terms of the different ways in which the state approached the problem of formulating a new accumulation strategy and hegemonic project.

THE FIRST PHASE OF REFORM

'Reform' was inaugurated by the appointment of the Wiehahn and Riekert Commissions at the end of the 1970s to investigate the place of Africans in the industrial conciliation machinery and urban residential policy. These 'reform' agendas, concerned solely with industrial relations and urbanisation, were confined to developing a new accumulation strategy and hence did not address the requirement of a corresponding resolution of the national question. The hegemonic project of Verwoerdian apartheid still remained unchallenged.

The language and the conceptualisation underlying the 'reforms' thus bore the stamp of their historical origins in Verwoerdian apartheid. They were permeated with apartheid terminology. The Riekert and Wiehahn strategies adopted in 1979 were based on the premise that a clear distinction could be drawn within the African population between urban 'insiders' and rural 'outsiders'. Urban 'insiders' with residential rights would supply industry's need for a more skilled and productive workforce. Rural 'outsiders' would be sealed out of urban labour markets and confined to the bantustans. Union rights would be extended to the 'insiders' only, which would allow them to increase their bargaining position relative to 'outsiders' and provide a further material basis for the differentiation. Economic gains plus upward occupational mobility would lead to social stratification within the urban African population, and labour market controls would differentiate more sharply between 'insiders' as a whole and 'outsiders'.

The campaign of shop floor resistance and legal battles by the emergent independent unions destroyed this form of implementing the Wiehahn reforms. Consequently the extension of trade union rights rapidly moved beyond the parameters set by this first phase of reform. From this time on there was always a dislocation between the reforms operating at the level of collective bargaining and other spheres of state policy.

The first phase of reform was limited to restructuring the dominant accumulation strategy and represented a 'racially selective Fordist' response with no attempt whatsoever to address the issue of the national question. It attempted to incorporate economically a protected 'insider' urban black population into the mass production economy, whilst compartmentalising the marginalised, unemployed, unemployable population as 'outsiders' in the bantustans. The reforms aimed at the 'insiders' would (it was hoped) stimulate new demand, arising out of their increased wages, at least part of which would be offset by increases in productivity.

This first phase of reform did not go sufficiently beyond the premises of territorial apartheid. Indeed the aim of this first phase of reform was to underpin, on a new basis, territorial segregation by legislatively strengthening the division between urban and rural Africans. It was believed that urban Africans would accept marginally strengthened local government structures while casting their votes as citizens in one or other of the bantustans. Rural 'outsiders' were still to be accommodated (economically and politically) within the bantustans. Thus Riekert's division between 'insiders' and 'outsiders' still operated within the framework of influx control, and notwithstanding its commitment to differentiation of the African population, it ran counter to the processes of class differentiation occurring within the economy. Consequently it succeeded neither in fostering class differentiation nor in coopting middle-class Africans.

Riekert, as Hindson (1987) argues, still conceived of South Africa as being structured by the historic divisions between precapitalist and capitalist sectors. This reform strategy therefore assumed that the 'insider–outsider' divisions of his reform policy could be based on sealing off the bantustans from urban areas. However, the bantustans were no longer simply subsistence repositories of surplus labour. They were integrated into the national economy through a massive system of commuter migration to the metropolitan areas. Furthermore, the urban areas were themselves being constantly reformed and reshaped by the massive proliferation of squatter settlements feeding into reconstituted metropolitan areas extending far beyond the legislative or traditional municipal boundaries. The classic patterns of labour supply and reproduction based on the migrant–settled, rural–urban dichotomies were being superseded by the restructured urban regional economies around the industrialised metropolitan areas which encompassed traditional rural bantustan, rural white, urban industrial and urban black areas.

In addition to these structural tendencies undercutting Riekert's agenda, the state reforms generated mass resistance. The period witnessed major resistance from squatter communities, mushrooming on the perimeters of metropolitan areas, to the 'insider–outsider' strategy. Furthermore, the extension of trade-union rights, instead of acting to facilitate the regulated rise in market demand of the urban 'insiders', served as one of the major defensive mechanisms of the working class as the economic crisis bit deeper. On the political level, instead of allowing for the cooption of the 'insider' working class, as Wiehahn intended, they rather provided a vehicle of political protest; instead of facilitating the depoliticisation of industrial relations, they fostered the radical and overt politicisation of working-class economic struggles.

Riekert's shift away from direct central state intervention was designed to go hand in hand with the decentralisation of administrative control to local black township councils. Collective consumption requirements in the townships catered for previously by direct state involvement were to shift, and thereby be depoliticised, to these township councils. But this 'privatisation' of housing and other forms of township collective consumption required that these councils were able to take on the financing and regulation role that the central state had previously applied. This required a major expansion of the local revenue base, occasioning severe rises in rents, rates and township service charges. The net result was that the state facilitated a wave of serious resistance from ordinary township dwellers to this process. The state's reform initiative to depoliticise collective consumption in the townships produced its direct opposite – the massive politicisation of struggles over township collective consumption.

Verwoerdian apartheid had depended heavily on the relationship between town and country reproducing labour power in the form of migrant labour. The inability to maintain this dovetailed with the state's inability to maintain coercion in the urban areas. Although the Wiehahn and Riekert reforms were not primarily concerned with the national question, by posing the status of the urban working class and urban residential rights, their resolutions spilled over into the national question and threw into turmoil the old hegemonic project. In attempting to avoid the national question, Wiehahn and Riekert only created slippery slopes on all sides around the issues of trade-union rights, permanent residency and political rights.

This is not surprising if one takes a comparative perspective on the problems that exceptional states have experienced in undertaking reform of their authoritarian structures. As Poulantzas pointed out with respect to the dictatorships of Southern Europe, 'any opening of "controlled liberalisation" on the part of the state rapidly becomes a gaping hole through which the popular movement rushes in. How can the state authorise the creation of "relatively representative" trade unions, for instance, in order to permit the power bloc to "negotiate" with them, when this very breach in the dyke leads to the unions being rapidly occupied by the genuine representatives of the popular masses...?' (1976: 94)

By 1982 the first reform phase was visibly fraying round the edges. Riekert could not solve the urbanisation problems and was unable to sustain the legislative division between 'insiders' and 'outsiders'. The attempt to exclude contract workers from belonging to trade unions had failed miserably both in practice and in law, and trade-union rights were accorded on a non-racial, democratic basis. The Wiehahn reforms had been pushed way beyond their initially proposed limits, and the state had been forced to map out a non-racial and democratically equal terrain of industrial collective bargaining. Black workers were granted the status of being equivalently identical (as opposed to separate but equal) to white workers. Viewed through the limited perspective of their union windows, black workers were wholly incorporated into the nation-people. Yet the state had no equivalent hegemonic project to address the national question.

This reform phase failed because it ignored the fact that the material basis of territorial segregation had been fundamentally eroded by structural changes in the socio-economic terrain of the society; and because it posed no solution to the national question. The crisis had thrown into disarray the previous hegemonic project and forced upon the dominant classes the need to formulate their own new resolution of the national question. Economic solutions that failed to address this were doomed to failure.

The National Party responded by formulating a new hegemonic project, the 'tricameral parliamentary system', to restructure the national question, which Verwoerdian apartheid no longer could deal with. From the governing National Party's point of view this was the high point of the first reform phase. It purportedly now had a hegemonic project to resolve the national question and thereby unite white, coloured and Indian citizens behind its policy. With no hegemonic project of its own, capital seized with near desperation on this new policy and supported the National Party in the November 1983 referendum. An overwhelming majority of the whites voted to amend the constitution and extend the boundaries of the 'nation' to include coloureds and Indians through the tricameral parliament. It deracialised by opening up parliament to previously excluded groups, while simultaneously re-racialising that system with its contortions into 'own' and 'general' affairs.[16]

Whilst the referendum acted to consolidate support from amongst the dominant classes for the regime's new hegemonic project, it only served to sharpen the problems that the popular classes faced. By maintaining the underlying basis of territorial segregation for Africans, insisting on their exclusion from any central organs of power, and creating representation only at local township level through community councils, whilst simultaneously parading tricameralism as a liberalisation of apartheid, the state effectively exacerbated the contradictions it faced in creating a new basis of social stability and economic growth. Instead of inaugurating a new period of political stability, as the state and capital fervently hoped it would, tricameralism brought the national question as the principal political issue to the fore. The very process of election for the new political dispensation provided renewed arenas for the growth of opposition to the regime's proposed solution to the national question. It was an immense irony that, instead of bringing peace, the highlight of the first reform phase served mainly to bring the national question to the boil amongst the popular classes, spilling over into insurrection in 1984, to the startled surprise of capital and the state.

THE COLLAPSE OF THE FIRST REFORM PHASE

The popular classes were able to take advantage of the partial opening up of ideological and political life which was one of the central elements of the state's 'reform' programme. This process of 'democratisation' was limited but, nonetheless, significant. It was seen as an integral and necessary part of the shift away from Verwoerdian apartheid by capital and *verligte* Nationalists. Furthermore, the lessons of trade union struggles and the example of the stabilisation of this arena were not lost on capital and were often used as a comparable lesson to be applied outside of industrial relations. Space was opened up for political organisations of the popular classes to emerge openly and engage in a series of high-profile mobilisation campaigns. The process of opening up such space was significantly influenced by mass struggles themselves, which extended the liberalisation beyond the parameters that many in power had intended.

Once again this is a problem that all authoritarian regimes face when attempting to liberalise under pressure from below. Such regimes 'are faced with the need to undertake a change at a point when they can no longer manage to control the popular movement by force, and precisely because they cannot manage to do so; this in itself means that they cannot in any way control and direct their own transformation. The regimes find themselves faced with the age-old dilemma: either they give too little, and then the changes they have in mind will in no way meet the needs of the situation; or these changes act as an incentive for more, and then the regimes appear almost automatically to have given too much.' (Poulantzas 1976: 95)[17]

The insurrection of 1984–6 was the primary cause of the collapse of this first phase of reform. The institution of the tricameral parliament gave the popular classes a focus against which to channel anger. The process of limited democratisation that had accompanied this first reform phase provided the opportunity and the political space to organise. In the black townships the failure of Riekert's urbanisation strategy, and the attempts to institute revamped community councils, coupled with the manifest crisis in black school education, fuelled anger on the ground.

Caught in its 'distance–dependency' relationship with the state, capital again displayed its political incoherence, retreating into a corporatist socio-economic intervention. By the mid-1980s, the more internationally oriented fractions of big capital, exemplified by Anglo American, had abandoned their hope in the Riekert reform initiatives and were calling for a more thoroughgoing reform of state policy towards the reproduction of labour power. This would entail the abolition of influx control measures, the acceptance of African urbanisation, expansion of the urban metropolitan reserve army of labour, and hence decreasing the pressure from unionised workers for rising wage levels. The Urban Foundation coordinated and led a well-directed campaign to abolish the pass laws. It argued strongly for an alternative accumulation strategy, which it called 'positive urbanisation', but made little attempt to follow this through politically by articulating an alternative hegemonic project to resolve the national question.[18]

For a brief historical moment this fraction of capital seemed able and willing to grasp the nettle. Relly from Anglo American led a delegation to see the ANC to begin exploratory talks about what they had in common. The

FCI (Federated Chambers of Industry) let Johan van Zyl have his head and put out the Business Charter – the most liberal and coherent political document of reform that had ever emanated from the ranks of organised business. It was not, however, sustained. Big capital revolted and had Van Zyl's head instead,[19] and Relly lost his political will over the central African savanna. Capital retreated in its fear of the excesses of the township struggles into the comfort of the state's protective security barrier.

The popular forces, on the other hand, overestimated their own power to overthrow the state. It was one thing to be able to resist the state's attempt to impose a resolution of the national question. However, it was quite another to present a coherent and viable political alternative which could win over the rest of the society to an alternative political solution to that proposed by the National Party. Riding on the crest of the wave of mass militancy, both the UDF and the ANC[20] were seduced by the slogans of 'ungovernability', 'insurrectionism' and 'people's power'. They placed inordinate stress on an insurrectionary strategy, even though it often contradicted other strategies designed to split the ruling bloc and win over significant sectors of the white population to the political ideals of a non-racial, democratic and unitary state. The strategies of mass mobilisation, international isolation, and high-profile meetings with major political and economic groupings from inside the country, projected the ANC as the political organisation able to save the nation from the economic and political crisis. However, the degeneration of the mass struggles into the organisationally ungovernable undermined this alternative hegemonic project and served instead to unite the ruling bloc firmly behind a state promising the restoration of law and order.

The insurrectionary strategy ignored the strength of the adversary, and dismissed the restructuring processes taking place within state policy as irrelevant and signifying no change at all. Instead of attempting to separate out, at least for their own purposes, those elements of reform, such as 'democratisation' and 'deracialisation', that were integral to their own struggles and required defending, the popular organisations lumped all these elements together and declared that the whole process of reform was merely 'apartheid in drag'.

Underlying this misplaced euphoria was an assumption that South African society was undergoing a period of 'dual power'.[21] Hence an insurrectionist strategy seemed most appropriate. However, this strategy fundamentally overestimated the strength of the popular classes and underestimated the strength of the state. The mass of the population had recently embarked on the process of spontaneously gaining an angry consciousness of their potential power. The insurrectionist strategy mistook this for a period when a disorganised state and capitalist class, unable to rule, were confronted with nationally consolidating real organs of alternative and countervailing popular power.

Having put forward an insurrectionist strategy based on 'dual power', the popular organisations found themselves in an impasse. In many, but not all, areas they were strong enough to challenge the local, delegated organs of state power, yet were fundamentally unable to even begin to challenge and overthrow the central organs of state power. The state, on the other hand, may have been unable to proceed successfully with its policy of localised 'co-optive domination' but it was by no means shaking on its very foundations.

By mid-1986 the state had clamped down heavily. Through the successive states of emergency it abandoned the limited process of democratisation which the first phase of reform had inaugurated and moved to crush the uprising. The first phase of reform had been shown to be inadequate. However, capital's inability to break its own political bonds of 'distance–dependency', and the popular movement's inability to formulate a credible alternative hegemonic project, left the reformulation of the state's reform project squarely in the hands of the National Party.

THE SECOND REFORM PHASE

The insurrectionary period resulted in a twofold response from the state. This was the consequence of a struggle between two antithetical positions within the state over how the next phase of reform should be structured. The struggle was essentially between two different state apparatuses – the Department of Constitutional Development and Planning (DCDP) under Heunis which until then had been leading the reform process and the Departments responsible for state security under Magnus Malan.[22]

The DCDP realised that it had to escape the premises of territorial segregation which underlay and trapped the first reform phase. It did so by shifting gear and moving away from a resolution of the national question towards restructuring the reproduction aspects of Verwoerdian apartheid's accumulation strategy. The DCDP decisively turned its back on the Riekert strategy and backed the pressures from the Urban Foundation for a new urban policy for blacks. This crystallised in the publication of the President's Council report, An Urbanisation Strategy for the Republic of South Africa (1985), which abandoned the old underlying premises of territorial segregation and accepted the interdependent, and interconnected, nature of South Africa's political economy. Its premiss was that South Africa be administered via eight interlocking, functional regional units which included or cut across bantustan borders. The purpose of state planning was to form functional economic, social and political administrative units.[23]

The report's major conclusion was that urbanisation of the African population was not only inevitable but also desirable. Hence future reform measures should operate to maximise this process for stability rather than attempting to undermine it. State policy towards influx control, housing, and employment therefore required radical revision. It was no surprise that it therefore recommended the abolition of legislative influx control measures and the pass laws – the cornerstone of Verwoerdian apartheid's accumulation strategy with regard to reproduction, and the basis on which the apartheid resolution of the national question was predicated.

The overturning of the economic and social basis of traditional apartheid held major political implications for the path that reform had been following. Now it was abstractly possible to put forward a hegemonic project which could include a place for Africans in the conception of the 'nation'. The first phase of reform had always foundered on the structural impossibility of including Africans in a political settlement since they were structurally excluded from the South African 'nation-people'. Even in circles sympathetic to the National Party's hegemonic project this was an obstacle that made nonsense of the tricameral parliament – no matter how much these ideologues contemplated various options around a fourth chamber, it was impossible to find a political place for Africans even within the framework of 'racial groups'. 'Own affairs' made no sense if there was nothing to 'own' outside of the bantustans.

In July 1986 the state set out a new policy called 'orderly urbanisation' and abolished the key Acts denying urban residential rights to Africans. These had been the basis on which influx control, the pass laws and territorial segregation had been founded. Instead of influx control legislation, it recommended using squatting legislation and state intervention concentrating on urban social engineering to control urbanisation. The point was to encourage informal settlements for the poorest layers of the working class and thereby differentiate them from those layers of the working class which could afford ordinary or upgraded township housing.

This urbanisation policy was an integral part of the accumulation strategy that the Urban Foundation, under the dominance of externally oriented capital, had been touting for some time. It fitted neatly into the policy put forward by all fractions of capital for free-market dominance in the spheres of production and reproduction. Both inwardly oriented and externally oriented capital concurred on restructuring the accumulation strategy towards government withdrawal from these spheres, agreeing on deregulation, privatisation and 'positive or orderly urbanisation'. In opposition to the Riekert strategy which had attempted to protect urban workers from rural workers, this new policy essentially proposed exacerbating intra-urban class divisions. Rather than protecting 'insiders' from 'outsiders', it proposed using urbanisation to encourage class differentiation amongst Africans: upgraded existing townships for those who could afford them, new middle-class suburbs to allow the black middle class to establish their own social milieu, and finally, squatter settlements on the peripheries of the cities to accommodate the lower strata of the black working class and the urban unemployed. By accelerating the process of class differentiation and ordering it within a racially revised social and spatial framework, the state sought to re-establish order and to lay the groundwork for economic revival.

In line with this shift, the state simultaneously initiated a contradictory process of restructuring the racially hierarchical social boundaries that so clearly constrained and characterised apartheid South Africa. Some aspects of social life, mostly revolving around racially discriminatory social amenities (petty apartheid), were clearly 'deracialised'. Black people were allowed access to a variety of social amenities that had been hitherto denied them – e.g. parks, cinemas, hotels, restaurants, pubs and beaches. Furthermore such previously sacrosanct pillars of apartheid as the Immorality and Prohibition of Mixed Marriages Acts were also abolished. At the same time, however, it embarked on a process of 're-racialisation' of other more sensitive aspects of social life, e.g. health and education, which fell under the respective 'own affairs' that tricameralism assigned to them.

Verwoerdian apartheid was based on the differentiated provision of social welfare services, with race as the basis of inclusion or exclusion. With 'orderly urbanisation' as the new basis of an accumulation strategy, the state attempted to change the terms of inclusion or exclusion to class-related criteria. Privatisation allows for the incorporation of those racially excluded or discriminated against by downgrading the state-provided social welfare functions for the poor sections of the 'nation' whilst simultaneously redirecting all other more affluent strata into privatised social services. In this way the incorporation of blacks into the same urbanised social sphere as whites could occur without presenting the state with requirements for massive expenditure increases in order to meet the new demands on state-provided collective consumption. This is fairly clear in the field of medical care where the state has been restructuring, on a new class basis, the national health-care system. The state is using privatisation as a means of excluding, through redirection to private-sector medical aids, clinics and hospitals, those citizens previously catered for by state hospital services, whilst simultaneously including blacks into a significantly downgraded hospital and primary health-care system.

Although 'orderly urbanisation' made possible a more politically inclusive strategy towards Africans, this second phase was ironically dominated by a political conception of reform which basically avoided the formulation of a hegemonic project to resolve the national question. Capital, already hesitant about putting forward an alternative resolution to the national question to that posed by the tricameral parliament, and threatened by the economic, social and political disorder which the slogan of 'ungovernability' seemed to promise, retreated into the protection of a state which offered the return of law and order.[24] Furthermore, although the shift towards 'orderly urbanisation' signified a closer link between 'free marketeerism' and the Heunis faction within the DCPD, this did not lead to a corresponding tackling of the political issues of the day. Ironically this major shift away from the accumulation strategy of Verwoerdian apartheid in this phase of reform remained politically stillborn. For the successful implementation of the state of emergency had the effect of shifting the balance of power within the state towards Magnus Malan and the security apparatus. Most businessmen, with the exception of a small grouping of domestically oriented capital in the Consultative Business Movement, avoided the political issues of the day and placed their trust in the state being able to contain the political demands for 'people's power'.

In this phase, state reform policy was increasingly determined by the 'securocrats' under Malan through the National Security Management System. The NSMS vested enormous administrative power in the hands of the security forces and operated to ensure a coordinated security and redistributive welfare intervention. The JMCs identified problems in a community and deployed expertise to upgrade township conditions in an effort to defuse their political consequences.

This system gave the military apparatus an important say in decision-making over social policy.[25] Redistribution of socio-economic infrastructural resources became the primary objective, not the political process of negotiation. In a classic 'corporatist' solution, the underlying theme was that the provision of social services would result in the legitimacy required, as long as all other forces were controlled. Hence the dual stress on security and township upgrading and the absence of any attempt to reformulate a hegemonic project in line with the shifting accumulation strategy set out in the policy of 'orderly urbanisation'.[26] The 'corporatists' within the state argued that it was unnecessary to resolve the national question through reconstituting the 'nation-people' since political demands could be deflected by satisfying socio-economic demands. In Malan's words, 'there is presently only a limited section [of black people] which is really interested in political participation. I think for the masses in South Africa democracy is not a relevant factor.'[27]

In terms of state policy, the first reform phase was characterised by an inadequate accumulation strategy and the inability to formulate a coherent hegemonic project to resolve the national question, whilst the second phase was characterised by a radical revision of the accumulation strategy underlying Verwoerdian apartheid, coupled with the total denial of the need to put forward a corresponding restructuring of the national question.

THE THIRD REFORM PHASE

Although Malan's corporatism was intended to be a wide-ranging redistributionist strategy, in practice it operated essentially as a security strategy. In this respect it worked effectively for the limited period of time it was implemented. By 1989 the state had not only crushed the insurrectionary mood of the mid-1980s, it had also successfully eliminated insurrectionary conceptions from the discourse of the popular forces, substantially weakened their organisational strength and placed the trade-union movement on the defensive. However, this policy's short-term success in containing the insurrection was also its long-term downfall. With the fear of 'ungovernability' no longer a serious political issue, it became patently clear to capital (and many other whites) that the strategic vision of this security-centred corporatism had no answer for the two central problems of the crisis. It posed no resolution of the national question, since it denied the existence of the problem; and, in the absence of a strategy to overcome South African capital's increasing international economic isolation, it had no clear long-term solution to the economic crisis that bedevilled capital.

By the end of the decade it was clear that capital's panic over the immediate effects of international isolation in the form of sanctions had been overhasty. After all, the economy had not collapsed, it had survived the debt crisis, masses of black workers had not been rendered unemployed, exports were still possible albeit at a premium, and imports were available even if at escalated prices. The hysteria over the short-term effects lifted, but this only served to bring home the long-term reality of the severe economic crisis facing the country. Although the internationally oriented sections of capital were most severely threatened, all capitals suffered from the economic crisis, and none could easily claim exemption from the bleakness of a future dominated by increasing international economic isolation.

This crisis for capital in general produced a deep, almost subliminal, unity centering around the need for long-term political stability. However, the old problem of capital's inability to create a mass political party to exert influence in a direct way still existed. Consequently, even before De Klerk became president, indeed even when he was still 'crown prince in waiting', organised capital began pinning their colours to his mast as the only viable alternative. Whilst Anglo American backed Zach de Beer within the Democratic Party, they seemed to do so more as a means to pull De Klerk closer to their preferred options than with any great faith in the Democratic Party as an electoral alternative.[28]

De Klerk responded to these overtures by reorienting the National Party closer to capital and starting to espouse the political equivalents of 'free marketeerism'. For the first time in a decade of reform, the governing Party started to pose both an alternative accumulation strategy and a credible resolution of the national question. Faced with the choice of remaining forever locked in the stasis of P. W. Botha or shrugging off the ideological weight of the past, De Klerk surprised everyone by grasping the nettle. To the horror of his erstwhile allies and the confusion of his own supporters, the discourse of National Party policy was suddenly shrouded in an entirely new language: 'negotiation', 'democracy', 'non-racialism', 'free choice', 'one nation', 'bills of rights', 'minority rights', 'freedom of association', 'elimination of ethnicity and race as the basis of group protection', etc. De Klerk caught the popular movement off balance by allowing marches and protests, curbing the police, scrapping the NSMS, and releasing most of the Rivonia trialists. Then before they had recovered, he changed the configuration of South African politics forever by unbanning the ANC and SACP, releasing Mandela, committing the government to unconditional negotiations over a future constitution embodying democracy and equality of the franchise, whilst at the same time reaffirming the economic strategy of privatisation, deregulation and defence of free marketeerism.

In a relatively short space of time De Klerk has acted to overcome capital's historical incoherence by reorganising its interests and fears from within the confines and ideology of the National Party. As a result, the National Party has become the organising and structuring political mechanism for capital, posing a resolution of the national question totally out of character with Afrikanerdom's ideological and political past and certainly at odds with apartheid. For most observers, not to mention those involved in the heat of political struggle, such a radical shift in the National Party policy was almost unthinkable. After all, most commentators had based their analysis of the South African state on its rigidity, militarism, the overwhelming dominance of the 'securocrats', and the controlling power of a security apparatus operating as 'a state within a state'.

Those who saw a 'securocrat' behind every state intervention were startled by De Klerk's rapid jettisoning of the central military core of the state. Such a rapid structural shift in the form of the state could not be accounted for by the militarisation thesis.[29] It overemphasised the dominance of the military and ignored the institutional flexibility inherent in the racially exclusive bourgeois democratic state. In its zeal to declare a 'coup by subterfuge', the militarisation thesis underemphasised the theoretical significance of what lay behind, and made possible, the dominance of the NSMS, i.e. the structural change that had occurred in the form of the state and the governing party during the 1980s.

P. W. Botha, whose rise within the ranks was primarily due to his being the ultimate party man, ironically presided over the demise of the democratic power vested in the federal structure of the National Party. Instead he replaced it with a political party which gave lip service to all the rituals and ideological trappings inherited from the past but which in fact was totally dominated by the centralised exercise of executive power. The National Party was transformed from a democratic, federally based party to a heavily centralised executive party. Likewise with respect to the state, Botha used the new tricameral constitution, the insurrectionist threat and the state of emergency to shift power away from the parliamentary apparatus towards the executive controlled by the president.[30] This resulted in an inordinate degree of party and state power being concentrated in the office of the State President.

De Klerk has used the transformation of the National Party into an executive party and the centralisation of power within the executive to bring about radical changes in the name of a party and state that were historically structured to oppose those very changes.[31] In this way De Klerk was able to be of the National Party but not to represent the historical interests of that party at one and the same moment. Journalistic orthodoxy has counterposed the securocrats with party democracy and thus concluded that De Klerk has returned power to the caucus and parliament. This is very far from the truth of the matter. Instead, he has shifted power from one executive mechanism to another, whilst still retaining the office of the State President as the fundamental decision-making centre. Instead of generals, the key inner decision-making group is now a small coterie of ministers and officials. If indeed he had returned power to the caucus and parliament he would never have been able to do what he has done. The parliamentary caucus and constituency branches of the National Party could never have operated in total defiance of its own historical legacy, which is why De Klerk never informed the National Party parliamentary caucus of the 2 February decisions until immediately after his speech at a special emergency caucus meeting. Indeed the cabinet as a whole was only informed of the contents of the speech the day before.[31] He could only adopt the interests of capital in general and appropriate the policies of the Democratic Party by using his executive power to seemingly stand above political groupings and yet institute these appropriated changes in the name (albeit hypocritically) of the party of Afrikaner nationalism and apartheid.

CONCLUSION

De Klerk's restructuring of the political agenda has meant that 'the central question will become not whether South Africa is going to be a one nation state but what kind of one-nation state are we going to become' (Van Zyl Slabbert, 1989).

The government's commitment to a negotiated settlement and the restructuring of South African political life have laid the basis for a number of different paths of economic development and political stabilisation. This makes possible what was always very difficult to achieve under the previous regimes of state reform – the adoption of a coherent accumulation strategy and resolution of the national question. However, the existence of a unitary nation state does not necessarily imply the pursuit of policies of equitable treatment for all its citizens. For hegemony can be secured via 'one-nation' strategies which attempt to unify and materially accommodate the mass of the nation-people or via 'two-nation' strategies which fracture society and are materially aimed only at the most privileged strata.

The crisis has changed the face of South Africa's political economy irrevocably. At present the organisations of the dominant classes are responding to the changing needs of the game more rapidly and flexibly than those of the popular classes. The former have finally, after a decade of experimentation, formulated the basics of an accumulation strategy and hegemonic project to resolve the national question and restore growth on terms most favourable to themselves. The onus now rests heavily on the organisations of the dominated classes, in particular the African National Congress, to respond flexibly to the new political conditions. For decades, the ANC have insisted that their struggle is one of national liberation of the oppressed black majority. That was perfectly true but the problem they face now is how to formulate an appropriate alternative social vision that is perceived to encompass the totality of South African society. The ANC have to resolve the national question by projecting themselves as the saviour and creator of the whole nation, and not, as they will be portrayed, as merely the principal representatives of the black majority at the negotiating table. The nation that has to be built now out of the divisiveness of apartheid is a truly non-racial one, encompassing large sections of the white community that previously would not normally have gravitated towards the ANC.

The growth model that is currently dominant within capital, as well as the National and Democratic parties, is one that we have termed the '50 per cent solution'.[33] The basic terms of this model are political settlement and growth. It proposes to solve the economic crisis by focusing on the issues of growth rather than those of redistribution. It specifically eschews radical state intervention to redistribute social and economic resources on a non-economic basis, and puts its reliance instead solely on the operations of the free market. According to this growth model, a political settlement will result in a rapid inflow of foreign investment, and, if coupled with minimal state intervention and radical free-marketeerism, the consequent effect will be high rates of growth. Redistribution will follow as a trickle-down effect of accelerated growth so that, in the long run, if we let the rich get richer the poor will get richer also, which is why it can best be summed up as 'redistribution through growth'.

The social effects of this model will be to accelerate the tendencies already taking place in the occupational division of labour[34] and economically incorporate those blacks at the upper end of the class ladder. These are the black bourgeoisie primarily based in commerce, the professional and managerial middle class, the small traders, the new skilled, clerical and supervisory black working class, and finally, although to a lesser extent, the better-paid strata of the semi-skilled black working class. The rest of the black population living on the periphery of the mass consumption economy are thus to be left to fend for themselves.

This would go hand in hand with a political strategy that resolves the national question with the maximum accommodation to demands for majority rule but with the minimum of accommodation to demands for major restructuring of the economic and social fabric of the society. The thrust of this strategy would be to establish hegemonic consent by politically appealing to the material interests of those classes amongst the black population who are to become the greatest beneficiaries of a '50 per cent solution'. Those on the other side of the divide would be symbolically accommodated but their material needs would not be systematically catered for.

This strategy is designed to exacerbate class differentiation amongst blacks and to polarise the upper strata away from the mass of employed workers (semi-skilled and unskilled) as well as the unemployed and marginalised. It is essentially still a 'two-nations' strategy but one where race is no longer the basis of differentiation.

An alternative growth model to the '50 per cent solution' will have to set out policies that tackle both the economic and the political crisis. This entails putting forward a 'one-nation' strategy (in effect, a 100 per cent solution) which sets out an accumulation strategy that aims to deal simultaneously with the problems of growth and redistribution as well as a hegemonic project aimed at resolving the national question. Solving the problem of growth would thus be directly linked to the significant redistribution of resources towards the majority of the population. Such a policy places the emphasis on redistribution as the mechanism to regenerate economic growth. Economic growth would be stimulated through the redistribution of productive capacity and capability towards the more marginalised sectors of the population. This would entail a systematic political commitment to welfare policies, redistributive mechanisms and urban–rural reconstruction, such as residential infrastructure, housing, electrification and telecommunication programmes.

A 'one-nation' strategy to resolve the crisis and inaugurate a period of social stability and renewed accumulation will therefore have to couple the need for state-stimulated economic restructuring with a negotiated political settlement. To focus only on resolving the national question is to ignore the factor that so forcibly threw that question onto the stage of history in the 1970s – the economic crisis. On the other hand, to deal only with the growth aspects of the economic crisis and not the plight of the urban poor, unemployed and marginalised is to abandon the hope of successfully resolving the national question through the realisation of the slogan 'one nation, one people'.

3

The crisis and South Africa's balance of payments

BRIAN KAHN

Because the South African economy is closely linked to the international economy through trade, finance and investment, the current crisis in the economy is mirrored in the balance of payments, which itself is an index of South Africa's insertion in the international division of labour. The internal structure of the balance of payments (that is, the nature and movements of both the current and capital accounts) responds to changes and crises in both the international and the domestic modes of regulation. Thus the extent to which the domestic crisis has affected the balance of payments is affected in turn by the changing nature of the international monetary system.

This chapter analyses developments in the balance of payments since 1960 and highlights not only the domestic and international processes which affect the balance of payments but also the effects that the balance of payments adjustment process has had on the economy. The plan of the chapter is as follows. Section 1 outlines the changing structure of the international mode of regulation and its effects on the balance of payments. The changing nature of the adjustment process is also examined. Section 2 analyses the overall structure of South Africa's balance of payments and emphasises its vulnerability to both exogenous and endogenous shocks. Section 3 examines in greater detail the different components of the balance of payments accounts, whilst developments in South Africa's exchange-rate policies are analysed in section 4.

THE CHANGING INTERNATIONAL MODE OF REGULATION

The beginning of the 1970s saw the end of the reproductive boom in South Africa. This coincided with the period of 'mutational crisis'[1] in the international economy after the breakdown of the Bretton Woods system. Under this system each country pegged its currency to the dollar, which in turn was convertible into gold at a fixed price of $35 per ounce. This meant that the international economy operated on a fixed exchange-rate system, and exchange-rate policy was directed at maintaining a fixed link with the dollar. The role of central banks was to buy up dollars when there were excess supplies and sell dollars during periods of excess demands. The dollar was the major reserve currency by means of which international debts were settled.

During the 1960s, the supposedly 'clear-cut and stabilising rules' of the international monetary game were progressively disrupted, and US industrial hegemony, which formed the basis of the stability of the international order after the Second World War, also disintegrated. In addition, the financing of the war in Vietnam created a massive increase in the world supply of dollars, and this made the fixed exchange-rate system increasingly fragile. This monetary order was replaced in the 1970s by a more ambiguous and less stabilising system. Not only had US industrial supremacy declined, but the international monetary system had changed in two fundamental ways. Firstly, flexible exchange rates were introduced and, secondly, there emerged what Aglietta calls the 'international debt economy'.[2] Together these factors altered the nature of the 'international monetary constraint', that is, the constraint imposed by the international monetary system on national economies.

The international monetary constraint refers to the fact that current account deficits have to be settled over time. The current account is the net flow of imports and exports of goods and services. By definition, a current account deficit indicates an excess of domestic expenditure over domestic output, or equivalently the excess of domestic investment over domestic savings. In other words, current account deficits are an indication of increased investment or increased consumption (decreased savings) or both. The financing of a deficit can be achieved either through reducing international reserves or through the inflow of foreign capital, which in effect supplements domestic savings. The latter option allows for a delay in current account corrections and therefore relaxes the constraint on growth. A correction of a current account deficit requires deflationary policies, which then reduce consumption or investment expenditure or both in the domestic economy.

Under the Bretton Woods system the international monetary constraint operated in a 'very straightforward ... way through adjustments in balances of payments. Central Banks had to settle deficits in dollars by drawing from their dollar reserves. The subsequent diminution in the reserves restricted internal macro-economic policy, which led in turn to absorption of the deficit' (De Vroey, 1984: 61). However, with the emergence of the private international credit market in the form of the Eurocurrency markets, the discipline imposed by the previous regime was relaxed as adjustments could be made to balance of payments disequilibria by means of debt accumulation, which allowed current account deficits to persist for a longer time. As a result, the international monetary system has become more fragile, as evidenced by general debt consolidation and the consequent international debt crises. A general financial crisis has only been averted because of the flexibility shown by the private banks in debt rescheduling. But, as we argue below, the lesser degree of flexibility shown by the banks to the South African debt situation in effect tightened the constraint on the economy at a time when expenditure and investment decisions were being made under the revised rules of the game.

In addition to providing greater access to foreign capital, the new system also changed the nature of the adjustment process as a result of the adoption of flexible exchange rates. Under such a regime, a depreciating currency is supposed to bring about the required deflation through its expenditure-reducing and expenditure-switching effects. The former refers to the fact that a depreciation reduces real incomes through its effects on the price level, and the latter to the fact that because domestic goods are now relatively cheaper than foreign goods, more domestic goods will be consumed both domestically and internationally. The resulting increase in exports and decline in imports helps to bring about an improvement in the current account.

The move to flexible exchange rates has also had implications for primary commodity prices. Of particular importance from the point of view of South Africa's balance of payments was the fact that under the new system, gold was no longer fixed in price. Whilst this represented a boost for the South African mining industry in particular and the economy in general, the volatile nature of the gold price ensured that the economy and the balance of payments had to adjust occasionally to the unpredictable and sometimes sharp swings in the gold price. At the same time, increasing domestic militance and repression led periodically to capital outflows and forced the

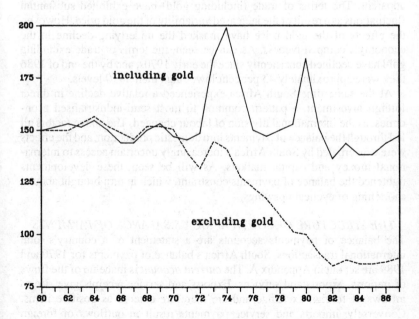

Fig. 1. South Africa's terms of trade

economy into policies that required urgent balance of payments adjustment different from those typical of the new international monetary structure.

Historically, South Africa became inserted into the international division of labour as an exporter of precious metals and raw materials and as an importer of capital, developing little technological capacity of its own. Moreover, a policy of import substitution, attractive profits from mineral resources, low wages and high profitability during the apartheid boom period encouraged high levels of direct foreign investment in the country. Such was the pattern of adjustment to cycles that an acceleration of investment expenditure and the concomitant increase in capital goods imports would eventually hit the balance of payments constraint. Concern for the level of foreign exchange reserves led to monetary restraint and higher interest rates. The consequent increase in the cost of capital slowed the rate of investment and consumption, and thereby reduced imports. The need to protect the level of reserves under fixed exchange rates was the basis of the balance of payments or international monetary constraint.

During the 1970s, at the same time that the international monetary system was undergoing changes, South Africa experienced a decline in its terms of trade as a result of higher import prices, particularly capital equipment, and fluctuating prices of primary commodity exports. As can be seen in Figure 1, minor fluctuations in South Africa's terms of trade occurred during the 1960s, but from the beginning of the 1970s significant changes became apparent. The terms of trade (including gold) have exhibited substantial fluctuations as a result of the increased variability of the gold price. However, the effects of the gold price have masked the underlying decline in the economy's competitiveness. As can be seen, the terms of trade excluding gold have declined consistently since the early 1970s, and by the end of 1986 they were approximately 43 per cent lower than their 1970 levels.

At the same time South Africa experienced a relative decline in direct foreign investment, a pattern common to most semi-industrialised economies, as the international division of labour changed. The above factors all fed through the balance of payments into domestic production, and the effects were exacerbated by South Africa's increasingly uncertain access to international money and capital markets. As will be seen, these developments tightened the balance of payments constraint, which in turn brought about a shortening of cyclical upswings.

THE STRUCTURE OF SOUTH AFRICA'S BALANCE OF PAYMENTS

The balance of payments accounts are a statement of a country's total international transactions. South Africa's balance of payments for 1970 and 1989 are set out in Appendix A. The *current account* is made up of the flows of imports, exports and services. Exports and service receipts give rise to inflows of foreign exchange and are therefore entered as positive items. Conversely, imports and service payments result in outflows of foreign exchange and are entered as negative items. The balance on the current

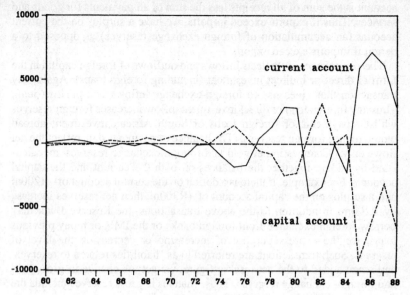

Fig. 2. Current and capital accounts of the balance of payments (Rm)

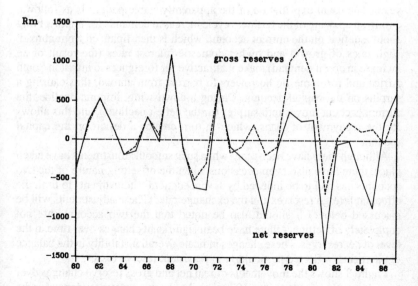

Fig. 3. Net and gross reserve changes (constant 1975 prices)

account is the sum of all receipts less the sum of all payments for goods and services. Thus if exports exceed imports, we have a surplus on the current account (an accumulation of foreign-exchange reserves) as opposed to a deficit if imports exceed exports.

The *capital account* reflects inflows and outflows of foreign capital in the form of direct or indirect investment (including foreign loans). Again, any transaction that gives rise to foreign-exchange inflows is a positive item. Thus any foreign loan or direct investment inflow increases foreign reserves whilst repayments of foreign loans or South African investment abroad would reduce foreign reserves. (The balance of payments data reflect the net flows only.) Changes in net gold and foreign-exchange reserves are calculated by adding together the balances on both the current and the capital accounts. For example, if there is a deficit on the current account of -R200m and a surplus on the capital account of +R300m, then net reserves increase by R100m. In addition to the above transactions, the Reserve Bank may borrow foreign exchange from foreign banks or the IMF, or repay previous loans, for the express purpose of increasing or decreasing the level of reserves. Such transactions are referred to as 'liabilities related to reserves' and when added to the net reserve figure, we get the change in gross gold and foreign- exchange reserves. Thus, changes in gross reserves include the Reserve Bank's policy reaction to net reserve changes.

South Africa's balance of payments has been characterised by cyclical features, which exhibit an apparently stable structure over time. Figure 2 shows the movements of the current and capital accounts over the past 20 years. The usual explanation of the apparently stable pattern is as follows: during an upswing of the business cycle increased demands for imports bring about a deficit on the current account, which is then financed from abroad. High rates of growth and higher domestic interest rates (the result of an increase in credit demand) make it attractive for foreigners to invest in South Africa and for domestic borrowers to borrow from abroad, thus causing a surplus on the capital account. During a downswing, imports decline, the current account moves into surplus (or the deficit declines), and this allows for the repayment of loans, which in turn creates a deficit on the capital account.

Although there have been times when such smooth adjustments have taken place, there have also been occasions when the offsetting nature of the two accounts has had to be induced by severe economic adjustments to limit the effect on foreign reserves and the exchange rate. (These adjustments will be discussed below.) It should also be noted that the two accounts are not completely offsetting – there have been significant changes over time in the level of net reserves. These changes indicate overall instability in the balance of payments, reflected particularly in persistent reserve losses.

Figure 3 shows the movements of real net and gross reserve changes over time. It can be seen that since 1970 there has been a greater tendency for the changes in net reserves to be negative, and the extent of these changes has been much greater than in the 1960s. It is clear that domestic developments brought about a secular decline in reserves and a greater tendency for crises to occur in the balance of payments, reflecting the general crisis in the economy. It can also be seen from the movements of gross reserves that whereas there was little need in the 1960s for the authorities to intervene directly through borrowing to bolster the level of reserves, such actions were needed consistently throughout the flexible exchange-rate period, despite the fact that exchange-rate flexibility itself is supposed to create lower fluctuations of reserves.

As mentioned, there were times when the apparently stable pattern had to be induced. These adjustments were necessary because of the susceptibility of South Africa's balance of payments to external 'shocks' (unexpected events over which the monetary authorities have no control) that have their roots in both the economic and the political structure of the economy. On the economic front, the economy's dependence on gold makes the current account prone to shocks that arise from the volatile nature of the gold price (which is determined in the international markets). The political dimension also renders the capital account sensitive to shocks, as outbreaks of violence and increased repression are usually followed by withdrawal of foreign loans and investment. These shocks to the capital account in turn have implications for the financing of current account deficits.

Adjustment to shocks

It is important at this stage to discuss how the balance of payments adjusts to the several shocks to which it is prone. The current account shocks are somewhat easier to adjust to than capital account shocks. Consider initially a current account that has been in balance over the medium term. A sudden fall in the gold price can be dealt with in one of the following ways.

(1) If the exchange rate is not flexible, adjustment can be made to the ensuing current account deficit by reducing imports directly through deflationary policies which reduce consumption and investment. This adjustment often occurs automatically because to the extent that the decline in the gold price is expected to be of long duration, there will be a decline in investment and hence a decline in imports. Increased availability of foreign loan capital in the 1970s lessened the immediate need for deflationary adjustments, but such an option is only available if there is easy access to foreign capital markets.

(2) The need for foreign borrowing can be also reduced through depreciation of the rand which, as described earlier, would effectively cut down on imports by making them more expensive in rand terms and at the same time act as a stimulus to exports by making them cheaper in foreign currency terms.[3]

Both of these techniques – deflation and depreciation – are designed to reduce a country's international expenditures. Deflation results in increased unemployment and downward pressure on real wages whilst depreciation also reduces real wages through its effect on inflation. In short, the adjustment works through its effect on the level of employment and the level of income.

The sensitivity of capital flows to political crises has made the capital account extremely vulnerable to shocks, and these tend to require even more severe adjustment than in the case of the current account. Such shocks occurred in the aftermath of Sharpeville (1960), Soweto (1976) and again in 1985 following events in the Vaal Triangle. Events like these had the effect of restricting South Africa's access to international money markets and at the same time resulted in 'capital flight' (large outflows of capital in response to political or economic uncertainty), which could not be offset by recourse to increased borrowing. When a capital account shock involving restricted access to new loans occurs, the deficit on the current account cannot be financed by new loans. Rather, it will have to be overcome by increasing domestic savings, that is, bringing about a current account surplus through reductions in consumption and investment. This requires a policy of deflation that reduces incomes and expenditure, employment and imports.

In 1977 not only was there a large outflow of capital but no new loans were forthcoming. This provoked strenuous deflationary policies on the part of the monetary and fiscal authorities, and led to real domestic fixed investment declining by over 10 per cent between 1976 and 1979. Real consumption expenditure fell by less than 1 per cent in 1977. As a result, a cumulative current account surplus of almost R5 billion was generated during 1977–9, which helped to finance the capital account deficit of R3.1 billion over that period. During this period the exchange rate was fixed to the dollar and thus the entire adjustment had to take place through adjustment to the current account via deflationary policies. The effects of the recession would have been even more long-lasting had the gold price not begun fortuitously to increase in a dramatic way in 1979. This took the pressure off the current account, and the increased investor confidence which it brought in its train helped re-open the doors to South African borrowers abroad.

The 1985 shock was more severe in that not only were no new loans forthcoming but existing loans were recalled on a large scale and suddenly. In this case the adjustment was brought about partly through deflation – although it was induced more so by the collapse of local investor confidence than by deliberate government policy – and partly through depreciation, which saw the rand collapsing to 35 US cents. Despite the precipitous drop, it was insufficient to protect the level of foreign-exchange reserves, and the fiscal authorities declared a moratorium on all debt repayment in order to stem the outflow of capital. The effects of a currency depreciation will be discussed more fully below, but it should be noted in this context that although a depreciation may improve the current account, it may also have a detrimental effect on the economy. This is partly because of the contractionary effects of depreciations,[4] and also because depreciation does not reduce the dollar value of outstanding debt but does increase the rand value of the debt denominated in dollars. What this all implies is that more resources have to be surrendered by the economy in order to repay a given amount of debt. The combined effect of depreciation and recession caused a dramatic turn-around on the current account: after a cumulative deficit of R9 732m in 1981–4, it experienced surpluses of R5 925m, R7 196m and R6 152m in 1985–7. During this period real domestic fixed investment fell by almost 24 per cent, whilst real consumption expenditure fell by 3 per cent between 1985 and 1986, before starting to rise again in 1987.

Our analysis underlines several important observations about South Africa's balance of payments. Firstly, both the current and capital accounts are extremely vulnerable to sudden shocks, and the adjustment to these shocks can impose severe hardships on the economy. Secondly, we can see the dilemma that faces the authorities at a time of capital account shocks. The need to repay capital requires a current account surplus but at the same time the desire to stimulate the economy and improve the unemployment position would tend to contradict this, for if such policies were successful they would bring about increased imports. Thirdly, these severe shocks have shaped the pattern of the current and capital accounts moving in opposite directions. The only periods in which South Africa experienced sustained current account surpluses have been in the aftermath of political crises. In other words, the general tendency to run sustained current account deficits is reversed following capital account shocks.

Finally, it should be emphasised that because a current account surplus reflects an excess of saving over investment, in the absence of foreign capital inflows such a surplus can be achieved either through an increase in domestic savings (a decline in consumption) or through decreased investment. In South Africa's case it is clear that the major determinant of current account surpluses from the mid-1970s has been declines in real gross domestic fixed investment. Real consumption expenditure has only fallen marginally in the wake of political shocks (although the rate of increase has declined). This relationship will be analysed in the next section.

THE COMPONENTS OF THE BALANCE OF PAYMENTS

The current account

Imports

As mentioned above, imports move strongly pro-cyclically. As incomes and economic activity increase, expansion of output requires the importation of capital equipment, because of South Africa's dependence on imported machinery. There has been a marked tendency over the past few decades for the capital intensity of production to increase, signalling a growing reliance on imported capital goods. The Report of the Study Group on Industrial Development Strategy (Kleu Report) has highlighted this capital deepening of production, noting that the capital–labour ratio has been rising since 1946, particularly in the manufacturing sector. At the same time, the inability of the manufacturing sector to expand capital goods production increases the dependence on imported capital goods. This dependence is clearly illustrated in Table 1, which shows the import penetration ratios (i.p.r.) for different categories of manufactured goods. The i.p.r. is the ratio of total imports to domestic consumption, and is calculated as M/(Pr+M−X), where M and X are the rand values of imports and exports respectively and Pr is the value of gross domestic sales in the different sectors.

Table 1. Import penetration ratios (SIC categories)

	1965	1970	1975	1980	1985
Food	9.7	11.3	12.7	6.0	7.7
Beverages & tobacco	4.5	5.3	4.0	2.4	4.9
Textiles	37.8	30.2	20.8	15.8	15.8
Clothing	10.8	14.6	10.1	6.7	7.2
Footwear	3.4	8.4	10.5	8.6	10.4
Wood & wood products	25.0	19.7	18.7	12.0	9.3
Paper & paper prods	23.4	24.3	17.9	16.4	13.6
Chemicals	25.0	25.2	16.5	15.1	15.1
Metals & metal prods	21.1	17.1	16.5	7.0	11.1
Non-metal mineral prods	22.8	17.7	12.6	6.3	20.0
Rubber products	21.4	20.2	19.3	22.8	20.6
Machinery	50.3	57.0	52.3	50.1	52.1
Motor vehicles & transport eqpt	37.1	39.2	34.5	31.4	30.0

Source: Kahn (1987b).

The most significant feature revealed by the data is that import penetration of machinery is at a high level and has shown no tendency to decline over the past two decades. According to these figures, over half of South Africa's domestic expenditure on machinery is devoted to imported machinery. It should also be borne in mind that this figure is an understatement of the true picture as the value of sales reflects the final selling price of goods whereas much of the locally produced machinery could contain a large proportion of imported inputs. The true value of domestic production or sales would then be overstated. Moreover, within the broad category of machinery, there is almost complete reliance on certain categories of more sophisticated machinery.

The other major category with a high import penetration ratio is motor vehicles and transport equipment. Here there has been a slight downward trend over the past two decades but, given the high degree of protection afforded this sector, a sharper fall could have been expected. Most other sectors have shown declining or fairly constant ratios, except for footwear, which rose from 3.4 per cent in 1965 to 10.4 per cent in 1985 after reaching 15.6 per cent in 1984. Other sectors with relatively high i.p.r.s are textiles, chemicals, non-metallic mineral products and rubber products.[5]

Table 2 shows the value of imports for the three International Standard

Industrial Classification (ISIC) sectors of agriculture, mining and quarrying, and manufacturing. Also included in the table is a category designated as 'other' by the Central Statistical Service. The major components of this category are petroleum and armaments imports.

Table 2.   Import values by ISIC classification (Rm)

	Mining	Manuf	Agric	Other
1968	106.0	1 699.4	45.2	29.8
1970	93.5	2 343.2	67.7	74.3
1972	181.4	2 546.3	56.7	242.8
1974	114.2	4 629.1	119.7	881.1
1976	136.6	5 547.8	139.1	1 613.8
1978	101.4	5 971.4	159.7	1 786.5
1979	130.9	6 647.1	197.6	2 929.1
1980	223.4	9 638.9	228.1	4 290.9
1981	212.9	13 242.6	324.0	4 650.3
1982	194.5	13 423.3	303.0	4 443.9
1983	197.0	12 260.0	550.3	3 216.7
1984	305.2	17 167.6	997.3	3 165.7
1985	469.1	18 214.0	578.4	3 429.1
1986	460.6	20 397.8	98.6	5 306.5
1987	365.4	23 221.8	613.3	4 472.1

Source: *South African Quarterly Bulletin of Statistics*, various issues.

It is clear from the table that the manufacturing sector is the most import-intensive of the sectors, but the unclassified category of arms and oil has risen significantly since the oil crisis of 1973 and increased arms build-up. Whereas this category accounted for 2.9 per cent of imports in 1970, it had risen to its peak of 29.8 per cent in 1980 and stood at 19.8 per cent in 1986. The relative decline of this category is probably attributable to the coming on stream of Sasol 2 in the early 1980s. The effects of the drought in 1983–4 can be seen in the increase of agricultural imports, from 1.6 per cent of imports in 1982 to 4.6 per cent in 1984. In 1983 and 1984, maize imports alone accounted for 1.3 per cent and 2.5 per cent of total imports respectively. The largest import category is machinery which in 1985 accounted for 30 per cent or 49 per cent when taken with oil and arms. In 1980 this figure was 56 per cent. The other major import categories are motor vehicles and transport equipment, chemicals and textiles.

Table 3 shows the percentage of imports by broad economic categories for 1980–5.

Import demand elasticities have also been calculated for manufacturing, agriculture and the sub-sectors of chemicals, and machinery and transport equipment. The methodology behind these calculations is set out in Kahn, 1987b. Elasticities could only be calculated for these limited categories because indices of import unit volumes and values were not available. Table 4 summarises the results.

Table 3. Imports by broad economic categories (% of total imports)

	1980	1981	1982	1983	1984	1985
Food and drink	2.5	3.0	2.5	4.2	4.8	4.6
Industrial supplies:						
primary	2.8	2.4	2.6	3.9	5.4	3.9
processed	20.3	20.0	20.4	22.1	23.2	24.0
Capital goods (excl. motor)						
cap goods	20.2	21.4	20.9	20.5	21.8	20.7
parts & access.	5.9	6.3	7.0	6.9	7.5	9.0
Motor vehicles & transport equipment	12.2	14.4	15.3	14.0	13.6	11.5
Consumer goods	6.3	7.4	7.5	8.7	9.2	7.6
Unclassified (arms & oil)	29.7	25.0	23.7	19.8	14.5	18.7

Source: South African Quarterly Bulletin of Statistics, 1988.

Table 4. Price and expenditure elasticities of import demands

	Price elasticity	Expenditure elasticity
Manufacturing	1.15	2.16
Agriculture	0.79	0.19
Chemicals	1.37	0.70
Machinery and transport equip.	0.14	2.96

Source: Kahn (1987b).

Table 4 shows that the responsiveness of manufacturing imports as a whole to relative price changes is elastic – in this case a 1 per cent increase in South African prices relative to import prices will result in a 1.15 per cent increase in the quantity of manufacturing imports. The expenditure elasticity describes the responsiveness of imports to domestic expenditure changes. Thus, for every 1 per cent increase in domestic expenditure, there will be a 2.16 per cent increase in manufacturing imports. This result is consistent with the cyclical nature of imports described above.

Agriculture, on the other hand, has a low price elasticity and an even lower expenditure elasticity, since most agricultural imports are necessities (e.g. maize) and not determined by the state of the business cycle. The results for machinery and transport equipment again confirm South Africa's dependence on imported technology and the cyclical nature of such demands. The income elasticity is high at 2.96 whereas the price elasticity is a low 0.14.

The importance of capital goods and intermediate goods imports explains why adjustments to the current account are transmitted predominantly through changes in domestic fixed investment. This is illustrated clearly in Figure 4, which plots the real value of imports against real consumption expenditure and real gross domestic fixed investment. Real consumption

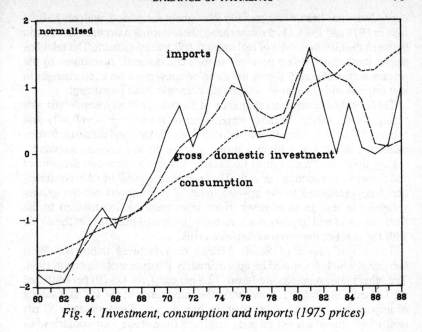

Fig. 4. Investment, consumption and imports (1975 prices)

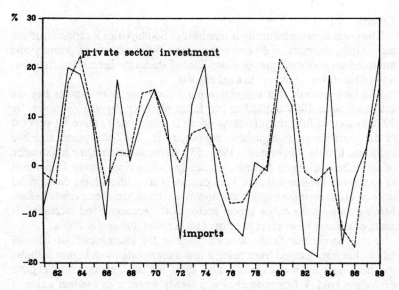

Fig. 5. Changes in imports and private investment (1975 prices)

expenditure has been rising steadily throughout the period, with only slight falls in 1978 and 1984. On the other hand, there is a much stronger correlation between the turning-points of real imports and real investment. The relationship is particularly strong between imports and domestic investment by the private sector. Figure 5 shows the close co-movement between changes in real imports and changes in real private domestic fixed investment.

This clearly demonstrates the nature of the balance of payments constraint facing South Africa. Because imports move strongly pro-cyclically and because there is a strong relationship between imports and domestic investment, any attempt to achieve a strong positive current account surplus by cutting down on imports must necessarily involve a decline in investment and therefore in economic growth. The consistent upward trend in consumption has contributed to the upward trend of real imports but the greater volatility in real gross domestic fixed investment has contributed to the fluctuations in real imports. Such volatility increased markedly in the 1970s with the onset of the current economic crisis.

The major source of South African manufactured imports is West Germany, which accounted for approximately 19 per cent of imports in 1984. The other major suppliers are Japan (12.9 per cent), the US (10 per cent) and the United Kingdom (11.2 per cent). The importance of the UK as a source of imports has been declining over the past decade. Approximately 52 per cent of all manufactured imports originate from these four countries, although Asia is becoming an increasingly significant source, having risen in importance from 3 per cent in 1976 to 8 per cent in 1988.

Exports

The crisis in manufacturing is manifest in South Africa's export structure and despite attempts to diversify the export base, exports of primary and intermediate goods still predominate. Table 5 shows the percentage contribution of the different sectors to total exports.

The table reveals that manufacturing's contribution to exports has not increased since 1968 and in fact had fallen till 1985. As Van Zyl notes, 'in 1970 exports of fully manufactured goods ... constituted a mere 3.1 per cent of the gross value of manufacturing output. By 1982 this percentage had increased to only 3.3 per cent' (1984: 53). At the same time, the importance of agriculture has been declining markedly, falling from almost 15 per cent of exports in 1968 to just over 3 per cent in 1986. Although the drought did have an effect on these exports, the downward trend had been evident before. Mining remains the major export sector of the economy and increased its share of exports to an average of over 60 per cent during the 1980s.

It is apparent that South Africa's place in the international division of labour has not changed from being a raw material supplier. Figure 6 shows the behaviour of real exports since 1960 for total exports and exports excluding gold. Whereas there was a steady increase in the real value of exports in the 1960s, the real value of exports began to fluctuate more

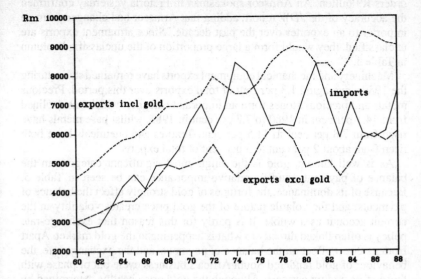

Figure 6. Imports and exports (1975 prices)

markedly in the 1970s, and non-gold exports have exhibited a downward trend since the mid-1970s.

Table 5. Percentage export contributions by major ISIC sectors

	Mining	Gold	Total mining	Agric.	Manuf.	Unclass.
1968	13.0	33.7	46.7	14.8	37.3	1.1
1972	10.1	34.3	44.3	11.8	37.2	6.3
1976	14.0	32.4	46.4	8.4	35.2	9.9
1980	12.4	50.9	63.3	5.2	28.4	3.0
1981	13.8	45.8	59.6	6.4	29.7	4.1
1982	15.6	45.0	60.6	6.0	29.0	4.4
1983	13.8	48.2	62.0	4.0	31.1	2.9
1984	15.5	44.6	60.1	2.9	33.8	3.2
1985	18.2	42.3	60.5	2.9	33.1	3.2
1986	17.3	39.8	57.1	3.2	37.4	2.4
1987	15.8	41.2	57.0	3.6	37.2	2.2

Source: Adapted from *Quarterly Bulletin of Statistics* and *SARB Quarterly Bulletins*. As these are ISIC categories, certain processed agricultural goods are included in the manufacturing sector.

An expanding area of exports is armaments. As the *Cape Times* (21 January 1988) quoting *Jane's Defence Weekly* reported, 'Armscor is now the largest single exporter of manufactured goods in South Africa, with sales to 23 countries valued in 1987 at R1,8 billion ... and the current backlog of orders R9 billion. An Armscor spokesman in Pretoria yesterday confirmed the accuracy of the *JDW* report, adding that Armscor had changed from an importer to an exporter over the past decade.' Since armament exports are unclassified, they would form a large proportion of the unclassified column in Table 5.

Machinery and mechanical equipment exports have remained static during the 1980s, averaging 1.3 per cent of total exports over this period. Precious metals and precious stones form an important category, which has declined from 14.1 per cent in 1980 to 7.2 per cent in 1985, whilst base metals have risen from 7.8 per cent to 11.5 per cent. Textiles and chemicals have both risen from about 2 per cent to 3 per cent of total exports.

As is well known, gold is the single most significant category in the balance of payments, and its relative importance can be seen in Table 5. Because of its dominance, the fortunes of gold strongly affect the balance of payments; and the volatile nature of the gold price creates volatility in the current account as a whole. It is partly for this reason that exchange-rate policy is often linked directly to what is happening in the gold market. Apart from its direct effect on the balance of payments and the exchange rate, the behaviour of gold rendered South Africa's business cycle out of phase with those of its trading partners. Although the gold price itself is determined by a variety of factors, a major variable is the direction of the dollar and the state of the US economy. A decline in the dollar has generally resulted in a stronger dollar gold price, which has fed into the South African economy through a healthier balance of payments position and increased economic activity. Because a weaker dollar generally coincides with a downswing in the US economy, South Africa's cycles are in consequence out of phase. (As mentioned before, the stronger gold price does not result in a longer-term improvement in the current account as the upswing leads to higher imports.) However, being out of phase with the US and Western European economies does generate problems for export expansion when the South African business cycle is in an upswing.

Table 6. Relative share of SA trade in imports of selected African countries

	Malawi	Zambia	Zimbabwe
1982	36.4	14.5	23.2
1983	38.5	n.a.	24.2
1984	40.3	22.8	19.3
1985	38.8	19.7	18.9
1986	30.1	23.0	21.4
1987	36.1	25.0	20.8
1988	33.6	23.4	24.9

Source: *Direction of Trade Statistics Yearbook* (1989), International Monetary Fund.

According to the International Monetary Fund's Direction of Trade Statistics, by 1988 Asia (excluding Japan) had become a major destination of South African non-gold exports, accounting for over 9 per cent of exports. The bulk of these exports go to Taiwan and Hong Kong. Other major export markets in 1988 were Japan (8 per cent), the US and West Germany (7 per cent each) and the UK (6 per cent). Although only 5 per cent of South Africa's exports went to Africa, South Africa remains an important source of imports for Malawi, Zambia and Zimbabwe, as shown in Table 6.

Factor service receipts and payments

Most analyses of the balance of payments neglect the distinction between the current account and the trade account. The latter is a component of the current account, and refers to the imports and exports of goods and non-factor services (e.g. insurance, freight charges, etc.). The current account is the trade account plus factor service receipts and payments. These services include migrant labour remittances, dividend remittances by foreign companies operating in South Africa, and payments of interest on foreign loans. Although for most developed countries the distinction is not important, because South Africa is a major importer of capital and labour these payments make the current account differ markedly from the trade account. This can be seen clearly from Figure 7, which compares the total current account with the trade account. Whereas the total current account experienced persistent deficits between 1965 and 1976, and again between 1981 and 1984 (the surplus periods having been induced in order to finance the shocks to the capital account), the trade account experienced deficits only in 1970–1, 1974–6 and again in 1981. These factor service payments are therefore substantial enough to bring about current account deficits, despite almost persistent surpluses in the trade account.

Factor payments to foreign capital (that is, excluding migrants' remittances) are shown in Figure 8. It can be seen that interest payments on indirect investment have risen dramatically in recent years and at a far greater rate than direct investment remittances or dividend payments. One should note that as the pace of disinvestment increases, the rate of dividend payments on direct investment should decline. Figure 9 reveals that the proportion of these service payments to GDP had risen to almost 5 per cent in 1986. Figure 10 demonstrates the increasing proportion of these payments to both imports and exports, reaching 21.4 per cent and 14.7 per cent respectively in 1986.

What is noticeable is that these payments have tightened the balance of payments constraint by placing a substantial additional burden on the current account. Even at the height of the debt crisis in 1985–6, South Africa maintained its commitment to allow free flows of interest and dividend payments. In this way the capital shortage came to be compounded, for net exports have to finance not only capital repayments but also net debt servicing and dividend payments.

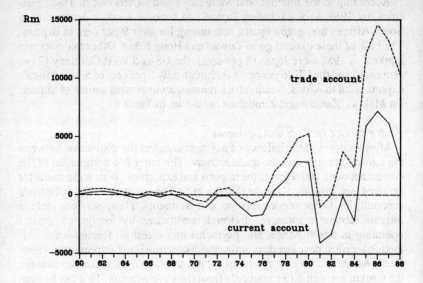

Fig. 7. Current account and trade account

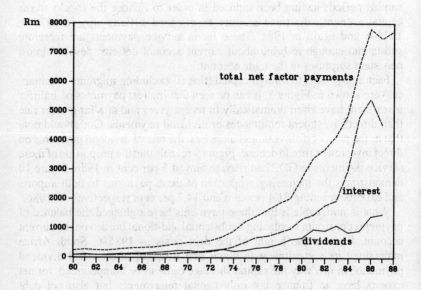

Fig. 8. South Africa's factor payments

BALANCE OF PAYMENTS

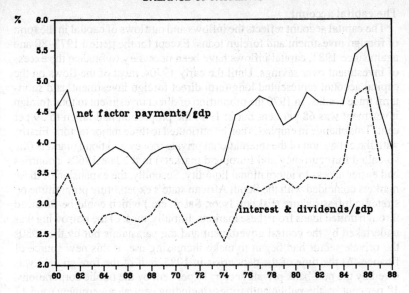

Fig. 9. Factor payments as a percentage of GDP

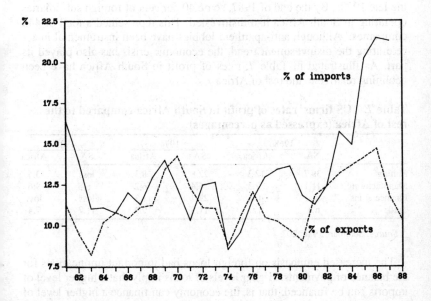

Fig. 10. Interest and dividends as percentage of imports and exports

The capital account

The capital account reflects the inflows and outflows of capital in the form of foreign investment and foreign loans. Except for the period 1977–80 and again since 1983, capital inflows have been necessary to finance the excess of investment over savings. Until the early 1970s, most of the flows on the capital account represented long-term direct foreign investment[6] and short-term trade credits. In 1970 the proportion of direct investment to total foreign investment was 68 per cent but in 1984 the proportion had fallen to 39 per cent. This change in emphasis can be attributed to three major factors. Firstly, with the expansion of the international private money and bond markets (the so-called Eurocurrency and Eurobond markets) in the late 1960s, countries had easier access to international liquidity. Secondly, the expansion of these markets coincided with the South African state's expenditure programme on strategic infrastructure (Eskom, Iscor, Sasol, etc.) which could be financed through untied loans from these markets. Initially, most of the borrowing was undertaken by the central government and the parastatals but by the 1980s the private sector had began to make increasing use of this new source of finance. At the time of the debt crisis in 1985, half of the foreign debt was owed by the non-bank private sector, 15 per cent by the public corporations, 18 per cent by the public authorities (including central government) and 17 per cent by the banking sector.

The third reason for the decline in the relative importance of direct investment was that the rate of disinvestment increased significantly from the late 1970s. By the end of 1987, some 40 per cent of foreign subsidiaries operating in South Africa had disinvested. This represented a total of 445 companies.[7] Although anti-apartheid lobbies have been instrumental in accelerating the disinvestment trend, the economic crisis has also played its part. As illustrated in Table 7, rates of profit in South Africa have been declining relative to the rest of Africa.

Table 7. US firms' rates of profit in South Africa compared to the rest of Africa (expressed as percentages)

	1968		1974		1982	
	SA	Africa	SA	Africa	SA	Africa
Mining	39.7	12.3	22.8	8.3	loss	3.5
Manufacturing	11.1	7.3	15.4	6.2	6.5	9.6
Finance & ins	n.a.	n.a.	29.7	11.3	loss	loss
Trade	n.a.	n.a.	11.5	19.2	11.7	7.4

Source: Seidman, 1986: 202.

The increased emphasis on foreign loans had important implications for the balance of payments. Higher levels of loans imply that a higher level of imports can be financed, that is, the economy can finance a higher level of planned expenditure over time. However, loans are not indefinite and have to be repaid eventually. For this reason the loans must attract investments which generate sufficient foreign exchange to repay the debt, or else imports (and therefore economic activity) will eventually have to decline. The need for generating foreign exchange has, in the case of most other LDCs, imposed a policy of export orientation on the economy. Whilst this in itself is not a problem, even greater reliance has to be placed on the export of primary commodities, given the inability of the manufacturing sector to expand its export base significantly.

The increased reliance on foreign loans has also made the balance of payments increasingly vulnerable to the political shocks that directly affect the capital account. When investment and expenditure plans are made on the basis of future expected loans, a sudden cessation or withdrawal of loans requires the severe adjustments described above. In addition, the capital account was rendered even more vulnerable by the fact that the maturity structure of foreign debt declined significantly in the 1980s. As can be seen in Table 8 short-term debt represented 49.1 per cent of foreign debt in 1980. By 1985 this proportion had risen to 72 per cent. On account of the shorter maturity structure, continuous access to foreign markets is required in order to maintain or roll over a given level of debt. If access is suddenly denied, as happened in 1985, the impact is far greater than if there had been a longer maturity structure (as was the case in 1977 when the average structure was about 4–5 years).

There are a number of reasons for the declining maturity periods. In the early 1970s maturity periods applicable to countries were much longer, loans of 10–15 years maturity not being unusual. At this time there was much recycling of petrodollars and the newly emerging Eurobanks had large amounts of funds for lending. By the mid-1970s maturities in general began to shorten as fears of overlending and borrower default grew. However, South Africa's maturities began to shorten even more than in the case of most other countries after the Portuguese coup in 1974, which spread political uncertainty over the southern African region. Maturities shortened even further after Soweto 1976 and only began to lengthen again in 1979. However, in the early 1980s South African borrowers were again being increasingly limited to short-term loans as economic and political conditions in the country deteriorated.

In addition to the views of international bankers about South Africa's credit-worthiness, the liberalisation policies of the Reserve Bank encouraged almost uncontrolled foreign short-term borrowing by private companies in South Africa. Moreover, because of the high level of interest rates that prevailed in the economy during 1983, and the expectation that the rand would appreciate against the dollar (which did not come about), most trade financing, which is by nature short-term, was done externally. 'The Reserve Bank encouraged overseas borrowing without keeping tabs on the amount. "The Reserve Bank's reporting systems were inadequate. Its free market philosophy seemed to extend to not requiring information," said the managing director of one South African bank.... They had no idea of corporate foreign debt' (Grant, 1985: 71).

Table 8. South Africa's foreign debt, 1980–1987

	1981	1982	1983	1984	1985	1986	1987	1988	1989
Dollar value of debt ($ billion)	18.7	22.4	23.9	25.5	27.0	22.6	22.6	21.2	20.6
Rand value of debt (R billion)	18.1	24.3	29.1	48.2	60.1	49.5	43.6	50.4	52.5
Short-term debt (% of total)	57.9	56.5	65.8	68.0	72.0	72.3	n.a.	n.a.	n.a.
Rand value of debt as % of GDP	25.4	30.4	32.6	45.7	50.1	35.2	26.5	25.3	22.6

Source: South African Reserve Bank Quarterly Bulletins.

Apart from loan capital, there are other categories of capital inflows that are of importance but do not impart the degree of volatility to the capital account which loans do. These flows include share transactions and direct investment.

Although share or stock-exchange transactions are by their nature volatile, the balance of payments has been insulated from such transactions through the financial rand mechanism. Because these transactions are subject to the financial rand, they are sales from one foreigner to another and therefore do not represent net inflows or outflows of foreign exchange. It was only during the period 1983–5 when all controls on foreigners were lifted that such transactions resulted in net movements.

Given the refusal of foreign banks to extend long-term loans to South Africa and South Africa's commitment to repay previous loans at the rate of approximately $2 billion per annum, the capital account is likely to remain under severe pressure for the foreseeable future. The only type of capital inflow still open to South Africa is in the form of trade credits. As these are generally short-term and directly related to foreign trade, they do not provide long-term relief from the balance of payments constraint.

EXCHANGE-RATE AND EXCHANGE-CONTROL POLICIES

Exchange-rate policy

As explained earlier, the nature of the balance of payments adjustment process and the adjustment to shocks depend in part on the nature of the exchange-rate regime. Under the Bretton Woods system, South African policy was directed at maintaining the exchange rate at a predetermined level, and the Reserve Bank bought and sold foreign currency to accommodate any excess supplies and demands. Since the breakdown of the system in 1971, countries have had to decide on the degree of flexibility of their exchange rate. In a world of floating exchange rates, even if a country decides to peg its own currency to a key currency it will effectively be floating against all third-party currencies.

In the 1970s South African exchange-rate policy was characterised by indecision. From August 1971 the rand was pegged to the dollar, then to sterling and back to the dollar. After a brief period of 'independent managed floating', the rand was repegged to the dollar in June 1975. If one excludes the 17.9 per cent devaluation against the dollar in 1975, this peg was maintained until the publication of the Interim Report of the Commission of Inquiry into the Monetary System and Monetary Policy (the De Kock Commission) in 1978. On the basis of its recommendations a form of variable dollar pegging was adopted, and the authorities came to adjust the exchange rate more freely. In August 1983 the Reserve Bank ceased its policy of quoting the spot exchange-rate and instead began to allow a more market-determined rate in which they intervened, only when considered desirable, by buying or selling foreign currencies.

The direction of exchange-rate movements has been dictated to a large extent by the gold price and political developments. However, in following the gold price the authorities have adopted a definite downward bias in determining the exchange rate, for reasons to be explained. In general, sustained falls in the gold price are followed by marked and rapidly falling exchange rates; whereas during periods of a rising gold price the authorities have tended to prevent the rand from appreciating as much as it might have under a freely floating regime. Such policies have important *real effects*,[8] which we shall now consider.

In analysing the effects of exchange-rate changes we should distinguish between real and nominal exchange rates. The nominal rate is the relative price of domestic and foreign moneys, whereas the real rate is the nominal exchange rate adjusted for inflation differentials between the two countries. Thus, if inflation in South Africa is 10 per cent higher than in the US and the rand depreciates by 10 per cent against the dollar, the real rate remains unchanged. Even though Americans have to pay fewer dollars for each rand, South Africa's competitive advantage is offset by the higher domestic inflation. If, on the other hand, the 10 per cent inflation was accompanied by a 6 per cent depreciation, then the real rate depreciates by 4 per cent. It follows that exchange-rate changes have real effects, in terms of export competitiveness for example, only if a depreciation is not completely offset by rising costs due to inflation. Real changes arise when domestic commodity and factor prices adjust slowly relative to exchange-rate changes, or when there are changes in real variables such as productivity, labour costs, consumer tastes, as well as discoveries of new mineral resources or alterations in the price of a major export commodity such as gold. Figure 11 shows the behaviour of the nominal and real trade-weighted exchange rate over the past 15 years. It can be seen that both rates appreciated in response to the rising gold price in 1980 and 1982, and depreciated significantly after the political events of 1984.

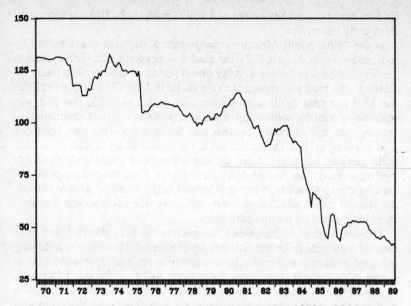

Fig. 11. *Nominal effective rand exchange rate (Jan 1979 = 100)*

The extent to which different sectors of the economy benefit from a real depreciation depends on whether they are export-oriented or import-competing, and on the degree of their dependence on imported inputs. A real depreciation makes South African exports cheaper in foreign currency terms and imports more expensive in rand terms. However, the competitive advantage gained will be reduced to the extent that these industries depend on imported technology and intermediate inputs, which will also have increased in price. The Kleu Report demonstrated that manufacturing, being far more capital-intensive than both mining and agriculture, is the highest net user of foreign exchange and has the highest import content in its exports.[9]

Changes in the exchange rate affect mineral exporters differently from manufacturing exporters. Because mineral prices are quoted in foreign currency terms and because these prices are set on world markets, a depreciation does not reduce the foreign currency price of these products and thereby increase competitiveness. The effect is to increase the rand price and thus directly increase the profitability of the industry. For this reason fluctuations do not affect the export possibilities of the mining industry. Indeed, because exchange-rate fluctuations are determined to a large degree by changes in the gold price, the gold mines are protected by these fluctuations, which tend to keep the rand gold price stable.

Although the manufacturing sector would benefit from a long-run depreciation of the real exchange rate because of its effects on international competitiveness, it is adversely affected by the *volatility* of the real exchange rate. Apart from the problem of the degree of imported inputs, fluctuations in the real exchange rate cause problems for manufacturing exporters because of their concern for market shares. Establishing export markets is risky and expensive, and the greater the degree of fluctuation in exchange rate, the greater the risk involved. Given such an uncertain environment, the manufacturing sector is less likely to embark on export expansion if there is a strong possibility that the exchange rate will change to their disadvantage. By contrast, an important feature in the success of the Asian NICs in expanding industrial exports was their policy of maintaining stable real exchange rates, which provided a significant degree of certainty to manufacturers. South Africa differs structurally from the NICs in that these countries do not have primary resource bases. The dilemma facing the South African authorities is whether to allow the exchange rate to adjust to accommodate changes in the dollar gold price (thereby stabilising the rand gold price) or to provide a stable environment for stimulating manufacturing exports. This would involve maintaining a stable real exchange rate by ensuring that the nominal exchange rate adjusts to offset inflation differentials.

On account of the importance of mining and mineral exports to the economy in terms of employment, foreign-exchange earnings and tax revenues for the government, it is not surprising that exchange-rate policy has implicitly been used to protect the fortunes of these sectors. The effect of the downward bias in exchange-rate policy has been to maintain or even increase mining profits in rand terms, given that the rand is allowed to depreciate markedly following a fall in the gold price. The concern for the mining sector has been well expressed by a senior advisor to the Governor of the Reserve Bank, who argued that 'it is desirable that the mining sector be allowed to reap higher profits stemming from currency depreciation' (Gidlow, 1988: 145). He further argued that 'the floating rand has been helping to insulate these [gold and platinum] exporters from fluctuations in the dollar prices of these commodities. The volatility in the rand may well be detrimental to numerous exporters in the manufacturing sectors whose operations are somewhat marginal in nature. Nevertheless, the onset of sanctions is likely to render the economy more dependent on fungible mineral exports like gold and platinum. The cushion provided to such industries by the floating rand could therefore assume even greater importance' (1988: 151).

Since real exchange-rate changes come about when wage and price increases in the domestic economy do not match a nominal depreciation, the initial distributive effect of such a currency depreciation is to reduce real wages of workers. However, the precise length of such a wage lag will vary from situation to situation and will in addition be effected at any point in time by such factors as the level of unemployment and the power of trade unions. At the same time, the currency depreciation implies higher profits for exporters. Moreover, Diaz-Alejandro (1965) has shown that regardless of the effect of a depreciation on total profits, the profit *share* in the national income will necessarily improve after devaluation. Given the bias towards depreciation rather than appreciation in the conduct of South African exchange-rate policy, such a stance tends to discriminate against wages in favour of profits.

A further effect of real exchange-rate changes is felt on foreign currency debts. A depreciation increases the local currency value of the debt and debt-servicing. For the economy as a whole this means that more resources have to be given up in order to repay the existing debt. Between 1983 and 1984 South Africa's foreign debt rose by 6.6 per cent in dollars, but because of the currency depreciation the rand value of this debt increased by 65.6 per cent and the proportion of total debt to GDP rose from 32.6 per cent to 45.7 per cent (Table 8).

Exchange-control policy

An integral part of South Africa's exchange-rate policy is exchange-control policy. Foreign-exchange controls were first instituted in 1961 after the Sharpeville shootings, which precipitated a large outflow of capital. Controls were imposed on both residents and non-residents, the latter having to use the blocked rand mechanism if they wished to repatriate capital. Essentially, this meant that the capital account was insulated from volatile flights of equity capital because sales of blocked rand could be sold only to other non-residents. There was thus effectively a pool of investment currency, the price being determined by foreigners' supply of and demand for South African securities. If foreigners were pessimistic about the future, the blocked-rand rate would fall, raising the discount against the commercial-rand rate. This would increase the return on South African securities as shares would be bought at a discount, but dividends could be repatriated at the commercial-rand rate. This system remained more or less intact until 1983, apart from a few technical changes which widened the scope of blocked-rand transactions with the change to securities rand (1976) and the financial rand (1979). An important feature of the latter change has been that foreign direct investment takes place through the financial rand mechanism, thereby giving the foreign direct investor more rands for every dollar. It also means that disinvestment has no direct implications for the balance of payments, as disinvesting companies are required to repatriate their capital through the financial rand mechanism.

The liberalisation trends that occurred in the financial and foreign-exchange markets in the late 1970s and early 1980s also affected exchange control. The Franszen Commission Report of 1970 argued strongly in favour of maintaining exchange controls, specifically in view of political factors that could cause capital flight and thereby put strain on the level of foreign-exchange reserves. By 1978, as a result of changing attitudes towards the market mechanism and the increasing influence of Dr Gerhard de Kock in the Reserve Bank, this approach had changed. The De Kock Commission proposed the complete elimination of exchange controls on non-residents and only limited controls for residents.

The Commission's report was written at a time when there were balance of payments surpluses, and when pressure was being brought to bear on the government by local conglomerates wishing to diversify investments abroad. Declining rates of profit and political uncertainty have made it increasingly attractive for South African companies and individuals to invest in other countries, thereby increasing pressure for relaxation of exchange controls. By proposing their relaxation, the Commission clearly put its faith in the ability of the market to eliminate capital flight through currency depreciation.

Exchange controls on non-residents were lifted in 1983, but had to be reimposed in August 1985 because of the debt crisis. With regard to residents, the Commission had recommended a gradual relaxation of controls, since a sudden lifting of such controls might have a large impact on real-estate and stock-market prices and interest rates. Some adjustments were made but, in general, exchange control remained intact, except that a far more liberal attitude was adopted towards applications by local firms to borrow overseas as well as towards certain types of investments abroad. However, when granted, such permission is still subject to *ad hoc* rules by the Reserve Bank.

A major problem with opening up the capital account is that it can change the composition of inflows and outflows, which might at a later stage lead to a potential debt problem. The phenomenon of simultaneous borrowing and investing in international capital markets has been analysed by Khan and Ul Haque (1985). They have shown that capital flight, in a number of countries, has caused a build-up of gross foreign debt, an erosion of the tax base and, to the extent that there has been a net real resource transfer out of the country, a reduction in domestic investment. Typically, when controls are lifted, residents invest their own money abroad and invest at home with borrowed foreign capital because they perceive a larger risk at home than abroad. Because of the uncertainty surrounding private property in a post-apartheid economy, the incentives for capital flight are high and rising as political instability increases, whereas the risk on assets held abroad is negligible.

Thus, although opening up capital flows may encourage a net inflow into the country, domestic private capital may be flowing out. A sudden political crisis would then result in an outflow of the borrowed capital, with no likely return of the locally owned capital from abroad. Growing foreign debt and the simultaneous rise in investment abroad indicate that this pattern holds for South Africa.[10]

Although exchange controls can never function perfectly, full liberalisation merely serves the interests of those residents and local companies that wish to hold their assets abroad. Given the necessity hitherto of supplementing domestic savings with foreign capital, it also seems highly questionable that scarce investment funds should be allowed to be invested elsewhere, particularly at a time when access to foreign savings is limited. Experience with liberalisation in LDCs and semi-industrialised countries has not been successful: the Southern Cone countries (Brazil, Argentina and Chile), Israel and South Africa have all reversed these policies in time. In the wake of the reimposition of the financial rand and the continuing debt crisis, liberalisation does not seem to be an option for the foreseeable future in South Africa either.

CONCLUSION

The balance of payments constraint facing any country derives from the fact that current account deficits have to be financed in some way. With the expansion of the Eurocurrency markets, countries were able to avoid immediate correction of current account deficits by borrowing. This lengthened the period over which growth could take place. But because South Africa no longer has easy access to foreign capital, and because of its commitment to repay previous loans, the country's capital account deficits are likely to persist. What this means is that these outflows will have to be financed through current account surpluses, which are already under pressure because of the substantial interest and dividend payments that have to be made. Surpluses can be achieved either through increasing exports or through decreasing imports. The former option is problematic because of the underlying crisis in the manufacturing sector. In addition, this sector faces increased threats of sanctions and has to battle with an exchange-rate policy that is not conducive to promoting exports. For these reasons, greater reliance will have to be placed on primary exports, yet the latter face the problems of price volatility and capacity constraints. Because of South Africa's dependence on imported capital equipment, a policy of reducing imports requires a reduction in gross domestic investment, with negative implications for employment and growth.

Recent events have illustrated how quickly the balance of payments constraint is felt. The large current account surpluses experienced since 1985 arose as a result of the collapse of domestic investment. In 1986 investment as a proportion of GDP was similar to that following the Sharpeville crisis of 1960. The moderate economic growth that occurred in the latter half of 1987 was sufficient to convert the current account surplus into a deficit by the middle of 1988. In response certain policies to curb investment and consumption growth were set in motion, such as the increase in interest rates and the imposition of import duties in the latter half of 1988.

South Africa's balance of payments constraint differs from that of other countries which have had to reschedule their debts, in that most rescheduling has revolved around the question of interest repayments. South Africa has the additional burden of having to repay a fairly substantial foreign currency debt in a relatively short period of time without access to further borrowing. Thus although the economy needs economic growth to generate employment, balance of payments considerations ensure that such growth cannot be sustained.

APPENDIX A

South Africa's balance of payments 1970 and 1989 (Rm)

Current account	1970	1989
Merchandise Exports	1 453	39 085
Gold Exports	837	19 228
Service Receipts	581	9 754
Merchandise Imports	−2 582	44 322
Service Payments	−1 206	20 857
Transfers	49	2 888
Balance on Current Account	−868	3 108

Capital account	1970	1989
Long-term Capital Movements	427	−1 230
Short-term Capital Movements (not related to reserves)	129	−3 115
Balance on Capital Account	556	−4 345
Change in Net Gold and Foreign Exchange Reserves	−312	−1 237
Liabilities Related to Reserves	26	2 626
SDR Allocations and Adjustments	24	−1 190
Change in Gross Gold and Foreign Exchange Reserves	−262	199

Source: South African Reserve Bank Quarterly Bulletins

4
The politics of South Africa's international financial relations, 1970–1990

VISHNU PADAYACHEE

This chapter examines the nature and structure of, as well as changes in, South Africa's international financial relations, concentrating mainly on this country's relations with the International Monetary Fund (IMF) and private international banks, in the period 1970–1989. It should be pointed out that the chapter does not represent an analysis of capital movements or South Africa's balance of payments *per se*. Rather it examines the history of, and basis for the changes in, the structure and character of the relations between South Africa and the two central 'non-state institutions' in the international financial arena – the IMF and private banks. Moreover, it briefly deals with the implications and effects of changes in these relations for South Africa, especially from the mid-1980s. Two points of clarification about the approach employed in this chapter need to be dispensed with at this stage.

Developments in the world economy since the early 1970s should convince all but the most arcane economist or state-centred international relations specialist of the inseparability of economics and politics in the study of the international political economy. For this chapter is indeed located within the framework of an international political economy approach, in which both economic and political developments at the local and international level, and changing power relations between states, institutions and classes, are all called upon in order to better explain the state of, and changes in, the world economy (Ladd-Hollist and LaMond Tullis, 1985).

The chapter is also located within a general analysis of the crisis in the South African economy, that is based on theories of 'capitalist regulation' and 'challenges to capitalist control'. More especially it takes as its starting point those aspects of this approach that relate to the international causes and dimensions of the crisis. According to this view, the growing problems in the international financial system from the late 1960s left the IMF for a time without a clearly defined lending or supervisory role, and created the space for the uncontrolled growth of private bank lending. According to De Vroey (1984: 63) one result of these developments was that 'deficits no longer exert[ed] pressure for their own elimination, since they [could] be turned into increased indebtedness'.

Furthermore, with the shift in the US basic balance from surpluses to erratic deficits from the late 1960s, there was a distinct fragmentation in the cohesive factors provided by American hegemony that held together the post-war international economy. In short, these developments threw into disarray the rules of the international trading and monetary system, so crucial to the social structure of accumulation established at the end of the Second World War. Through the 1970s and 1980s this created further uncertainties, shocks and instability in the international institutional framework within which accumulation proceeded in both advanced capitalist countries as well as middle-income and developing countries.

As far as South Africa is concerned, it is our view that these international developments and trends interacted with local economic and political forces and certain policy mistakes in such a way as to reinforce and exacerbate the growing crisis in the South African economy. The period after 1970 was punctuated by political instability, a fluctuating and uncertain gold price, and rising import bills for oil, armaments and capital goods. Under these conditions, the financial support of the IMF and of private international banks was to become more and more crucial.

The chapter begins by examining South Africa's relations with the IMF, before proceeding to investigate the character of its relations with private international (creditor) banks in the period 1970–1989.

SOUTH AFRICA AND THE IMF, 1970–1983

The overall aim of the IMF, as set out in its Articles of Agreement, is to facilitate the growth of world trade by extending loans to member countries which experience short-term balance of payments difficulties. However, it was only in the 1970s, as the world economy began to show signs of crisis, that the IMF came to the forefront of public attention. Although the funds available to the IMF to assist member countries did not increase significantly, the IMF was more and more called upon to lend foreign currencies to members. In order to qualify for such stabilisation assistance, most borrowers had to accept an IMF programme requiring changes in their economic policy, mainly in respect of exchange-rate, monetary and fiscal policy.

South Africa, a founder member of the fund, received its first credit or loan from the IMF in 1957/8. However, over the period 1947–69, South Africa's purchases of IMF currencies amounted to only SDR211.7m, or 10,8 per cent of its total purchases in the period 1947–85. In other words, the great proportion of IMF loans to South Africa were made after 1970, and this was largely concentrated in the loan of SDR643.2m in the mid-1970s and the 1982 loan of SDR902.2m. South Africa's share of IMF purchases amounted to nearly 50 per cent of all purchases by southern African states up to 1984; South Africa was also the second largest purchaser of IMF funds, after Zaire, among all African countries; and until 1983 ranked second behind Yugoslavia, as the largest borrower from the fund in its IMF-classified sub-group, that of 'major exporters of manufactures' among 'non-oil developing countries' (Padayachee, 1986).

By most standards, therefore, it would appear that the IMF represented a critical source of funds to South Africa for offsetting this country's balance of payments deficits especially in the period after 1970. In the two major recessions experienced by South Africa at this time – the one that lasted from September 1974 until the end of 1977, and that from the end of 1981 to date – South Africa borrowed heavily from the fund to supplement its foreign-exchange reserves. In the first of these, triggered by the oil-price shock of 1973, Gelb (1987) has shown that South Africa's foreign reserves declined in US dollar terms, despite two devaluations of the rand in 1975. This evidence he cites as one reason for characterising that cyclical downswing as 'non-reproductive', as opposed to a reproductive cyclical downswing in which, amongst other things, foreign-exchange reserves rise, so creating the stability required for more sustained accumulation.

The IMF provided South Africa with credits of SDR91.2m in 1975, SDR390m in 1976 and SDR162m in 1977. The IMF assistance to South Africa in the two years 1976–77 was in fact greater than the combined assistance to all other African countries for the same period. In those two years only two other countries, Britain and Mexico, were bigger beneficiaries of IMF assistance. These mid-1970s borrowings were made, according to South African and IMF sources, in support of the government's programme to strengthen the country's balance of payments. In our view, however, international and internal developments, both economic and political in nature, were largely responsible for the parlous state of South Africa's balance of payments in the mid-1970s, compelling the country to turn to the IMF. The effect on exports of a falling gold price was exacerbated by still rising prices of imports of both capital goods and intermediate inputs, especially oil. The SDR160m purchase from the fund was announced on 15 November 1976, under the fund's Compensatory Financing Facility (CFF).

Yet though the implications of the developing international crisis for a country such as South Africa, with its narrow primary export base and high propensity for, but low elasticity of, import demand, were undoubtedly crucial, internal pressures arising from the need to defend its undemocratic political order also impacted negatively, in both direct and indirect ways, on the country's balance of payments.

Most of the fundamental problems facing the world economy and South Africa remained in place throughout the 1970s and 1980s. In particular, the Soweto uprising of 1976, the nation-wide schools boycotts of the early 1980s, and from 1984 the unprecedented mass resistance to apartheid, ensured that South Africa continued to sustain pressure from within. Internationally, the years 1977 to 1981 witnessed an increasing deterioration in the state of the world economy. At the same time, a series of internal and external developments partially insulated South Africa and eased the pressure on the balance of payments sufficiently to reduce South Africa's dependence on the IMF.

Nevertheless, these events acted as no more than a palliative, and the underlying imbalances and vulnerabilities of the South African political economy constantly remained close to the surface.

However, between 1981 and 1984, with the gold price declining sharply from over $800 an ounce and imports surging, South Africa once again ran up a deficit on the current account of its balance of payments. By mid-1982 it appeared that there was little likelihood of an improvement on the current account. Despite numerous denials at first by top financial officials (*FM*, 10 September 1982), South Africa approached the IMF for a standby loan of $1.1 billion. South Africa's approach to the Fund was made in mid-1982 (*FT*, 25 January 1983) but, on the US State Department's advice, was not announced until October 1982, a few weeks before an IMF executive board meeting. The expectation that the application would be controversial was to prove correct. Opinion ranged from those who urged that the loan should not be approved under any circumstances because of South Africa's racial policies, (United Nations, 1982a, 1982b, 1982c) to those who stated that the loan should be approved on condition that South Africa abolished 'apartheid imposed' rigidities (*Economist*, 16 October 1982; *NYT*, 20 October 1982). However, despite the objections of the United Nations, human rights groups and others in the US (based on political grounds) and the opposition (on economic grounds) of at least five IMF executive directors voiced at the confidential IMF executive board meeting of 3 November 1982, the South African request was approved. As far as can be ascertained, approval was given without conditions related to the country's political system or the impact of this system on the efficiency of its economy. Like the loans of the mid-1970s which were also approved against objections from many quarters, South Africa received the crucial and numerically vital voting support of the advanced capitalist countries, most notably the UK and US.

Yet the international outcry did have some effect, for both the IMF and the US shifted their stance vis-à-vis future South African applications to the IMF. At a 20 June 1983 meeting of the IMF, its staff presented a report critical of apartheid on economic grounds – the first in the IMF's history. The report noted that apartheid impedes the optimal utilisation of South Africa's resources, especially by creating labour-supply bottlenecks. It went on to argue that apartheid does not have a short-term effect on the balance of payments, but that its effects are felt in the medium and long term. The Center for International Policy (1983), commenting on the report, stated that in terms of its recommendations, South Africa would no longer be able to count on automatic access to some of the IMF's conditional facilities unless it dismantled apartheid. Until then, South Africa would have available only the low conditionality facilities it used in November 1982. At the time such facilities would still have provided Pretoria with $1.5 billion.

After much debate in the US Senate and House of Representatives, and its respective banking committees and sub-committees, an anti-apartheid amendment to a bill authorising an increase in the US quota to the Fund was agreed upon. American law has since required that the US executive director of the Fund oppose South African applications for loans unless the Secretary of the Treasury certifies in person before the banking committees that the loan conditions would reduce the distortions caused by apartheid. Furthermore, the Treasury would have to give a three-week public notice of another IMF loan application from South Africa (CIP, 1984).

The amendment left the US with enough manoeuvrability, so that its support for IMF loans to South Africa was not entirely prevented. It was definitely restricted, but restricted in ways that suited the purposes of 'reform-minded' individuals and institutions in South Africa and abroad. The scrapping of influx control legislation and other reforms since then has probably eased the South African position although this has not apparently been tested by a loan application to the Fund. Nevertheless, in contrast to the deafening silence over loans to South Africa in previous decades, the actions of the IMF and US are not insignificant.

Elsewhere I argue more fully that South Africa, especially in relation to other comparable countries, was particularly well received and accommodated at the IMF despite its pursuance of economic policies which are generally anathema to the IMF's view of the world and its ideological stance on questions of development (Padayachee, 1986 and 1988). This chapter has presented only the bare bones of this case. Nevertheless, two issues needs to be addressed. Why, and on what basis, did South Africa receive such favourable treatment? And what was the rationale behind the IMF and US rebuke of South Africa in 1983?

In 1978 President Carter's National Security Advisor identified Iran (then under the Shah), Brazil and South Africa as 'regionally influential nations', noting that changes within these countries would have profound consequences throughout the areas in which they are situated and in the major centres of world capitalism (Petras and Morley, 1981). For one thing, the West's reluctance for so long to impose tougher trade and financial sanctions against the apartheid state can be partly explained by its strategic concerns for regional stability in southern Africa. To secure these concerns, support for a regional power like South Africa was crucial. Among other channels, it is through Western-dominated agencies like the IMF that this support is provided. Such support becomes especially crucial when the regional power is itself, like South Africa has been since the early 1970s, threatened from within and without its borders.

But these strategic concerns are not the only reasons explaining the West's long-standing support for South Africa through agencies like the IMF. Western nations have, in addition, financial and economic interests in this country, which make South Africa's fate of no small significance to them. These investments have generated a substantial proportion of earnings for specific firms and countries in the core areas of the capitalist world economy. South Africa's raw material, potential markets and politically sympathetic government are all important in maintaining successful and efficient production in the Western industrialised nations. In other words, countries like South Africa have developed within the sphere of international capitalism in such a way that they are vital to the advanced capitalist countries in stalling or mediating crises that arise from their own specific forms of accumulation.

However, it has become clear to the West that its interests cannot be served in the longer run under an old-style apartheid form of state. Unless some settlement of the South African conflict is reached, ideally by the election of a non-racial, though pro-Western government, it is now recognised in the US and Western Europe that there is the danger of South Africa slipping out of the Western sphere of influence altogether.

In our view, this recognition lies behind recent efforts at nudging the apparently intransigent South African regime towards negotiation. The IMF and US rebuke of South Africa's racial policies in 1983, which followed the adverse reaction to the 1982 IMF loan to South Africa, must be seen as among early Western attempts towards this end. IMF assistance in the mid-1970s and again in 1982 was important to South Africa, both in providing much-needed liquidity and in giving banks and the international financial community the signal that the IMF supported and endorsed the policies of this country. In these ways, the IMF played an important role in staving off the imminent crisis in the South African political economy. The 1983 decisions of the US and IMF in regard to future loans were meant to signal this to South Africa. They were also, like the 1986 limited US sanctions programme, part of an effort to persuade South Africa to reform its policies, pre-empt more radical change, and so become 'respectable' again within the world capitalist system.

SOUTH AFRICA AND PRIVATE INTERNATIONAL BANKS

Whereas during the period 1946–73, the main sources of foreign capital investment in South Africa were in the form of direct investment (risk capital, which was profit-yielding) and indirect investment (dividend-yielding share purchases and short-term credits), there has since been an increasing reliance on loan capital, including syndicated loans and credits on the Euromarket and public bond issues on the Eurobond market. The portion of international credit represented by private international bank lending to South Africa increased from 15 per cent of total foreign investment in 1974 to 32 per cent in 1976. (These figures exclude bond issues managed or guaranteed by these banks.) These loans and bonds to South Africa have been made, managed or underwritten by some of the largest international banks, which have extended large quantities of credit to South Africa since the early 1970s, especially between 1973 and 1976, and again between the end of 1981 and late 1984.

One point should be noted that relates to the value of international credit to South Africa, though it is also relevant to other borrowers. For South African borrowers, especially the state and public corporations, private international credit has distinct advantages over both direct foreign investment and public international credit, such as IMF loans, in the sense that it is usually 'untied'. That is, it does not imply any ownership or control of any political leverage. IMF loans usually come with conditions attached – although this was not always true of South Africa's IMF borrowings – and direct foreign investment occurs at the discretion of foreign investors motivated by their own views on the profitability of different sectors or industries in the economy.

In contrast, private international bank loans and funds raised from bond issues in the international capital market can be determined and utilised according to the economic and political priorities and judgements of borrowers themselves – provided their credit rating holds, of course. For the very essence of direct foreign investment is that the investor maintains control over the investment: in contrast, private international bankers have most often neither the capacity nor the inclination to monitor the use of loans provided. However, investment necessitating international bank finance, of the sort undertaken by the South African state and parastatals in the 1970s, may itself create a new type of dependency: such borrowers may in turn 'become mortgaged to the Eurobanks' (Stalling, 1985; Lipietz, 1987). For South Africa as a whole, this ambiguous state of 'dependence upon, yet relative autonomy from' the diktat of the international financial community at large intensified from the early 1970s, as the relative importance of direct foreign investment declined at the expense of foreign capital inflows of the loan variety.

In general, the inflow of loan capital to South Africa between 1973 and 1976 reflected the greater availability of recycled petrodollars after the 1973 oil price hike at a time of low investment demand in the industrialised economies. As a result of the political uncertainty generated by South Africa's abortive Angolan venture and the Soweto student revolt of 1976, medium-to-long-term international credit from private bank sources dried up completely in 1977 and generally eased off between 1977 and 1979. The 1979 oil price increase, expectations of boom sparked by the increase in the gold price (1979–80), and growing fears of debt repudiation elsewhere, gave a fillip to bank lending to South Africa, especially from the second half of 1981.

The slow-down in direct foreign investment does, however, indicate that foreign investors, motivated by South Africa's poor economic performance virtually throughout the period 1973–85 and by political instability for much of the time, have taken a fairly negative view of South Africa's long-term economic and political prospects. This slow-down in risk and equity finance placed upward pressure on local interest rates, in turn forcing domestic borrowers to seek foreign funds. Especially between 1973 and 1976 and again between 1982 and 1984, the demand was easily met by foreign bankers anxious to place money in South Africa, which seemed a good risk – for this type of investment anyway – in view of its impeccable debt-servicing record up to then. During the early 1980s, this was in stark contrast to the performance of major debtors such as Mexico and Brazil, which were hit very hard by rising interest rates in the international money markets.

From late 1981 and clearly from 1982, South African borrowers in both the public and notably the private sector went on what may best be described as an orgy of borrowing from private international banks directly and from the international capital market by means of bond issues. This massive upsurge in foreign borrowing culminated in the dramatic developments of August 1985. This frenzied period of South African international financial market activity was, however, preceded by a term of consolidation in which South Africa rebuilt its credit rating after the uncertainties following the Soweto uprising and the simultaneous downturn in the economy.

By 1982 the gold-led boom of 1979–80 was clearly over. The South African economy had moved into an almost continuous recessionary phase from 1981, except for a brief consumption-led surge in activity between the end of 1983 and the second quarter of 1984. Inflation ran at double-digit levels exceeding 13 per cent in 1984 and 17 per cent in 1985. The rand was under almost uninterrupted pressure from the beginning of 1982. The money supply had been expanding at annualised rates of between 25 per cent and 50 per cent between 1981 and 1984 (*FAR*, 1985). Government borrowing and the budget deficit increased, and there appeared little scope for reductions in spending because of the state's requirements for external defence, internal 'law and order', and 'reform'. The gold price averaged $376 an ounce in 1982, rose to $424 in 1983 but fell to $361 in 1984 and $317 in 1985 (De Kock, 1987). The current account of the balance of payments was in deficit between 1981 and 1984, changing to a surplus in 1985 only because of the sharp increase in the value of gold and other exports consequent upon the dramatic fall in the value of the rand in that year. Personal savings dropped in the early 1980s to extremely low levels, seriously affecting capital formation from domestic resources.

On the international front, the world economy was itself in a state of decline for much of the period. Unemployment remained a major problem in most Western capitalist economies, though many of them had greater success in bringing down the inflation rate through the use of fairly tough monetary measures. International capital markets were in a parlous condition because of the South American debt crisis and the erratic movement of the US dollar. New lending to most developing countries by private international banks came to a virtual end between 1982 and 1984, forcing the construction of new initiatives to address the growing international debt problem.

At the beginning of this period South Africa's credit rating remained good. In a survey conducted by the international banking magazine *Euromoney* in September 1982, South Africa's credit rating was upgraded to no. 20 in the table of sovereign borrowers, compared to no. 36 in September 1981 and a lowly no. 60 in 1980. The upgrading occurred largely because of the debt crisis in some middle-income developing and East European countries. The *Rand Daily Mail* (7 September 1982) announced proudly that 'the main reason for South Africa's change – in spite of the deterioration in the balance of payments – is that the average international banker has realized that South

African borrowers not only honour their obligations but sometimes repay loans ahead of time.' Conditions for South African activity on the international financial market were undoubtedly favourable.

Early in 1982 the South African Reserve Bank, by lowering marginally the premium for forward cover on loans of three months or longer, encouraged foreign borrowing (*FM*, 19 March 1982). The intention was to help the current account and to limit domestic bank lending and hence the money supply and interest rates. With dollar interest rates often 10–12 per cent below rand rates, private-sector borrowers needed little prompting to switch to short-term foreign currency borrowing. However, as *Euromoney* (December 1985) pointed out, 'the Reserve Bank encouraged doubters by manipulating rates in the forward foreign exchange markets which it controlled. It offered forward cover at cheaper than pure market rates, so that the total cost of foreign currency borrowing plus forward cover was lower than the domestic rand cost.' While the Reserve Bank encouraged foreign borrowing, it appeared to do this without improving its reporting and information system on the amounts borrowed. The managing director of one South African bank told *Euromoney* (December 1985) that 'the Reserve Bank's reporting systems were inadequate. Its free market philosophy seemed to extend to not requiring information.' The result was that when South Africa's debt crisis finally came to a head, the Bank did not know how much short-term debt there was owing.

An increasing proportion of international loans in this period were made by the private sector, and loans from all types of borrowers were increasingly short-term. On both counts these represented significant changes in the nature and structure of South Africa's international borrowing, when compared to borrowings in the mid-1970s.

Private-sector investments were financed largely from retained earnings in 1980–1 but the scale of projects undertaken, notably in pulp, paper and chemical industries, under conditions of tightening cash flows, made international borrowing essential thereafter (*The Banker*, May 1982). The provision of off-shore finance, especially short-term funding, had become a significant feature in the borrowing plans of the major South African corporations, including Premier and South African Breweries (SAB) by 1983–4. In 1983 particularly, both South African and private international banks stepped up their off-shore funding for domestic (South African) companies eager to take advantage of lower interest rates outside the country. Private-sector borrowers appeared to have little difficulty in raising foreign- bank financing on favourable terms.

But the public sector, and Eskom especially, were not insignificant borrowers themselves. In 1982, Eskom was in the process of building four large coal-fired power stations, with a further two plants in the pipeline. Eskom had become the biggest borrower of medium-term funds abroad, its 1982 borrowing (including export credits) being close to R2 billion. Other major public-sector borrowers in this period included the Post Office, the

Industrial Development Corporation, and the South African Transport Services (SATS).

In general, over this period there was a rapid growth in the short-term component of South Africa's borrowing, as a result of both the nature of South Africa's borrowing requirements – which, given the balance of payments problems, were mainly trade and accommodation credits – and the decision of US banks in particular to reduce the terms of their South African loans. This became especially apparent after the township unrest began in September 1984, and enabled a quick retreat in the case of adverse political or economic events. By mid-1981 short-term debts amounted to some 46.4 per cent, according to the Bank for International Settlements, rising to 53.5 per cent at the end of 1981, 59.6 per cent in mid-1982 (*FM*, 14 January 1983) and 66 per cent in 1984 (*FAR*, March 1985). As the rand fell against the dollar, the country's debt burden increased – as most of the debt was denominated in dollars – and the short-term component of the debt increased further as more and more importers held off repayment in the vain hope that the rand would recover (*FM*, 14 December 1984).

Having no shortage of private international loan sources, South Africa's R1.24 billion IMF loan in November 1982 appeared to be directed at reassuring its private international creditors. 'International bankers', commented the *Financial Mail* (8 October 1982), 'will be happy because there will be less pressure on South Africa to borrow abroad and they will be reassured that South Africa's economic progress is being monitored by the IMF.' More than this, however, a UN study on South Africa's IMF loan application charged that the loan to South Africa 'demonstrated to the world financial community that, despite increasingly violent confrontations between the majority black population and the white minority government in South Africa, the IMF would continue to support the status quo and the white authorities' (*NYT*, 21 October 1982). Militz (1985) observed that in 1983, after the IMF loan was approved, private international banks provided South Africa with an unprecedented amount in credits (mainly short-term), showing clearly the importance of the IMF in the new conditions of the international economy.

On 7 February 1983 South Africa abolished exchange controls (and with it the financial rand) over non-residents, and received plaudits from the IMF and some of its major international creditor banks. It was only after the August 1985 crisis broke, however, that criticism of the liberalisation of foreign-exchange policy began to surface. Volkskas's managing director, Danie Cronje, and Trustbank's Chris van Wyk came out publicly in opposition to such liberalisation and especially to its timing. Despite this criticism from the two major indigenous ('non-British') banks and other mainly Afrikaner businessmen, the key Reserve Bank officials remained committed to the progressive liberalisation of exchange regulations. Yet at the time of implementation (1979–83), their policies had been enthusiastically endorsed by the international financial community. The latter must bear at least some of the responsibility for the uncontrolled and damaging capital outflow which immediately followed the abolition of exchange controls, as well as to the debt crisis of 1985 itself.

While South Africa's standing in the international financial market and in financial centres was repaired after the abolition of the financial rand, the country's monetary authorities were by 1984 becoming concerned about the capital losses and diverted resources flowing from investment by South African companies abroad.

Another development which ought to have caused South Africa's monetary authorities some concern – but which, until it was too late, did not appear to do so – was the unsound international financial dealings and practices of some of the large South African banks. These banks had from about 1982 been borrowing heavily abroad. Charles Grant (*Euromoney*, December 1985) gave the following foreign currency exposure for the main South African banks: Standard Bank $2.5–$3 billion; Barclays National $2 billion; and Nedbank $1.6 billion. According to the *Euromoney* Bank Report (October 1985) of the major South African banks, Nedbank as a wholly owned South African bank 'was freer to expand overseas in contrast to its bigger rivals, Barclays National Bank and Standard Bank of South Africa, whose British controlling parents took charge of their overseas operations' (Lind and Espaldon, 1986). The problem, as the *Financial Mail* (13 September 1985) noted, was that 'Nedbank which has been borrowing short-term in New York at lower inter-bank rates and turning these credits into long-term interest loans in South Africa (often to the government and parastatals) had been thought by US banks to be travelling in dangerous financial waters for some time'.

By July 1984, at the time of the collapse of the rand, the South African economy was in the midst of the worst and longest-running recession in its history. The economy was hit by high interest rates, excessive growth in the money supply, rising inflation, falling output and investment, as well as increasing unemployment and bankruptcies, a large balance of payments deficit, 'excessive' government spending, low domestic savings and a rising tax burden. Foreign debt was increasing and becoming ever more short-term in nature. But because the international financial market from 1982 was characterised by reasonably high liquidity and a shortage of 'good' borrowers, and because South Africa enjoyed an exceptional and enviable record of debt repayment, talk of a debt crisis, of debt rescheduling or debt moratoriums was noticeably absent.

It took the upheavals in South Africa's townships and factories in September 1984 to concentrate the minds of international bankers and local monetary authorities alike. Spring 1984 brought with it massive protests against the elections to the tricameral parliament, organised by the United Democratic Front (UDF), National Forum Committee (NFC), and the independent trade unions. 1984 also saw a sharp increase in strike activity, the number of working days lost being more than treble 1983 figures (Militz, 1985: 2). In general, there was a noticeable rise of unprecedented proportions in organised political protest by civic, student, youth and worker organisations. In July 1985 the state responded by declaring a partial state of emergency, in itself an admission of the seriousness of the situation.

The underlying tensions and crisis in the South African political economy could no longer be hidden under talk of 'political stability'. The risks involved in investment and loans to South Africa became strikingly apparent, and South Africa's foreign loan exposure at last began to appear problematic: well over 50 per cent of its debt was scheduled for payment within a year, and large amounts were due for repayment between 1985 and 1987. A financial report published in March 1985 wrote of South Africa's external finances as being in 'absolute chaos'. Worse still, the report went on, 'no regard whatsoever has evidently been paid to the maturity structure of this debt – with the startling result that, according to the Bank for International Settlements, no less than 66 per cent of the debt is due to be repaid within one year' (*FAR*, 1985).

Anti-apartheid and disinvestment pressure on US corporations and banks increased sharply. US banks in particular, from the last quarter of 1984, began to reduce their exposure. At this time, a proportionally higher percentage of the total lending by US banks to South Africa was in the form of short-term credits (85 per cent, compared to 57 per cent and 31 per cent for UK and West German banks, respectively). 'Because of this the liquidity of the South African economy was very sensitive to the decision of US banks regarding the roll-over of this debt' (Lind and Espaldon, 1986: 3).

South Africa was also heavily dependent for short-, medium- and long-term syndicated loans and for bond mangement and underwriting to a small group of West German and Swiss banks. On account of South Africa's heavy borrowing during the years 1982–4, even these banks who were very supportive of the financial needs of the apartheid state, were beginning to bump up against their internal credit limits and controls on exposure to the country. This new mood towards South African issues was noticeable in the response to a $100m bond issue for Eskom in July 1985. Although the yield was 11.5 per cent – almost 2 per cent higher than a US Ford Motor Company issue in the same week – the issue was poorly received and sold slowly (*South*, August 1985).

Although Swiss, German and Japanese banks initially filled the gaps left by the reduced US exposure, the continuing violence in the townships, the disinvestment campaign, and the underlying weakness of the South African economy made them ever more reluctant to raise their exposure to the country. It is important to stress that the view held by banks and other investors of the future potential of the South African economy had changed. As the *Financial Mail* (6 September 1985) pointed out, the economy had come to be seen as ill-equipped to cope with the world of the 1980s and 1990s, in contrast to the 1970s when there had been an explosion of foreign borrowing. Whereas that burst of loans had appeared to contribute to productive and infrastructural investment and technology inflows and increased productivity, and in these ways enhanced the strength, resilience and potential performance of the economy, the loans of the early-to-mid 1980s were overwhelmingly short-term credits related to loans and balance of payments difficulties, and reflected if anything the weakness and vulnerability of the economy rather than its strength.

When inflation was high and another energy crisis was expected, an economy which produced the greatest proportion of the world's output of gold as well as vast quantities of coal, was bound to do well. So it was believed in the 1970s. But at a time of low world inflation, a depressed oil market and an increasingly hi-tech-led growth syndrome, South Africa's traditional comparative advantage and exports appeared much less valuable. As the *Financial Mail* (6 September 1985) almost ruefully noted, 'All the old faithful standbys no longer apply.' The same report continued, 'The unrest is now so serious that any decisions to lend or to invest in South Africa are not just based on moral scruples – they are hard to justify by any measure of risk and reward.' Holden (1989: 26–8) has shown that at this time South Africa's debt-service ratio (104.0 in dollar terms and 118.1 in rand terms) was not significantly better than that of Brazil (132.6), Chile (153.3), Mexico (161.8) and Argentina (214.9). These figures suggest that South Africa, like other major debtors, was overborrowed in relation to exports, raising fears on the part of creditor banks about its ability to meet its current commitments out of its export revenue.

On 1 August 1985, a report appeared in the *New York Times* that Chase Manhattan Bank, the second largest US lender to South Africa, had ceased extending credit to South Africa. The report precipitated a rapid withdrawal by other US banks during August of about $400m – as part of their refusal to roll-over short-term debt due for payment. Some of this was taken up or offset by additional lending from other banks, including some Japanese banks, according to Lind and Espaldon (1986: 3). The value of the rand continued to fall (reflecting these tensions), dropping to a (then) record low of $0.353 on Tuesday 27 August 1985. Finally, on 1 September, as other foreign banks 'presumably indicated that they could not continue to counterbalance the US bank withdrawals, South Africa unilaterally proclaimed a moratorium on payment of the principal of the short-term debt' (Lind and Espaldon, 1986: 3). The moratorium was originally to last until 31 December 1985, but was later extended through the first quarter of 1986. South Africa's long and (most often) favoured relationship with private international banks and the international financial market in general had reached its nadir.

To summarise: The nature of capital accumulation in South Africa – in particular, the centrality of the gold-mining industry as a stabiliser and foreign-exchange earner, the heavy dependence on imports, especially of capital goods, and the racial form under which accumulation occurred – as well as developments in the international financial market, proved to be crucial in shaping South Africa's relationship with private international banks and with the international financial community in general during the period 1973–85. Between late 1978 and 1980 the soaring gold price brought in sufficient foreign exchange to finance imports and other requirements, including oil and armaments, in the wake of the curtailment of international bank credit after the Soweto uprising in 1976. And when the gold price declined after 1980, short-term international credit provided another palliative, so hiding the underlying weaknesses in the economy. 'We were faced with a big decline in exports in 1982 and 1983, and instead of adjusting we replaced lost real income with short-term [foreign] finance,' admitted Chris Stals, then deputy governor of the Reserve Bank (*Euromoney*, December 1985).

Private international bank lending and credit to South Africa's regime expanded rapidly in 1982–4. However, the panic brought upon the international financial community when South Africa's black townships exploded in sustained and organised resistance served to strip away, temporarily at least, some of South Africa's comparative advantages within the world economic system. Its enormous short-term debt could no longer be 'discounted' by virtue of its 'political stability', economic prospects or enviable debt-servicing and repayment record. The underlying weakness of the South African economy was exposed as was the gross financial mismanagement of its banking and monetary authorities. In the longer term, its growth and development prospects were perceived to be limited by its inability to match the newly industrialising economies. For foreign investors and international bankers, the liability of political instability and economic lethargy finally became too heavy to bear.

The factors used by international bankers in their assessment of a country's risk include political stability, national cohesiveness, adequacy and diversity of the resource base, the quality, effectiveness and efficiency of its economic and financial management, and its external financial position (Baughn and Mandich, 1983). By August 1985 none of these factors were running in South Africa's favour.

RECENT DEVELOPMENTS

Following the declaration of the standstill on debt repayment in September 1985, South Africa began the task of negotiating with the international banks to reach an agreement on the resumption of debt payments. This involved setting up a local negotiating team, the so-called Standstill Co-ordinating Committee (SCC) to put the South African case, and acquiring the services of an acceptable arbitrator. Dr Fritz Leutwiler, former head of the Swiss National Bank, was finally chosen for this job. The standstill was, as expected, extended to March 1986 as neither side was able to sort out even the most basic aspects of the problem, such as the total amount of South Africa's debt. At a meeting between the banks, the SCC and Leutwiler in London on 20 February 1986, South Africa and the major creditor banks reached an interim accord, so bringing a technical end to the repayment freeze imposed in September 1985. The South Africans were very satisfied with most aspects of the agreement – Stals, the head of the SCC, hailing it as a first step on the long road to normality for the country. South Africa agreed to pay 5 per cent or $500m in four phased instalments beginning in April 1986, on the approximately $10 billion in loans maturing by 1987. Interest rates were increased by 1 per cent and the major banks agreed to roll over the unpaid balance for a one-year period, by the end of which time it was hoped that a final agreement would be reached.

In March 1987, South Africa struck a three-year deal with its creditor banks, so further easing its foreign debt problems. In terms of the agreement, which ran until 30 June 1990, South Africa agreed to pay a total of 13 per cent of the outstanding capital amount – approximately $1.42 billion – of the debt trapped inside the standstill net, in six monthly instalments over the three years. The interest rate on capital still outstanding was raised by 1 per cent above the going London inter-bank rate. In addition a facility was introduced to enable creditors to convert short-term claims inside the net, to long-term claims outside the net, repayable over 10 years. This 'exit clause' became an important option, especially for US creditor banks such as Citibank and Manufacturers Hanover.

Business Day noted that the bankers were initially reluctant to commit themselves to a three-year agreement, but that 'confronted by a Third World debt problem that is threatening to run off the rails, international bankers, particularly those in the US, decided to knuckle under to the inevitable criticism which would follow a longer term agreement' (25 March 1987).

Significantly, neither this nor the February 1986 interim agreement compelled creditors to enforce repayments inside the net. Given the attractive rate of interest (1 to 1.25 per cent above LIBOR, the London inter-bank rate), banks would in any event have been tempted to roll over even these payments. As Hirsch points out, by October 1987, the Deutsche Bank, unlike most US and UK banks, had received no repayments at all, obviously choosing not to enforce their rights under the agreements (1988: 27). Furthermore, when South Africa announced its unilateral debt moratorium in mid-1985, it excluded trade credits from the standstill net; and when the banks decided earlier to refuse further loans to South Africa, the declaration specifically excluded trade credits from the ban. In other words, trade credits, critical in facilitating external trade, continued to be made available to South Africa. But in addition, as Jenkins (1989: 19) points out, trade credits have increasingly been used since 1985, as a substitute for other forms of lending such as bonds and syndicated loans. Thus, for example, Eskom has shifted some of its sources of funding for capital projects to trade credits (Jenkins, 1989: 19).

While South Africa remained economically weak and still politically vulnerable during the period of these debt negotiations, it was nowhere nearly as badly off by March 1987, as it was in the third quarter of 1985, when the country appeared to have reached close to the threshold of its vulnerability to international economic pressure. South Africa's status as a defaulter has been 'officially blurred' by the medium-term agreement of March 1987 (*Independent*, 25 March 1985). That, by any standard, was an improvement on its renegade debtor status of 1985.

Foreign bankers, who came to recognise that their bargaining power vis-à-vis South Africa was slipping away the longer it survived the initial trauma of the events of August 1985, were gradually forced to adopt a more conciliatory attitude in the debt-rescheduling talks. Their basic aim was to get their money back, and the desperation and urgency which characterised their actions in 1985 seemed to pass. The recovery of some degree of political stability in South Africa, albeit by extremely repressive means, facilitated this change in the banks' attitude. Any thoughts that may have been held in July–August 1985 that the South African regime was about to fall, so jeopardising the chances of the banks recovering their loans, were dispelled. This reduced the need to pressurise the apartheid state with the intensity (and consequent added risk of precipitating a default) that appeared to characterise their actions of mid-1985.

The actions of the banks were also affected by other considerations. There was, for example, some degree of concern among banks about the effects of a possible South African default on the stability of the international financial system. The threat of an international financial crisis did, however, begin to recede shortly after 1985. Secondly, the decision by the government to exclude the 1982 IMF loan repayments from the net, that is, to continue repaying the loan, also appears to have been important to bankers. Given the nature of the relationship between private international banks and the IMF, South Africa appears to have recognised the importance of not aggravating further its already difficult relationship with the Fund. By the end of 1987, South Africa had paid back the IMF loan in its entirety.

Perhaps most important among the factors influencing the banks was that South Africa had made the painful though necessary 'adjustment' on its balance of payments on current account. That this surplus was largely based on a rise in the gold price was of no consequence. International banks were apparently pleased at South Africa's success in generating the reserves to pay off at least part of its debt.

By early 1989 there was further evidence of an improvement in South Africa's relations with private international banks. This is clear from the following examples. First, in March, the South African government negotiated a R200m loan with an unnamed bank, reputedly to have been Swiss. Although the amount was small, it created optimism among financial officials and enabled them to budget for new foreign loans for the first time in three years (*Business Day*, 17 March 1989). In general there appears to have been a partial improvement in loan flows to South Africa after 1986, especially from Swiss banks. According to Ovenden and Cole (1989: 78) Swiss bank loans to South Africa, which fell from SFr300m in 1984 to a low point of SFr38m in 1986, increased thereafter, rising to SFr115m in 1988.

Secondly, two of South Africa's US creditor banks, Citibank and Manufacturers Hanover, came to an agreement with South Africa in April 1988 not to call in their loans of $670m and $230m, respectively, when the February 1987 interim agreement on debt rescheduling expired in June 1990, chosing to take the 10-year exit-option on conversion into long-term loans. These banks were the only two financial institutions which publicly admitted taking the exit option. Clearly, however, over the three years since the second interim agreement which offered this 'exit clause', more banks took the option than just these American banks. By October 1989 only $8 billion was left inside the net, which fell due for payment in June 1990, after the conversion of $4 billion into longer-term loans under this option (*Business Day*, 17, 18 October 1989). There exists some dispute about the extent of conversions by US banks under this scheme. The US Assistant Secretary of State for African Affairs claimed, in mid-October 1989, that most of South Africa's US banks loans had been rescheduled under the 10-year plan. Figures released by the US Federal Reserve Bank, however, showed that 'more than half of South Africa's debt to US banks falls due by the end of June 1990', and that debt repayable within a year or less stood at $1.27 billion out of the total short-term debt of $2.43 billion (*Business Day*, 17 October 1989).

The restoration of some degree of political stability and the diminished prospects of, what banks in the mid-1980s viewed as an 'unsettling' and rapid transformation to black majority rule, do not necessarily or automatically mean that South Africa can now expect unlimited access to new foreign loans in the years ahead. Partly this will depend on developments in the international financial markets as a whole and on the state of the world economy. An important factor here could be the emergence of Japanese banks as major lenders and bond managers in the 1980s. Japanese banks were responsible for 55 per cent of the increase in international bank activity between 1986 and 1987 (Flight and Lee-Swan, 1988: 157). There are also major new banking centres in South Korea, Hong Kong and Taiwan, countries with which the South African state has maintained good economic relations. While Japan has recently been somewhat embarrassed by its importance in South Africa's foreign economic relations, Far Eastern banks may in future prove to be sources of considerable international credit for this country.

One indicator relevant here is evidence of growing foreign direct investment in South Africa. The *Sunday Times* (28 February 1988) reported that 55 companies, of whom 42 were foreign, intended to invest more than R66m in South Africa's decentralised areas in the following year. The preponderance of Far Eastern companies among this number is striking. These companies were involved in production in a wide variety of goods, from clothing and textiles to office furniture and power tools. In addition to these direct investments and the bank loans mentioned above, some companies made equity investments in South Africa in 1987.

CONCLUSION

The institutions set up at the end of the Second World War, such as the IMF and World Bank, which facilitated economic growth as well as social prosperity in the advanced capitalist countries, were also to make their impact felt in peripheral and semi-peripheral countries like South Africa. At the same time, South Africa adjusted to the new conditions, demands and possibilities of the post-war international economy with some degree of success, through the rationalisation, concentration and centralisation of production, and the use of foreign technology and capital.

Similarly, when the Western industrial economies entered troubled times from the late 1960s, the international crisis affected countries like South Africa, in terms of the external shocks delivered (the two oil-price hikes, the decline in the gold price), as well as what Piore and Sabel (1979: 166) call the 'limits of development' (for example, the saturation of certain markets) in the post-war system. However, South Africa's inability to adjust and diversify its economic base, the dogmatism and lack of vision which characterised the responses of South Africa's economic policy-makers and planning agencies, and the instability and uncertainty generated by its racially based political system in the face of rising international pressure, especially after 1976, all acted in ways that exacerbated the effects of the oncoming international crisis.

Just as 'more enlightened crisis management' (Piore and Sabel, 1979: 166) may have spared the world of the 1970s and 1980s the worst of the crisis – though there would undoubtedly had to have been critical reforms of the system – we would argue that the adoption of more innovative and appropriate strategies by South African monetary, fiscal and exchange-rate policy-makers, as well as development planners, could have better insulated South Africa from the worst effects of these international developments and assisted growth as well. Had at least some of the more far-sighted development strategies and policies been pursued early enough, possibly in the 1960s at a time of prosperity, it is probable that South Africa's need for IMF assistance, its timing and amount, and the burden of repayment which this brought about, especially in the illiquid circumstances of 1976–8 and again in 1985–6, might have been very different. Similarly, the disastrous consequences of the uncontrolled forays into the international financial markets from the late 1970s – with the encouragement of the Reserve Bank, it must be added – may have been averted. Certainly, the mismanagement of the country's financial operations – 'justified' in terms of the free-market zeal of the monetary authorities – contributed in no small way to the financial chaos of mid-1985 and after. Under different and more enlightened economic and political circumstances, the nature and structure of South Africa's relations with the IMF and the international banks may have been somewhat different from the pattern they eventually assumed after 1970.

South Africa's relations with the international financial community became more involved, complex and costly after 1973 – though clearly not as costly as for countries like Jamaica – when compared to the entire period since the Second World War. This was the result of both international developments and the problems of accumulation in the world capitalist economy, and local economic and political developments and policies. These factors combined to produce a particular balance in the power relations between South Africa and the IMF and the international banks, a balance which tilted away from South Africa as the level and structure of the country's indebtedness changed dramatically. By mid-1985, with IMF assistance already circumscribed for two years and the international banks calling in their loans, South Africa had reached a position of extremity.

Following the interim debt-rescheduling agreements of 1986 and 1987, however, the continued availability of trade credits and the resumption of some, albeit small and irregular, amounts of bank loans to South Africa, the country's international financial relations have improved somewhat, especially when compared to the parlous state of this relationship in mid-1985. Despite this improvement, the restrictions on IMF balance of payments assistance and the limited access to international credit and capital exert a continuing pressure on foreign-exchange reserves and hence on the restoration of more robust and sustainable economic growth.

The degree and extent of pressure on the balance of payments, on foreign-exchange reserves and ultimately on growth prospects are in turn linked, among other factors, to the effectiveness of the international sanctions campaign. This campaign has in the years since 1987 come to be concentrated on financial sanctions. More especially, anti-apartheid activists have attempted to pressurise South Africa's creditor banks into extracting political concessions from the South African government as a condition for rescheduling the debt which fell due in June 1990.

Significantly, and indicative of the very changed political environment of 1989, the more recent call in this regard was less dramatic than the earlier demand that the South African government resign as a precondition for rescheduling. However, the decision by some banks to take the 'exit option' of converting claims inside the net which were due in June 1990 into long-term loans outside the net, eased the pressure on South Africa's external finances, and reduced the potential value of the debt-rescheduling issue as a political weapon.

However, in a recently published book commissioned by the Australian government, Ovenden and Cole argue that the conversion option taken by some banks signal that they were indicating 'once and for all that they were done with South Africa', and that they would 'have no more to do with the country' (Ovenden and Cole, 1989: 96). Ovenden and Cole make this generalisation on the basis of an interview with one official of an unnamed bank. This assessment ignores the historical evidence of the wide variations in the responses of private international banks to all aspects of their relationship with South Africa in the period since 1970. Furthermore, their conclusion is based on banks' perceived moral objections to apartheid and ignores wider economic and other forms of political influences. We would argue that private international banks are likely to assess their attitude to and exposure in South Africa as well as their stance on rescheduling, even in the period before apartheid is finally abolished, as their differential responses in the fluctuating economic and political environment of 1984–9 demonstrate. This on-going assessment is likely to take account of both changing global and domestic (South African) factors, and not only their 'opposition' to the apartheid regime, important as this factor may be for some banks in some countries.

For we would argue that the actions of banks in August 1985 were based on the normal criteria of country risk analysis – that is, political stability, current exposure levels and debt maturity structure, economic performance, future growth prospects and the like. Political strategies which rely on the belief that banks were and are genuinely concerned to make a positive contribution (as opposed to the self-interested gesture of August 1985) to the South African struggle by seriously and steadfastly posing the alternative of a non-racial democracy in this country as a precondition for the restoration of normal international financial relations, would, in our view, be based on a fundamentally misconceived and romantic understanding of the real dynamics of international banks. The same can no doubt be said for multinational corporations in South Africa. When they coincide with other strategies and in specific political conjunctures, financial sanctions may well prove to be a critical element in campaigns to bring about political change in countries such as South Africa. These possibilities clearly existed in 1984–6, and the actions of the banks in regard to rescheduling South Africa's debts might have been decisive then, had they been firm in demanding political change. But for reasons we have set out above, they were not.

By 1987 the political conjuncture made it less probable that creditor banks would insist, as a condition for rescheduling, on more significant political change towards a non-racial, democratic South Africa and less likely that any watered-down form of bank pressure or intervention would in fact succeed. Gelb (1988: 73) has recently argued on the general question of the timing and appropriateness of sanctions against the apartheid state that 'the South African state is not the Iranian or Philippine state: it will take more than US pressure, even from comprehensive sanctions, to bring it down. To break with this approach – to redefine the meaning of sanctions 'working' – requires locating sanctions very explicitly within the context of the wider liberation struggle. This means understanding sanctions in relation to the existing balance of forces, and the overall direction of strategy. What is crucial here is that this is always shifting.

POSTSCRIPT

On 18 October 1989, the South African authorities announced details of a third interim debt rescheduling arrangement. The announcement was clearly timed to undercut a Commonwealth financial sanctions campaign to be announced the same day.

In terms of the agreement South Africa will pay $1,5 billion or 20.5 per cent of the $8 billion inside the standstill net in varying amounts over an extended period of three and a half years from June 1990 to December 1993 at one percentage point over applicable base lending rates (*Business Day*, 19, 23 October 1989). Banking sources observed that this one point margin over the base lending rate was 'higher than expected for a country of South Africa's financial standing under current market conditions' (*Business Day*, 23 October 1989).

Creditor banks retain the option agreed upon in March 1987 of converting debt inside the net into longer-term loans outside the net, by which arrangement some $4 billion had already been so converted and rescheduled between March 1987 and October 1989. The first instalments of such conversions in terms of the latest arrangement only falls due at the beginning of 1998. In addition to the debt covered by this agreement, the country owes another $12 billion 'outside the net' (*FM*, 27 October 1989). These loans have to paid when each falls due. Nevertheless, the agreement announced in October 1989 appears to have been made with a view to synchronising debt repayments inside and outside the net: redemptions of debt inside the net will be lowest when redemptions of debt outside the net are highest, so relieving some pressure on the balance of payments and foreign reserves.

This third interim agreement presents a significant easing up in South Africa's debt situation, although the fundamental difficulties faced by the country's capital account since the mid-1970s remain. This point should be emphasised: the agreement represents a further easing of, rather than a long-term solution to, the problems and pressures on South Africa's international financial relations. The prospect of a long-term solution to these problems was created by the unbanning of the ANC and other organisations in February 1990, making possible a negotiated political settlement. Spokesmen for big business and the present government have called for a resumption of loans and other capital inflows, which they see as a prerequisite for raising the economic growth rate.

At the same time, however, political change has intensified the debate about economic policies for a future non-racial democracy. In this context, the role of international lending, and of lending institutions, remains unclear. On the one hand, the recommendations of the ANC–Cosatu workshop in Harare in April 1990 suggest that the main emphasis in the future should fall on domestic savings, with foreign capital as a supplement. By contrast, Nelson Mandela's address to British businessmen on 4 July 1990 emphasised that South Africa's 'critical need' for rapid growth 'cannot happen without large inflows of foreign capital' (*Business Day*, 5 July 1990).

The desire to meet the demands of foreign lending agencies may well influence the wider debate. Like South African capital and the state, the IMF, the World Bank and private international banks generally favour market-oriented policies with minimal state intervention. In contrast, there is strong pressure within the ANC alliance for the state to play a leading role, and for a relatively (though not exclusively) inward-looking strategy. This approach might well involve some degree of nationalisation and control over the activities of foreign investors.

During the 1980s, restricted access to international lending was an important source of pressure for change in South Africa's domestic politics. During the 1990s it might well become a source of pressure for change in economic policy. At the same time, however, broader developments in the international economy could also help to limit foreign capital inflows. These developments include more intense competition for funds (most prominently from Eastern Europe), changes in the nature of their supply (witness the recent decline in the volume of syndicated bank loans, on which South Africa relied heavily in 1979–84), continuing uncertainty over exchange-rate coordination, increasing financial deregulation and securitisation.

These developments have generated enormous tensions and policy dilemmas within the international economy. Increasingly they point to a vacuum created by the failure to formulate internationally agreed rules of conduct which take account of the changes in the structure of the world economy since the late 1980s. The search for a new mode of global regulation is proceeding in an *ad hoc* manner, by a process of trial and error. The role of international finance in South Africa's future will ultimately be determined by the success or failure of this process.

5
South African gold mining in transformation

BILL FREUND

The centrality of the gold mines to the South African economy hardly needs to be pointed out to an informed audience. The gold mines have been fundamental to and instrumental in the implantation of the capitalist mode of production in the subcontinent. They have shaped the character of society and economy more than any other enterprise. Any serious examination of the South African economy indubitably involves a careful monitoring of the industry.

However, this chapter seeks to go further, for two reasons. An important thrust of the literature on the South African economy suggests that the dominance of mining has been increasingly giving way over the past forty to fifty years to manufacturing. One vital point to make is that this trend has been halted, or even reversed, in recent years. In the mid-1980s, gold returned to representing more than half the value of South African goods exported. It provides perhaps one-tenth of the state budget in direct taxes. The stagnation in most sectors of secondary industry has meant that the profits in the Anglo American empire come more and more from mining, with industry providing under 20 per cent and gold alone up to 40 per cent (*FM*, 2 October 1987). In 1978, capital expenditure in mining represented some 20.4 per cent of the value of expenditure in manufacturing. That figure rose to 25.5 per cent in the high gold-price year of 1981 and to 41.2 per cent in the recession year of 1984. In 1985 and 1986, the nominal value of mining investment rose again by more than 60 per cent (projected 68 per cent) while it stagnated in manufacturing. Even in strife-torn 1985, the real increase was 11 per cent (Chamber of Mines Annual Report, 1985). Capital expenditure in gold and uranium mining touched R1911m (*SAM*, May 1986). In sum, gold mining is actually occupying a more and more important role in the economy again.

Secondly, and this is the heart of the argument of this chapter, the relations of production in gold mining are beginning to change quite substantially. From the time of the Rand Revolt in 1922 until the 1970s, South African gold mining was dominated by certain features that were slow to change. These included the nature of underground supervision and control, the ratio and relationship of black and white miners, and the massive recruitment operations that brought most black miners in from outside the borders of South Africa. In addition, of course, lay the major constraint imposed by the fixed price of gold on the world market, always a basic factor in profitability. Despite the argument in some of the management-derived literature, technology has at certain points developed substantially, in the interests of controlling workers and reducing worker numbers, notably in the highly profitable years after 1933, when South Africa went off the gold standard.[1] Otherwise these aspects of gold mining in South Africa are very well known from the literature. It is only logical that such writers as Francis Wilson and Frederick Johnstone felt that the great battles in the history of the industry belonged to the early years of this century, culminating in the strikes and political contests of the early 1920s (Wilson, 1972; Johnstone, 1976). However, the entire set of circumstances so long dominant has been subject to pressure of growing intensity for the past fifteen to twenty years. Labour discipline, labour recruitment, technology, and the economic constraints of the gold trade itself are all in a state of flux, and as a result the available literature on gold mining is of declining value as a means of understanding contemporary conditions.

A clear theoretical comprehension of the integrated nature of these factors is essential. One cannot simply isolate out 'cheap labour', or the price of gold, or the nature of gold-bearing ore to explain conditions of capitalist accumulation on the gold mines. In recent years, using Aglietta's totalising notion of *regulation* in relation to the historical process of accumulation (Aglietta, 1979), a school of thought has emerged that tries to refine marxist ideas about crisis and the cyclical nature of capitalism, as well as the relationship of the economy to society and politics. Particular regimes of accumulation with particular social and political parameters determining class relationships and workplace conditions succeed each other under the pressure of crisis, though crisis does not imply the threatened collapse of the entire capitalist economy. The sea change which the international economy began to experience at the beginning of the 1970s and which has far from entirely worked its way through, has obviously been the inspiration for this approach. What this chapter suggests is that the South African economy is experiencing its own version of this sea change, and the gold-mining industry in particular needs to be seen in terms of a qualitative change in the social conditions of accumulation that prevailed over a half-century.

GOLD MINING UP TO THE 1960s

It may be useful to summarise some of the main features of gold mining in the 1960s, a decade that now seems to represent the end of an historic era. In 1970, the production of gold in South Africa peaked at 1 000 tons. After the Second World War, expansion had occurred on the basis of major new discoveries in the northern Orange Free State, around Carletonville on the Far West Rand, and around Evander, south-east of Johannesburg. South

Africa produced up to 70 per cent of the 'free world's' gold supply at profit levels which were modest in proportional terms, but huge in absolute terms, given the fixed character of the gold price.

Labour relations were notoriously despotic, corresponding to the colonial conditions distinctively identified by Michael Burawoy on the Zambia copperbelt (Burawoy, 1985). Jobs were defined racially, whether by statute or custom. White miners received much of their wage in the form of production-related bonuses and had an interest in pushing up output and maximising profits. Between the white supervisors, the black bossboys and the mass of labourers, coercive conditions were modified by relations of clientelism and informal ties of many sorts (Gordon, 1977; Moodie, 1980). Only on the surface could one find an important number of skilled, mainly clerical, black workers.

White trade unionism was not strong or organisationally effective to any extent but white voters could count on the patronage of the South African state. The number of jobs available for whites was steadily increasing and the real wages of whites had expanded absolutely and proportionately when compared with those of blacks. In 1972 whites earned approximately two-thirds of the entire wage bill, although they only made up 10 per cent of the workforce. By contrast, as Wilson has shown, the real wage of black labourers was no higher than it had been before the First World War. Although somewhat larger in money terms than what the farmworker earned, the miner's wage, paid partly in kind, could not support an urban family, even in poverty. Despite the immense costs involved in organised recruitment, the majority of black miners came to the gold mines from outside South Africa proper and were housed in compounds. Capitalist exploitation of the workforce necessarily 'articulated' with precapitalist forms of production to ensure the survival of the workers. Urban dwellers were not significantly found in the ranks of the mineworkers.

Earlier in the century, the relationship between the state and the mining companies had been most problematic but this was no longer the case by the 1960s. One big mining house, Gencor, was born with the blessing of Anglo American but had strong links to the Afrikaner business establishment. Most mining shares were owned in South Africa by South Africans. Crude attempts at milking the mines (and in the 1940s Afrikaner nationalists still talked about nationalising them) had given way to sophisticated taxation policies designed to maximise mining employment, expenditure and articulation with the industrial economy that was then growing rapidly in South Africa. Between the mining houses and the National Party government there were no very serious disputes. Anglo American, the strongest and ostensibly most liberal mining house, did not even try to use the official quota of 3 per cent of exemptions from the rule barring the establishment of black family housing on the mines (*SAM*, January 1985, statement by Gavin Relly).

Conditions in the secondary industrial sector, which rapidly expanded in the generation after the Second World War, allowed for very substantial profitability. But it is important to stress that the industrialisation of South Africa rode on the back of the gold-mining industry. The gold mines provided foreign exchange, attracted foreign investment and were competitive on the world market, while industry rested on a protected local market or on supplying the mines.

THE DIVIDE OF THE EARLY 1970s

There is general agreement on the historical significance of the world-wide economic downturn at the beginning of the 1970s. These years were marked by the collapse of the dollar-based Bretton Woods monetary system, rapidly mounting inflation in most countries, and the sudden impact in 1973 of dramatic increases in the price of oil. The money value of gold shot up several fold and, while the value has continued to fluctuate, it has been of a different order than before. The historic constraints on gold mining imposed by the fixed price regime have been lifted.

The most obvious result is that gold-mining capitalists have been able to increase investment in mining in a very substantial way, with regard to exploration, technical development and wages. This has made qualitative change on the mines *possible*. However, at the same time the effect on gold mining in other parts of the world has been yet more dramatic; as a result, the proportion of the world's gold mined in South Africa has steadily declined. In fact, South African capital has itself made important investments in foreign gold mines.

Simultaneously, technology constraints and labour problems came to make change *necessary*. The relative coolness of the ground has been a factor in making it possible to extract gold ore at exceptionally deep levels in South Africa. By the 1970s, however, levels were beginning to be touched that posed problems for the established methods of cooling mineshafts. Further developments began to require exploring where rock temperatures rise to 60°C and more. The deeper the mine, the greater the possibility of seismic disturbances or rockbursts. Although the safety record of gold mines, measured by the fatality rate, has steadily improved over time, in the category of rockbursts it has proved hard to lower the number of accidents. Ultra-deep mining without further technological improvement is bound to lead to mounting injuries and deaths. At the same time, given the absence of any spectacular new gold discoveries, larger and larger amounts of ore mined with more and more labour (and greater labour costs) yield a declining amount of actual gold, thus propelling the effort to spend and innovate in order to provoke a countervailing trend.

At the same time, labour relations in the mines have entered a period of unprecedented turbulence and violent conflict, the full story of which needs to be explored and explained elsewhere. Regional politics threatened the smooth operation of labour recruitment from two major sources, Malawi and Mozambique (the latter country after independence enjoying very poor relations with South Africa). The mining companies responded by rapidly raising black workers' wages and looking around for new zones of recruitment. While a new stability in labour sourcing seemed to be taking hold towards the end of the 1970s, the mining companies have been reconsidering the entire system of subsidised recruitment and dependence on migrant labour. In any event, their labour force consists of returning workers, with few new entrants each year even amongst the migrants (Crush, 1987). At the same time, in an attempt to contain the crisis of control, Anglo American, the biggest mining house, permitted trade unions to organise amongst their workers and, in consequence, the National Union of Mineworkers, taking advantage of this opening, has been able to establish itself as a very large and militant force on the minefields. The continuity of the labour force has aided the organisational efforts of the NUM (Crush, 1987). From the point of view of capital (which still does contain important anti-trade union elements), this is part of a process of establishing a new structure of control, of 'industrial relations', on the minefields. As workers win higher wages, moreover, the pressure to increase productivity and trim the labour force mounts, and the mining houses are in this way impelled further on a course of technological change.

Broadly speaking, what follows is intended to explore the nature and prospects of the gold mines, in the light of the mounting pressure on capital for change. In this industry, the rate of change is quite slow. Large-scale investments require many years to bring results. From exploration to development of a new mine can take five to ten years or more. Although it is possible to sketch the outlines of change, it seems reasonable to suggest that a process has begun to unfold whose full outline is still only vaguely apparent and that this process will take many years to crystallise in detail.

GOLD MINE EXPLORATION

For many years after the Second World War, gold mining emphasised expenditure on development over exploration. Only in the late 1970s did this substantially start to change. From a peak in 1970, the amount of gold mined in South Africa proceeded to fall by more than 30 per cent to 685 tonnes in 1985 and further to 607 tonnes in 1987. The grade of ore mined fell from 13.28 grammes per tonne in 1970 to 6.1 grammes in 1985 and 5.28 grammes in 1987. Put another way, while the tonnage of ore mined increased from 75 million tonnes in 1970 to 107 million tonnes in 1987, the actual amount of gold produced still fell (*SAM*, May 1986: 61, Tom Main speech, Chamber of Mines of South Africa, 1987–88 review). These figures register the impact of South African law, which, in order to protect low-yield mines, makes it legally obligatory to mine lower grades in response to particular price levels. As will be discussed below in more detail, state fiscal policy is also directed at heavily encouraging the expenditure of profits on the exploration of new mines. None the less, it is clear that had the mining companies chosen to concentrate on higher-quality ore, the overall life of the mines would be declining more rapidly and the need for new development would be ever more pressing. Thus it is not only state policy which has led to unprecedented expenditure on exploration (*MAR*, 1987).

The days of gold exploration dominated by intrepid prospectors and blind luck are long gone in South Africa. Exploration is an increasingly capital-intensive operation on which some R300–R400m was spent in 1987 (*SAM*, August 1987: 43; *FM*, 27 November 1987). Computerisation of the exploration process has begun (*SAM*, June 1987: 43). So far, there have been no spectacular new finds. Instead, development is focused on the logical extension of existing mines, on the exploitation (based on new technology) of abandoned gold mines – such as the Princess Mine near Roodepoort on the West Rand which closed in 1920, or the Eersteling Mine, South Africa's oldest, discovered in 1871, near Pietersburg – and the probing of exceptionally deep levels.

Thus new sections of major mines such as Kloof (Gold Fields-owned, Far West Rand) and Randfontein Estates (JCI, West Rand) are scheduled to open up (*MAR*, 1987). After six years of work, shaft one at Western Deep Levels opened in 1986. This is the deepest mineshaft in the world, with a depth of 3 582 metres, where miners cope with a rock temperature of 55°C (3 793 metres according to Elliott, 1986: 104). On this and another new shaft at Western Deep Levels, Anglo American will be spending R1.4 billion by 1992, according to one estimate (*Economist*, 8 June 1985; *SAM*, September 1986).

Much of the exploration is being conducted in 'gaps' between the major extant minefields, such as the Bothaville Gap south of the Vaal River and north of the Orange Free State gold mines, and especially the Potchefstroom Gap between the Far West Rand and the mines around Klerksdorp. It has been estimated, however, that to reach the gold of Potchefstroom will require penetrating to levels of 4 000 metres below the earth's surface (*FM*, Mining Supplement, 13 November 1987; *SAM*, June 1987: 5). New mines are apt to be 'low-grade and deep' (Tom Main in *SAM*, May 1986). It may finally be noted that Anglo American seems likely to consolidate and expand its strong position in gold mining in South Africa on the basis of this pattern of new developments (*FM*, 27 November 1987: 84, Mathison and Hollidge report to stockbrokers). Gencor expenditure on expansion is, however, also very substantial, at over R100m per annum (*FM*, 25 September 1987).

According to a recent journalistic estimate, 'the industry badly needs to find new reserves and open new mines merely to retain its present position' (*FM*, 27 November 1987: 84–5). This is in fact occurring, and some estimates suggest that a slight increase in the amount of gold mined in South Africa can be foreseen by the start of the 1990s (*Economist*, 8 June 1985). However, the costs for maintaining production at the 600–700 tonne level become increasingly high.

TECHNOLOGICAL CHANGE

The furtherance of technological change in South African gold mines has much in common with the main characteristics of new exploration and development: it is an economic necessity that is, however, extremely costly. (For technological change, see Hermanus, 1987, and the thoughtful assessment in Lever and James, 1987: 1–13). Research and development sponsored by the Chamber of Mines took off from a rather modest base in the mid-1960s but has become more and more significant since. The watershed was marked by the new price structure that emerged from 1973 onwards. In so far as direct mining techniques are concerned, the most remarkable new tool is the hydraulic drill, which is twice as productive as the pneumatic drills used presently. The hydraulic drill is also perhaps ten times more energy-efficient but it is a more expensive device whose use is just becoming feasible (*AMM* Circular, 3/1986; *SAM*, May 1986: 38–42). In particular, it becomes more practicable with the intensified use of water-cooling systems integrated into hydro-power systems (*SAM*, October 1987: 17). In one new mine, Harmony, sits the 'largest ice plant in the world' (*AMM* Circular, 2/86), that uses 1 000 tonnes of ice per day (*SAM*, May 1986: 42).

Expensive experiments are being run on structural changes that will make ultra-deep gold mining possible. Thus some thirteen mines by the end of 1986 were experimenting with backfilling, which would, it was hoped, lead not only to safer mining (in minimising rockbursts) and to savings on wooden pillar construction but would also avoid the huge wastage created by everthicker rock pillars left unworked so as to preserve passageways (Chamber of Mines Annual Report, 1986, E. P. Gush address). It remains a controversial technique (*AMM* Circulars, 1/85, 3/85, 2/86, 1/87).

The largest single non-labour expense within new mines involves underground transport. Here the big shift is towards rubber-tyred trackless vehicles that can negotiate inclined ramps. In Western Areas, the application of trackless transport is fairly advanced. A revealing article by the Northern Division manager, A. C. Naudé, pointed to the advantages it produced: higher productivity due to lower manning levels; 'concentrated supervision'; ability to reach previously inaccessible ores; saving on support material needed underground; better environmental control and reduced fire hazard; its suitability for a more skilled, stable labour force (*AMM* Circular, 2/86). None the less, trackless mining is also not without problems. Difficult and expensive to maintain, it has been unsuccessful in some experiments (Lever and James, 1987: 10).

We shall return to the whole thrust of the Naudé article from the perspective of labour control and labour process, but it should also be stressed that technological change can be associated both with new shaft sinking and new mine development and with the consolidation of existing mines. The most spectacular instance of such consolidation took place with the creation of Freegold in 1986 out of the President Brand, President Steyn, Western Holdings, Video and F. S. Geduld mines in the Orange Free state, ostensibly to allow for maximum exploitation of the ore and of capital assets, as well as to strengthen the financial base of operations and to prolong the life of the whole complex. This, the largest gold mine in the world, employed an estimated 105 425 workers at the time of its foundation. The old mines were based on the boundaries established by specific farm leases; Freegold is intended to exploit the actual structure of geological space (*SAM*, February 1986). Unit working costs at Freegold rose markedly less than the rate of inflation in its first year of operation (*FM*, 2 October 1987).

The emphasis in new research is to consider mechanised change in a systemic way, not merely in terms of particular isolated innovations (Clive Knobbs in *SAM*, May 1986: 42). The gold mines have been very interested in the application of computers and information technology to exploration, in order, what is more, to identify the site of seismic stresses and to allow management a clearer and quicker picture of what is going on underground at any time (Chamber of Mines Research Organisation Annual Report, 1986: 23ff). 'The development of sophisticated computer mine design models' has been predicted (E. P. Gush address in Chamber of Mines Annual Report, 1986). By 1986, the mines were pioneering 'the most advanced use of computers in deep-level mining technology anywhere in the world' (*SAM*, July 1986: 86) and probably are today South Africa's (and Africa's) single major user of computers.

Technological innovation in the gold mines has two features in common with exploration: the rising capital costs that make it possible and the need to confront the absolute barriers to ultra-deep mining presented by the techniques historically in operation. However, there is another element of cardinal importance: the growing emphasis on increasing productivity. Horst Wagner, director general of the Research Organisation of the Chamber of Mines, stressed in 1986 that this must be the most important single feature in mine development henceforth (*SAM*, October 1986: 7). The new emphasis on productivity contains a growing urge to reduce the amount of rockface that has to be drilled in order to obtain given amounts of ore, and reconstructing the design of stopes (ultimately with computer assistance) to minimise waste in ore. This becomes more crucial as ultra-deep levels, on the basis of present technology, require more and more supports and narrower and narrower stopes. Planners hope that they can achieve a 50 per cent increase in face advance per month and thus can mine one-third less face for the same ore value (*FM*, 13 January 1987).

In addition, new technology is being put forward specifically with the idea of reducing manpower levels. Such novelties as trackless mining and hydraulic tools will reduce the demand for labour markedly, as the Naudé report on trackless mining suggested. One rationale is to reduce the human input at increasingly dangerous levels with potentially inhuman working conditions. However, it is as much or more a question of cutting costs and using lay-offs as a means of disciplining an increasingly class-conscious, militant and skilled community of workers (*Mining Journal*, August 1986). Just as technological change must be seen as part of the overall context of changing mining design, that design has as well a plan for labour organisation, control and costs imbedded within it. (For an example of this approach, see the comments of Gavin Relly about the 'regrettable' impact of mechanisation, *FM*, 2 October 1987.) In 1984, the Chamber of Mines signalled this approach by approving a wide-ranging, coordinated research scheme to concentrate on labour reduction and changes in the labour process (Salamon, 1986: 97-98).

THE CHANGING LABOUR MARKET

In order to explain how changing labour relations impact with the introduction of new levels of mechanisation and mine expansion, several factors need to be isolated. The one which has occasioned the most comment in the literature has been the shift in the sourcing of migrant labour. At the same time that migrant labour arrangements with Malawi were suddenly cancelled, severe snags developed in the Mozambican recruitment operations which mine management feared would herald the end of labour supplies from Mozambique. In practice, the Frelimo government never has felt able to take such a step; indeed, quite the contrary (James, 1987: 11-15). Jeeves and Yudelman have provided a convincing account of how this was experienced as a crisis which forced capital to cope with large numbers of poorly trained recruits, and rapid turnover, and to raise black labourers' wages with extreme rapidity (Jeeves and Yudelman, 1986).

The crisis has not yet been fully resolved by any means. At the simplest level, there seems to be a commitment to flexibility in labour sourcing (Yudelman and Jeeves, 1986: 123; James, 1987: 32). In practice, Malawi again sends the Chamber of Mines many workers at present, as does Mozambique, which would like to send far more. However, and for the first time in this century, mining capital has been less and less committed to migrant labour. In 1980 the social scientist Merle Lipton was commissioned by Anglo American to write a highly condemnatory piece on migrant labour for their in-house journal *Optima* (Lipton, 1980). The very divided corporate reaction to this article revealed the extent to which strong sentiment was split on the issue. Migrant labour was felt by some to be a cause of the increasingly violent atmosphere in many workplaces, disrupting work and spurring resistance to management (Lever and James, 1987: 21). Hostels have become union strongholds (James, 1987: 8). While black workers do not all object to migration, they do object, and increasingly articulately, to the associated controls and lack of choice. The National Union of Mineworkers has tried to fight for the right of migrants to come to South African mines and work as always, but they have equally condemned the compound system. In the past decade, short-term migrant labour has in any event almost entirely fallen away. Migrants today are largely men who sign up and re-recruit for their whole working lives, and they more and more possess work skills that are not immediately acquired by new arrivals with no experience.

The wage hikes of the mid-1970s, moreover, created the possibility of greater numbers of voluntary workers coming to the mines without elaborate

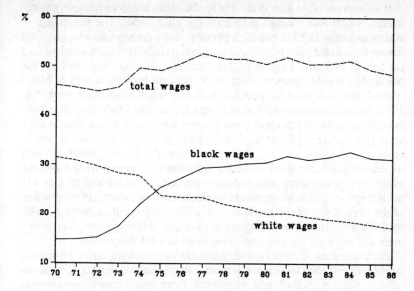

Fig. 1. Wages as a percentage of total costs

recruitment expenditure. On the whole, Teba recruitment stations in southern Africa are being rationalised and, in some regions, closed down (Crush, 1987). It has been suggested that labour recruitment in a distant place such as Malawi is simply becoming less and less worthwhile. In this case, exclusion of migrants on the suspicion of bearing the Aids virus may be a convenient excuse for dropping the recruitment system in time.

During the acute crisis years of the mid-1970s, mining companies experimented quite extensively with the employment of urban labour (James, 1987; Jeeves and Yudelman, 1985). While this has not entirely been abandoned as an option, it has been discredited as the main alternative for the mines. James has stressed that mining management finds urban labour too educated or too sophisticated, undisciplined and unprepared to accept the harsh circumstances of the industry easily. In addition, there is little reason to expect that wages, albeit much higher than they were in the 1960s, will be allowed to reach levels that would be competitive with jobs in secondary industry, unless accompanied by very much smaller workforces.

While there is general agreement as to the qualitative leap in wages levels after 1973, thereafter the opinion of organised labour and its antagonists differs. Thus, a professional journal estimated that real wages had risen about 40 per cent between 1975 and 1985 while a management source claimed that they had gone up 85 per cent in real terms between 1977 and 1987 (*Mining Journal*, August 1985; *FM*, 2 October 1987). By contrast, unionists would see real wages as going up little, if at all, after the late 1970s.

It is interesting to turn from this to the relationship estimated between overall wages and claimed profits on the gold mines. The percentage of wages to profits in 1972 was 52.9 per cent. Over the next decade, given the fluctuations in the price of gold, the proportion shifted between a high of 80.1 per cent in 1976 and a low of 21.2 per cent in 1980. But from 1982 to 1986 one finds a steady return to about the 50 per cent level (Chamber of Mines Annual Reports). What has changed markedly is that while black workers in 1972 earned about one-third of the wage bill, in 1985 they earned perhaps two-thirds and that ratio will have become even more favourable to them since (see Figure 1). There is good reason to think that the mines would be most unwilling to see the proportion of revenues going to wages substantially increased. Thus, the level of wages that mines are now paying represent, given the size and composition of the workforce, about the highest they will go, if they can help it. Management thinks, moreover, in terms of the problem of the overall size of the workforce. In 1975, the payroll included 370 000 men. This figure had risen within a decade to 514 000 workers, who mined more ore certainly but less gold (Tom Main in *SAM*, May 1986).

For this reason, it seems unlikely that labourers' wages on the mines will be easily permitted to rise much in real terms, barring further spectacular advances in the price of gold. At present, these wages may allow a man to make a major contribution to a subsistence family income. They enable the mines to recruit a labour force without 'articulating' with a viable peasant sector somewhere in the subcontinental rural periphery. However, they do not provide for a settled urban dweller with some education, except those in desperate straits. One account suggests that the mining companies will try to use the abolition of the colour bar on the mines to reduce the numbers of workers described as 'skilled', not to integrate the workforce (Rafel, 1987). Thus the main focus of recruitment will surely be in the urban periphery and in the South African bantustans, where dependence on the South African labour market is matched by compliant regimes unlikely to make the sort of precipitous, unfavourable decision about recruitment to the mines that President Banda of Malawi once did. More than 90 per cent of black gold miners from South Africa are domiciled in 'homelands' (Lever and James, 1987:17). Lesotho and, to a lesser extent, Botswana and Swaziland really fall into this category as well, despite their internationally recognised sovereignty.

The most considered judgements suggest that a minority of gold miners will be encouraged to settle down in an urban context near the mines, either in company-built, tied housing or, preferably, in purchased houses which the employer will help to finance, in this way creating the ideal blend of control and freedom. Wedela, the new Anglo American-sponsored township near Carletonville, is ultimately planned for a population of 35 000; Western Deep Levels, the feeder mine, employs 25 000 workers (Elliot, 1986:112). Freegold management estimated that some 15 per cent of their 100 000 workers will be able to take advantage of their home-owner scheme (Chamber of Mines Newsletter, 3/4, 1987). Wilmot James suggests that 20 per cent of workers will be able to 'stabilise' their way of life by the end of this century (James, 1987: 31).

The other way of looking at this picture is to emphasise the continued salience of migrancy for most workers. It may be that new compounds will be increasingly comfortable and relaxed in terms of social control; the new Joel mine in the Orange Free State, it is claimed, is supplied with hostels that are served by shopping centres, excellent sporting facilities, swimming pool, extensive storage facilities, greater privacy, etc. (*SAM*, June 1987). It is indicative that Rand Mines pushed very hard against local white opposition to get permission for massive new compound facilities that would enable East Rand Proprietary Mines in Boksburg to house an additional 6 500 workers (*SAM*, June 1987). As former Anglo American executive director Denis Etheredge said in 1985, migrant labour is intended to operate for a long time yet (*SAM*, December 1985; but see Lever and James, 1987:18–19, for a more equivocal view).

The questions of labour sourcing and wages need to be considered alongside what seems to be a fundamental change in the labour process itself. Two issues immediately come to mind here: differentiation and redundancy. In an important recent address, Chamber of Mines research director Horst Wagner pointed out that it will be less and less meaningful to talk about black gold miners as a single undifferentiated entity. More and more of that labour will be skilled and acknowledged as such. This, he argued, is essential for the technological changes on which the mining companies are banking. He called as well for an end to the 'existing hierarchical system of management', in favour of one that is cooperative and flexible precisely because technology will increasingly make mining operations more inflexible in some respects. Part of this process will highlight the growing importance of engineers as opposed to overseers (*SAM*, October 1986: 17). It is worth pointing out here that engineers are in short supply in the mines and that the South African gold mines remain, as they have been historically, in need of immigrants to provide most of their engineers. (More than 40 per cent of engineers on Chamber mines are graduates of British universities; see *SAM*, November 1986.)

With the disappearance of the legal colour bar and the progressive weakening of the position of the Mine Workers' Union as a bastion of white exclusiveness, mine management is poised to alter work routines, to replace white with black supervisors (already long true at the lowest level of supervision where black team leaders are in charge of daily operations at the mineface), perhaps at lower pay, and to intensify skills deployment in ways that tie into the new technology. To what extent will black men fill the shoes of the white miner, identifying consequently in some respects with management and being paid bonuses by result? How will the application to gold mining of job category scales borrowed from secondary industry, work (Lever and James, 1987: 66)? While this will only be resolved through the struggles of the NUM with the bosses, it must be an important goal for the union to press the issue. At the very least, the most sophisticated men in management realise the need to move towards a less violent, less despotic work-control order and away from one that can all too easily be identified in terms of racial antagonism by workers. (For control issues and changes already under observation, see Leger, 1986; Leger and Mothibeli, 1987.)

At the same time, as we have already seen in our discussion of the fundamental thrust of new technology, the mines are determined to reduce manning levels as substantially as possible, from the point of view both of costs and of labour control. Trackless mining, it is claimed for example, can reduce workforces by 30 per cent and more, while hydraulic drills will equally make important reductions possible. It is often said that given the nature of despotic workplace control and the cheapness of labour, productivity levels were of little importance historically to mine management. In an era of rising costs and with the need to find weapons to fight a developing, militant trade union, productivity becomes more and more crucial. The 1987 miners' strike resulted in massive dismissals (estimates range up to 50 000 men). These dismissals are not just a threat. They represent the discovery by Anglo American in particular that it is actually possible to work shafts with a much smaller workforce than was thought to be the case, even apart from the issue of technological change. (Information from M. Morris and V. Padayachee interviews with management.)

The August 1987 strike brought out at least 220 000 men by Chamber estimates, or up to 340 000 according to the NUM (Markham and Mothibeli, 1987). While the Chamber clearly won the strike, in that it succeeded in keeping to the initial wage increases it had unilaterally called for, and gained ground through the massive dismissals, this was a costly victory. The strike was marked by conflict, violence and destruction and the industry experienced significant production losses and profit cuts. Its occurrence should remind us that the kind of changes which the mining companies are introducing, the new order they are trying to create, is both contested by workers and reflects that very contestation, even when management remains in control. In a recent report on the mammoth Freegold complex in 1987, the year's events are said to have included the killing of two white managers following a workers' onslaught on their offices, the death of seven black miners at the hands of security, and the systematic execution of four black team leaders in front of a crowd of 3 000 workers after a 'workers' trial' for which it proved impossible to obtain any reliable witnesses. The killing of a team leader at one component part of Freegold, President Brand mine, was used to justify the withdrawal of recognition of the NUM (*FM*, Mining Supplement, 13 November 1987). The management of the Anglo-linked JCI claimed that more than twenty black miners who had refused to go on strike had been killed by militants in their workforce (ibid.: 54).

JCI has pioneered trackless mining in its Far West Rand mines, and the impact of mechanisation on the number of jobs has already been quite dramatic (*MAR*, 1987; *Economist*, 31 May 1986; *FM*, 2 October 1987).

Worker response has ranged between intense resistance to management on the one hand, and hostility to recruitment for trade-union membership for fear of losing jobs, on the other. It is not difficult to map out the broad lines of where Anglo American and, to a lesser extent, the other houses would like to be moving their workers: greater skill levels, involving smaller numbers, the most skilled being tied to management control by means of the carrot as much as the stick; black supervision up to much higher levels; and (much more tentatively) deracialisation in order to bring down costs at even managerial/engineering levels. The most obvious way in which the social reproduction of the new, relatively well-paid and stabilised black worker is assisted can be found in the housing schemes being planned for a minority. Some commentators have also suggested that it is essentially the South African workers who would be likely to be better-paid and more skilled (James, 1987; Ruiters, 1987). In order to establish this new order, capital is, in the words of the 1986 Chamber of Mines Annual Report, 'firmly committed to negotiation rather than co-optation' and to a relationship with a unionised work-force (Chamber of Mines Annual Report, E. P. Gush presidential speech: 5). However, in this atmosphere of intense contestation, it would be most rash to assume that capital will entirely succeed in achieving its goal.

THE INTERNATIONAL CONTEXT

It is at first sight quite easy to see the causes of transformation in South African gold mining as entirely local. However, the impulse provided by change in the international market is also crucial.[2] The problems facing accumulation in gold mining in South Africa cannot be distinguished from either those in the national economy or the international context of crisis and resolution. The increase in the price of gold after the early 1970s, but especially post-1979, correspondingly began to boost gold exploration in other parts of the world. The result has been a disproportionately large increase in gold production in other nations. In 1975, South Africa produced 75 per cent of 'free world' gold. In 1985, while total tonnage mined had risen from 945.7 tonnes to 1 212.8 tonnes, the South African share – partly influenced by the constraints imposed by the South African state on mining rich veins when the gold price is high – fell to 56 per cent (*Economist*, 14 June 1986: 72). In the strike year of 1987, this proportion fell to 44 per cent of an estimated 'free world' total of 1 373.4 tonnes (*Mining Annual Review*, 1988). The proportion will continue to fall, moreover, particularly to the advantage of the USA, Canada, Australia and Papua New Guinea. In addition, China is attempting to step up gold production very significantly (*Economist*, 23 August 1986: 67; *FM*, 16 September 1988).

For this there are a number of reasons (*SAM*, May 1986). New technology has enabled low-yield ores previously of little worth to be profitably exploited, and this has disproportionately benefited other producers than South Africa. Mining equipment, no longer needed because of the generalised

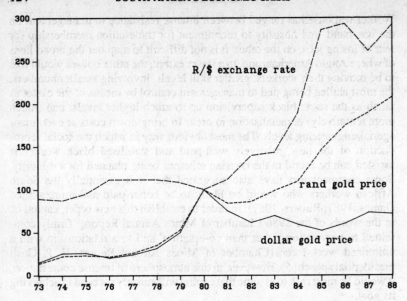

Fig. 2. Indices of rand and dollar gold prices and R/$ exchange rate (1980 = 100)

slump in base metals, can be relatively cheaply adapted to the only paying propositions – gold and platinum – that now dominate world mineral prospecting. However, it is also true that mining development tends to gravitate because of its fixed nature towards the most politically secure countries where capitalist production proceeds unhindered. The dominance of the USA, Canada and Australia here is noteworthy, and Papua New Guinea must be amongst the most docile and stable of Third World countries. What is more, investors from overseas have been uninterested in putting new money into South Africa. Total Canadian gold mine capitalisation is almost half that in South Africa, thus making the value of Canadian gold shares equal to 35 times annual earnings compared to only 6 times for South African shares: this reflects, at least in part, a political judgement (Chamber of Mines Newsletter, July–August 1986). According to one stock analyst, only 28.5 per cent of South African mining shares are now foreign-owned (*Mining Journal*, November 1985).

If South African investment has come to operate at a political discount, it is important to stress that competitive overseas gold mines, in which South African mining houses themselves invest, are by no means uneconomical. In rand values, the profitability of the South African gold mines is considerable but the situation is different when measured in other currencies (see Figure 2). The weak rand and the high inflation rate discourage foreign investors from South Africa also (*Economist*, 14 June 1986). Mines in North America and Australia use a minimum of labour and under new conditions are very cost-effective. 'According to calculation by leading London stockbroking firm James Capel, costs for 37 of South Africa's leading gold mines will average $232 an ounce this year against roughly $220 in Canada and Australia' (*Daily News*, Durban, 7 October 1987). For these reasons, pressure to mechanise and to limit wage gains of South African workers should be firmly set within the context of novel but intensifying international competition.

However, a couple of interesting additional questions also comes to mind in considering the political economy of South African gold from an international perspective. One is raised by the international campaign for sanctions against South Africa, which at times has involved talk about the role of gold. In the past, it has always been clear that to do without South African gold was to do without new gold coming onto the world market, by and large. This is less and less true and South Africa's biggest competitors are all involved already in forms of sanctions campaigns. According to another London financial luminary, Michael Coulson of Kitkat and Aitken, South African political leverage in the world declines as competing gold producers perform well. Coulson in fact suggests that South Africa would do better as a result were the real price of gold to fall again, because lower gold prices cause bigger problems for other producing economies than South Africa (*FM*, 27 November 1987). From the economic point of view, on the other hand, if one puts aside the possibilities of punitive gold sanctions, South Africa does far better if more marginal producing countries fight for a higher gold price, thus benefiting her. Even if cost structures are now becoming more favourable in other countries, their potential production still remains substantially less in the long term than South Africa's.

If we take the question quite seriously of the potential threat to capitalism in South Africa, it is worth asking whether mining capital has any contingency or, indeed, active plans to disengage from the region. There was good reason to think that the one big foreign-owned producer, Consolidated Gold Fields, which was probably the toughest firm in terms of refusing to acknowledge trade unions, worried increasingly in the 1980s along these lines, and its investment strategy became specifically designed to avoid too great a dependence on South African gold (*Economist*, 20 April 1985). This was true even though Gold Fields is spending substantially on new shafts and technology in South Africa (*Economist*, 8 June 1985). Throughout the turbulent mid-1980s, the local mining companies continued to make large and long-term investments in South Africa, notably in exploration. Nor have I seen any evidence to suggest that the lack of foreign investment has limited South African mine development. Anglo American appears to be a relatively unsuccessful investor outside South Africa and continues to make its main impress felt within the country, rather to the surprise of foreign commentators (*Mining Annual Review*, 1987; *Economist*, 8 June 1985: 69). During late 1988 and early 1989, Anglo unsuccessfully attempted to use its large overseas capital holdings to purchase Consolidated Gold Fields. Instead, Hanson Trust, a British conglomerate, was the successful buyer. The South African business world expects to see Consgold's South African interests sold to a local buyer. It is as though mining executives assume that the state, any South African state, will inevitably be obliged to consider their interests and operations as a top priority. An alternative view recently propounded suggests of Anglo that 'this secretive giant has shown little taste for competing in free markets. That explains why it has failed to become a global company despite its size and clout, and why it now finds itself trapped within the gilded cage of South Africa' (*Economist*, 1 July 1989: 63).

MINING AND THE STATE

The one really vital area requiring consideration in which there is very little talk of fundamental change lies in the relationship between the South African state and the mining companies. It has been suggested early in this chapter that a quite cordial, if complex, relationship regarding manpower, foreign relations, tariffs, state security and fiscal policy had evolved by the 1960s between these two forces, despite their problematic past. Some dispute exists over the taxation of the mines, especially in the light of the intensified profits of the good years of the 1970s and 1980s. The fiscal relations of state and capital have been dominated by the anxiety of the state to provoke the maximum investment and prolong the life of low-yielding mines, thereby benefiting employment. Although the tax structure of the mines is complicated, one can identify the key elements as the ease with which capital investments can be written off and the extent to which exploration expenses can be used to reduce the tax burden on very profitable existing mines. According to the recent Margo Commission, which has proposed the reconstruction of the South African tax system, 'It is still possible for a company to reduce the income of a profitable mine by deducting first its own capital redemption allowances and then the working losses of a developing mine, ending up with a very small taxable income or even a tax loss (Margo Commission, 1987: 245). Such a mine can also ensure that it continues to pay dividends to its shareholders. The tax issue is very significant, given the centrality of mining in the provision of revenue and in the process of accumulation in the national economy.

The state seems to be moving in the direction of trying to extend its revenues and reduce some of these benefits to the mines. Thus between 1985 and 1986, a new tax amendment was passed to make it more difficult for mines to deduct on taxes for new capital expenditure on prospective title leases, while assistance to marginal mines through tax relief was made conditional on application (Chamber of Mines Annual Report, 1985; *SAM*, February 1986; *Mining Journal*, May 1986). The year before, the surcharge on income tax from excess profits was raised (*MAR*, 1985). A corporate representative claims tax and lease payments rose from 13.9 per cent of profits in 1970 to 22 per cent in 1985 and were 'onerous' (*SAM*, May 1986:

61). In response, and in the wake of the great 1987 strike, closure of the most marginal mines and shafts has been taking place. The Margo Commission tentatively suggested going further to bring mining taxation in line with industrial taxation and end any efforts by the state to prop up mining as opposed to other enterprises (1987: 252). Perhaps because of the declining interest in supporting white labour, the historic tendency to try to keep low-grade mines in operation is decried. Instead, any special effort to assist gold mining should be directed, according to Margo, to supporting investment in new ultra-deep mines independent of the tax system. One suspects that the continued centrality of gold mining to all facets of the economy will make the Margo Commission suggestions unpalatable for the state.

To the extent that one can talk of a political view on the part of mining capital, one can note little indeed of the radical edge that occasionally was articulated before the legalisation of the ANC in February 1990 by representatives of secondary industrial interests or commerce, beyond the desirability of a political settlement (see Gavin Relly in *FM*, 2 October 1987). As Merle Lipton points out in *Capitalism and Apartheid*, mining is moving away from a dependence on crude apartheid. But this brings it broadly in line with the reform strategies of the National Party, with which it shares a total hostility to the prospect of radical change. Just after the proclamation of the national state of emergency, on 24 June 1986, the outgoing president of the Chamber of Mines, C. G. Knobbs, stressed in his address: 'The major crisis of confidence which developed late in 1985 and which overshadowed all else was South Africa's seeming inability to get to grips with its fundamental political and economic problems. This might have proved calamitous were it not for the excellent performance of the mining sector. Fortunately, the mining industry's earnings, aided by a depreciating, image-battered rand, increased by 36 per cent to a total record of R26 000 million (Chamber of Mines Annual Report, 1986: 5).

Knobbs considered South Africa to be 'one of the few viable areas in an otherwise tragic continent' (ibid.: 6), and he exemplified the sympathies of senior mining executives in supporting the National Party reform thrust, perhaps wanting it to show renewed vigour, and opposing any break in South African national history. Zach de Beer, a well-known figure from Anglo American and co-leader of the Democratic Party, sees the approach of capital to the NUM as a model for how the state should address itself to black militancy (*SAM*, November 1985: 55). On the whole, the mining industry is dominated by men who favour a close cooperative relationship between themselves and the state in considering the many areas where policies need to be ironed out on both sides. Undoubtedly they strongly support the political initiatives of F. W. de Klerk in recent days.

CONCLUSION

According to the *Financial Mail*, 'the South African gold mining industry is in decline' (16 September 1988). This chapter has tried to identify changing conditions in the world gold market, in the supply and work relations of labour and in mining technology, that is to say, in both the technical and social means of production which are increasingly essential for the maintenance of a profitable gold-mining industry. The search for relative, as opposed to absolute, surplus value, taking the form of an intensified concern for productivity in a traditionally labour-profligate industry, is becoming more and more important. The labour force is planned by capital to be divided and restructured in new ways that will increase differentiation along less rigidly defined racial or non-racial lines, expand skilled labour (which will perhaps not be paid as such) and modify the despotic character of workplace control. With the coming of the computer, the totalising nature of this change, organised in a regulatory manner as a system, is made more apparent and more possible. The future of the industry depends on the institution of a new regulatory regime.

Nonetheless, caution is needed in assessing the voices of management spokesmen, who have reason to publicise a technological revolution. So far as can be judged, the mining companies, like their political friends, are doing what they can to see that this era of change occurs in a conservative political context. While one recognises the structural character of change, it is finally important to point out that mineworkers on the goldfields have become militant and organised in recent years, that they would like to contest much of what management plans, and that the actual shape of future development cannot be gauged unless we know whether or not they will have some success in their struggles in a fluid political environment.

6
Coal mining: past profits, current crisis?

JEAN LEGER

Coal, for decades the Cinderella of the South African mining industry, rapidly rose in the 1970s to become, after gold, the second most important mineral in the country. Indeed, from 1970 to 1986 production more than tripled.

Coal is important for two reasons. Firstly, it increasingly provides the energy basis for the South African economy, particularly as cheap electricity. According to Hill (1987), the latest data available (1982) indicate that coal supplies 82 per cent of energy requirements, imported oil 17 per cent and hydro-electricity 1 per cent. More recently a Department of Mineral and Energy Affairs spokesman estimated that coal provides 88 per cent of current energy requirements. This suggests that about half of the diesel and petroleum used is now produced from coal, that is, between 10 and 15 per cent of total energy requirements. It appears to be government policy to maintain oil-from-coal production at about half of total oil consumption.

Secondly, coal has become one of the three most important earners of foreign exchange. Even more significant than the growth in total production was the growth in exports – from only 1.3 million tonnes (mt) in 1970 to 45 mt in 1986. Expressed in another way, in less than a decade South Africa captured a sizeable chunk of the Japanese market and became *the* major exporter of steam coal to the EEC, providing 37 per cent of EEC imports in 1986. The changes experienced in the industry since 1970 amount to a qualitative shift in its economic and technical base and mark a new era of accumulation.

The first part of this chapter discusses the spectacular growth in the scale and profitability of coal mining in the 1970s and early 1980s. The second part of the chapter focuses on labour issues. Despite the reorganisation and changing skill content of work necessitated by rapidly changing labour processes, despite worker militancy and, more recently, organisation of miners by the National Union of Mineworkers, developments since 1970 have involved limited wage increases and limited reforms in industrial and social relations. Of crucial importance, the low wages paid to black workers remain a pillar of profitability in the industry. Many of the exploitative features of the apartheid period persist, in particular the predominance of migrant labour and the low level of wages.

If the growth of the industry since 1970 has been meteoric, so too has been its reversal of fortunes. The industry's crisis in the late 1980s and future prospects are discussed in the third part. The oil price has collapsed following a glut in world oil supplies, while massive investments prompted by the energy shortages of the seventies have led to excess capacity and oversupply in the world coal market. The effects of these reversals were initially cushioned by the collapse of the rand but the profitability of local operations experienced a sharp turn-round in 1987. The modest strengthening of the rand against international currencies, new railway tariffs and the threat (and in some cases the effects) of sanctions have all exacerbated the industry's slump.

From the perspective of labour, coal employment is already in decline. The rate of growth in demand by Eskom, the largest consumer of coal, has declined considerably in recent years as overall production in the South African economy has diminished and new, more efficient power stations have come on stream. The spectacular growth of exports has for the moment reached a ceiling. Prices obtained in some cases have been insufficient to cover working costs. It remains to be seen whether the new structure of the industry, so thoroughly transformed in the 1970s to produce the world's

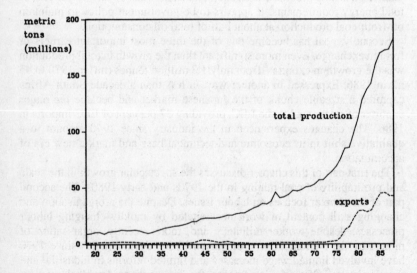

Figure 1. Total coal production and exports

cheapest coal, will be able to sustain the icy winds of international politics and an increasingly competitive market.

THE TRANSFORMATION OF COAL MINING

Until 1970 the coal industry was stagnant, with prices for coal being low and demand limited in growth. Figure 1 shows the trend in coal production since 1915. Low coal prices derived in part from geological factors – much of South Africa's coal is readily accessible, lying near the surface in thick, horizontal and infrequently faulted seams, making it comparatively cheap to mine. For example, the Petrick Commission (1975) noted that of the total reserves of bituminous coal, 23 per cent lay between 15 m and 50 m below the surface, and a further 73 per cent between 50 m and 200 m. Moreover almost half of the reserves were to be found in seams 4–6m thick, and a further third in seams 2–4 m thick. However, anthracite reserves are generally deeper and distinctly more difficult to mine as most occurrences are in narrow seams less than 2 m thick.

Apart from geological considerations, the South African coal price was kept amongst the lowest in the world by price controls introduced from 1951 onwards.[1] While the intention was to enable a return of 12.5 per cent on capital, rising inflation from the late 1960s raised capital replacement costs, pushing down the after-tax return on new investment to an estimated 2.5 per cent in 1976.

Three consequences followed from the coal pricing policy and declining profitability. Firstly, there was little significant investment in new collieries. Between 1950 and 1970 only two new mines were opened (Wassenaar, 1977). This led to frequent shortages of coal for the domestic market. Secondly, mining techniques were extremely wasteful. Low prices encouraged bord and pillar mining – pillars of coal were left behind to support the roof, thus eliminating the costs of purchasing and installing support equipment, as well as costs of surface rights or compensation for subsidence damage. According to a 1975 estimate, 'the total volumetric extraction of coal from all coal fields mined in South Africa to date is certainly well below 30 per cent' (Anon., 1977).

Finally, two distinct sectors in the bituminous coal industry emerged. On the one hand, there were 'tied' coal mines linked directly to power stations or metallurgical operations. Prices in these cases were negotiated, usually on a cost-plus basis, rather than controlled. On the other hand, 'commercial' collieries supplied the rest of the market, principally households, untied Eskom and municipal power stations, and the railways. Since these latter markets were stagnant or even shrinking, the 'commercial' collieries bore the burden of the limited market. In addition, they found it difficult to sell their 'fines' (low-grade, high-ash coal), which were suitable only for electricity generation. The consequent incentive to explore new avenues of profitability provided the mainspring for the transformation the industry entered upon from 1970.

The mine-owners' capacity to restructure the industry was given substantial impetus by the Transvaal Coal Owners' Association (TCOA). Efforts to restructure the industry initially involved both pressure to transform the domestic price and demand structure, and experimentation to develop processed coal of adequate quality for export markets.

A massive boost in export potential resulted from the OPEC oil crisis of 1973–4. It made coal a relatively favourable energy option while at the same time raising its world price significantly. The expansion of export volumes and revenues made it feasible to introduce more mechanised methods of extraction with far higher productivity yields and at the same time made it necessary to develop new transport facilities. Furthermore, South Africa's expanding export possibilities attracted the interest of the corporate oil 'majors', who assisted with finance, access to international markets and technologies.

Re-organisation of the internal market

The first and most obvious problem addressed by mine managements was domestic price control. Through the TCOA, mine-owners argued that inflation was eroding proposed price increases even before they were implemented. The demand that the system of price control be reviewed reached a crescendo in the annual reports of the mining houses and the Chamber of Mines. Management's vigorous lobbying on the issue met with increasing success.

A fundamental change in coal marketing was introduced in 1972. This was the recognition of four grades of coal, from A (high grade) to D (low grade). According to Berning, then chairman of Anglo American's colliery operations: 'Of far more potential importance [than a 30 cents increase in the coal price] is the introduction in the Transvaal of differential selling prices for varying grades of coal. In the past only two grades of steam coal, high and low, were recognised, with a difference in price of only five cents a ton. This resulted in consumers often demanding a higher grade of coal than was necessary for their needs. Secondly, to meet their needs, much usable coal was left unmined, to some extent irretrievably, and at many collieries there has also been excessive washery discard. It has also been difficult to place the output of low grade producers. Under the new arrangement there is a graded differential of 22.5 cents per ton between top and bottom qualities. While, in my view this is still insufficient, it is a step in the right direction' (Berning, 1972).

Over and above this, price increases granted after 1972 were much larger than the average of 5 per cent per annum conceded in the preceding 20 years. For example, for grade D coal, increases averaging 27 per cent per year were granted between 1972 and 1976 (Dutkiewicz and Bennett, 1976). In 1976 the Petrick commission of inquiry into coal reserves recommended 'that the whole system of price control for the coal mining industry be revised'. In the interim it suggested a substantial increase in the coal price in line with the

Chamber's submission, resulting in an increase of 50 per cent in 1977. Price control was finally abandoned from 1 April 1987.

Technological breakthroughs in coal beneficiation

Coal is classified by the government as a strategic commodity and exports are controlled by export quotas. As a result of lobbying by the coal owners, the government granted export permits to the Transvaal Coal Owners' Association (TCOA), the Anthracite Producers Association (APA), and Natal Associated Collieries (NAC) to export up to 12 mt of coal per annum from 1976. These permits eventually comprised phase I of the export programme, as it was expanded.

A more fundamental obstacle for exports when they were first mooted in the late 1960s was the relatively low quality of much South African coal, and hence the low prices that could be attained. Most world trade was in coking coal used for metallurgical processing and traded at a premium price. However, South Africa is short of coking coal and its export remains prohibited.'

The processing of coal to meet international requirements thus became a prerequisite for exporting. Anglo American engineers developed a two-stage washing process which economically produced both a low-ash coking coal and a high-grade steam coal from a single source. The price obtainable for the beneficiated product mix far exceeded that of coal processed by standard washing techniques. A 1970 analysis suggested that 100 tonnes of coal, normally sold for R150, could realise R240 if suitably treated (Anon., 1970).

In November 1969 discussions began with Japanese steel mill representatives for the supply of coking coal. The Japanese steel makers were prepared to blend the relatively low-cost South African coking coal with higher grades to reduce their costs and dependence on high-cost Australian coking coal.

In March 1971 the first major long-term contract, for 27 mt, was signed with seven Japanese steel mills and coke works. Exports started at 100 000 tonnes per year from November 1972, increasing to 2.7 mt per year from 1976 to 1986. As a by-product of this low-ash coal, large quantities of steam coal were produced, suitable for power stations, cement works, and the like. These quantities could not be absorbed on the domestic market and provided an additional incentive for further developing the export market.

The oil crisis of the early 1970s

The 1973–4 oil shortage provided the opening for South African coal exports, particularly into Europe, as coal suddenly became cheap relative to oil. Coal producers renegotiated the Japanese coking contract on substantially more favourable terms, and within months also clinched deals for 2 mt of steam coal to the US and smaller orders to West Germany and Belgium.

Given the high levels of profitability offered by exports, pressure on the authorities to raise export allocations became intense, and this from individual companies, rather than trade associations. In November 1974 it was agreed that phase II exports should begin in 1979, with total exports from Richards Bay of 20 mt a year. A third phase was envisaged and was confirmed in 1979, raising the annual ceiling of exports to 44 mt, which was reached in 1985.

A major bottleneck to increased exports remained in the shape of limited harbour capacity at both Maputo and Durban. New deliveries had to wait until the completion of the Richards Bay facilities in 1976 (Anon., 1974a).

The international oil crisis also had implications for the domestic coal market. Prior to the price rise, the low price of oil had encouraged some domestic consumers to begin using oil. For example, in 1972 the Cape Town power station converted from coal- to oil-burning boilers, resulting in a decline in demand from Transvaal collieries. With the rapid rise in the oil price and the uncertainty of supplies, many consumers converted back to coal or recommissioned and redesigned their plants to use coal (Sealey, 1974).

Apart from exports, the oil crisis led to a massive increase in the mining of coal for synfuels. The South African state, concerned about the strategic implications of world oil shortages, commissioned Sasol 2 and Sasol 3 at Secunda. Whereas Sasol 1 produced only 4 per cent of the country's liquid fuel needs in the early 1970s, the three Sasols currently consume 39 mt of coal to produce half the liquid fuel requirement (Oliver, 1982).

Mine mechanisation

While total sale tonnes of coal increased by an average of 6.5 per cent per annum between 1965 and 1985, employment rose from 77 244 in 1965 to 103 352 in 1983, an average of only 1.6 per cent per annum. In other words, there was rising productivity of labour, reflecting the introduction of new extraction techniques. These, particularly open-cast mining, improved the international competitiveness of South African coal.

As noted above, the controlled domestic price structure throughout the 1960s made intensive investment in mechanisation unprofitable. In this respect the profitability conditions for coal mines were analogous to those for gold, which also faced a fixed output price. Capital and working costs had to be minimised and levels of operating profit tied to extremely low wage levels for black miners. Technology was therefore labour-intensive – underground bord and pillar mining, and hand-loading of coal. This approach secured short-term profits, but in the long term the low extraction rates which resulted meant that individual seams were quickly exhausted.

The first major technical development was the mechanisation of hand-loading from the mid-1960s. There were reports by mine management in the late 1960s of increasing difficulties in recruiting workers for hand-loading and tramming, as well as complaints that bonus incentive schemes were not working. This pushed management to accelerate mechanisation of loading (Newman, 1972). The real wage increases of the early 1970s no doubt further propelled the process, which was by then well under way. By 1970 only 43 per cent of coal was hand-loaded, in 1978 as little as 10 per cent and today it accounts for less than 2 per cent of all coal production, having been abandoned in all but very narrow seams.

The second form of mechanisation was the introduction of open-cast mining. In 1969 Gencor announced their intention of commissioning the country's first walking dragline at Optimum mine. The dragline bucket had a design capacity of 50 m^3 – this single dragline produced a total output of 2.9 mt per year! Gencor's caution about adopting this new technology appears to have led them to offer a 40 per cent share in the operation to the dragline supplier, McAlpine, a UK corporation.

As a result of the spectacular success of the Optimum dragline a number of planned mines such as Arnot and Kriel, originally designed as underground operations, were redeveloped as open-cast mines with underground sections. By 1987 at least 17 major open-cast mines had been developed. Most of these mines have massive outputs – seven produce more than 4 mt each per year. In 1986 open-cast mining was responsible for 35 per cent of total production (Mehliss, 1986).

The major advantages of open-cast operations are, firstly, the very short lead time – a mine can come into operation within two years; and secondly, the high extraction rates that are possible – recovery of more than 90 per cent of the coal is feasible. Because open-cast mining is a capital-intensive technique and relies on imported machinery, it became economically attractive only as the coal price increased. A 1975 estimate put the level of capital investment in open-cast mining per annual tonne sold as approximately 50 per cent higher than bord and pillar operations (Anon., 1975). An open-cast mine can achieve similar production levels with about-one third the workforce of a mechanised bord and pillar operation.

The third leg of the mechanisation programme was the introduction in the mid-1970s of 'continuous miners', which do not require the use of explosives. Despite early technical problems, by 1977 nine continuous miners were in operation. By 1980 they accounted for 9 per cent of output, with 50 machines operating, and by 1985 some 16 per cent of total production. An important advantage of continuous miners is that they may be used for pillar extraction in bord and pillar mines. They are able to improve recoveries to 50–80 per cent of reserves – an extraction rate on a par with longwalling. Continuous miners require half the labour of bord and pillar operations at an increase in capital outlay of only 15–25 per cent. For new mines, higher capital costs of continuous miners are more than offset by savings on housing and other facilities for black workers.

The final major technology to be introduced was longwall mining. However, this has not had the same impact in South Africa as elsewhere. The high extraction rates possible can generally only justify the additional costs (estimated in 1979 to be 21 per cent more than mechanised bord and pillar methods) at mines tied to power stations or large process plants for steel or chemicals. Longwalling is therefore found at Sasol (Sigma mine), Secunda

Collieries and at certain Iscor- and Eskom-tied collieries. Longwalling accounted for only 4 per cent of total production in 1980 and 7.4 per cent in 1986.

Transport and bulk handling

A major obstacle to all coal sales is the costs of transportation, both to local and to international markets. In the 1960s and 1970s the costs of sea-freighting coal were 50–100 per cent higher than those of oil for the equivalent energy. In 1971 the TCOA's managing director estimated transport costs as 75 per cent of the final landed cost of coal in Europe: 31 per cent for railage and handling, and 44 per cent for ocean freight and insurance (Tew, 1971).

To make exports viable required the reduction of freight costs through the use of larger ships. Both ports available in the 1960s – Durban and Lourenço Marques (Maputo) – could accommodate only ships of a size for which the freight charges made South African coal uncompetitive in Europe, despite its low pit-head price. Therefore a suitable port with efficient bulk-handling facilities was essential. And so the Richards Bay 'super port' was constructed, and a railway line built from Witbank.

The catalyst for this development was the Japanese coke contract: even though the contract was not sufficiently large to justify the Richards Bay project, the Japanese made the latter a condition of its contract with the TCOA. This seems to have suited the TCOA well since it increased the pressure on the government to assist in developing transport facilities for coal exports. In 1971 the government agreed to proceed with the terminal and railway (due to be commissioned in April 1976). Both were planned for far greater capacities than the Japanese coke contract, which involved about 2.7 mt per year: the railway line was designed to transport 5 mt per year initially, and the port-handling facilities 9 mt per year (Sealey, 1974).

The TCOA exporters were required to guarantee the financing of the railway line. In addition, the coal-handling facilities of the Richards Bay Coal Terminal (RBCT) were paid for and are managed by the TCOA on land leased from South African Transport Services (SATS). Each participant in the contract invested in a proportion of the facility, which secured them the right to first use of that proportion of handling capacity in the future.[2] As a result, transport costs are no longer wholly in the hands of third parties and beyond the control of the mining companies.

Most of the operating costs of the RBCT derive from the capital invested, hence the charges levied to users depend on their participation in the various phases of coal export. The costing formula initially adopted for the Witbank–Richards Bay rail line was calculated on the basis of economic costs and divorced from the general tariff system.

By the time the Richards Bay facility was ready in 1976 at a cost of R730 million, export orders to France, West Germany, Japan and the USA totalling 9 mt per year had already been secured (Anon., 1976). To handle additional export tonnages, the terminal capacity was increased to 20 mt per year in

1978 and by 1987 had a capacity to handle 48–52 mt per annum. The railway has been upgraded to a capacity of 65 mt per year. The port was designed to accommodate ships of 150 000 dwt capacity, with provision for expansion to take ships of 250 000 dwt.

New linkages with international monopoly capital

In the aftermath of the oil crisis of 1973–4 the multinational oil companies began to take an intense interest in other sources of energy. The massive South African coal reserves caught their interest at a time when the state was particularly concerned about its vulnerability to sanctions against oil imports, especially as there were already severe world oil shortages. The state sought to secure its strategic interests by using the prospect of phase III exports to encourage the oil multinationals to enlarge their South African investments. In allocating export quotas, the state gave Shell, BP and Total Oil 34 per cent of the total exports of bituminous coal under phase III, totalling some 11.5 mt per year. Although this move was resented by South African coal companies, particularly those with no export allocation such as JCI, the state defended it on the grounds that the oil majors contributed to South African oil supplies.

To produce the coal for their export allocations, most of the oil companies entered into joint ventures with local mining houses. While the oil companies provided a certain amount of capital, no doubt more important was their access to international energy markets.

Shell entered into a 50–50 joint venture with Rand Mines to develop the Rietspruit export colliery. While Rand Mines was responsible for mining the coal at this massive open-cast mine, Shell was to market it. Rand Mines also had a minority stake of 5 per cent in the 89 per cent BP-owned open-cast Middelburg colliery, which produces over 5.5 mt per year for export. In December 1989, BP sold its interest in Middelburg to Rand Mines for R546m.

Gencor's coal subsidiary, TransNatal, entered into a joint venture with Total Oil and BP to develop the Ermelo Colliery. The initial estimated capital costs of R65 million were to be met by Total and BP, with TransNatal contributing its coal rights to the venture. The mine, a conventional bord and pillar operation using trackless coalcutters, loaders and shuttle cars, was planned and is managed by General Mining. BP and Total marketed the annual production of 3 mt, all intended for export (Coetzee, 1975).[3]

JCI, through its Arthur Taylor Colliery, mines about 1.6 mt per year of steam coal to provide coal for Total's export quota. Amcoal entered into a medium-term contract with Shell to market export coal from the Kleinkoppie mine.

As in other South African industries, new technologies were often obtained via joint ventures with foreign corporations. One of the first joint ventures in coal mining was Gencor's offer to McAlpine of a 40 per cent share in the Optimum Colliery, to ensure the success of the new open-cast technology. Initially McAlpine provided personnel and technical services for the operation of the walking dragline, though these were later taken over by Gencor.

Presumably to strengthen their marketing, reduce their reliance on the oil multinationals, and counter the sanctions campaign, local coal companies have also set up the 'Office of the South African Coal Industry' in London.

CHANGES IN LABOUR RELATIONS

The development of the technical capacity and infrastructure to exploit local and international coal markets was not accompanied by similarly dramatic changes in the labour situation in the coal mines. Notwithstanding the boost in black workers' real wages during the early 1970s, enabling the 'South Africanisation' of the labour force, neither employment's nor labour's share of the vastly greater quantity of value added in the industry increased to any significant degree. Profitability – and South African coal's competitive position in world markets – critically depended on the ultra-cheap labour system which the mining houses were able to reproduce through the migrant labour system. Unionisation beginning from the mid-1980s has not yet seriously dented management's control over labour relations. In this section, we examine changes in productivity, wages and patterns of migrancy.

Employment and productivity

Figure 2 illustrates the trends in the total number of employees and the number working below ground for all collieries. Since 1960 employment in coal mining has increased from 65 000 to about 95 000 workers.[4]

Chamber of Mines collieries, which represent a decreasing proportion of total coal production (59 per cent in 1987) and exclude the Sasol collieries as well as a number of open-cast export operations involving multinational oil majors, show an absolute decline of black miners employed and an increase in white miners. This suggests that black workers are not being given the opportunity to take up employment in the more skilled positions that colliery mechanisation has created.

Figure 3 illustrates productivity changes. The curve of output per underground employee (including open-cast pit workers) is more useful for assessing productivity improvements, for the reasons discussed in footnote 4. Firstly, the dramatic improvements in productivity should be noted – from 1 000 tonnes per year in the early 1960s to over 3 000 tonnes per year per underground worker by 1985.

Four periods in the productivity trend may be identified. Until 1966 there were only slight and almost inconsequential increases. From this point on there were steady and substantial increases in productivity until 1978, reflecting the mechanisation of hand-loading. The development of new open-cast mines and investment in continuous miners are marked by the dramatic, sustained productivity improvements from 1978 until 1984. In only seven years, productivity practically doubled. Since 1984 there has been a marked slow-down in the rate of productivity improvement because of the deterior-

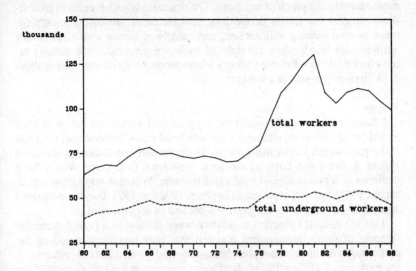

Fig. 2. Numbers at work in South African coal mines

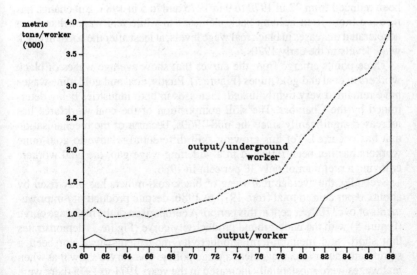

Fig. 3. Production per worker in all South African coal mines

ation in the world coal market, stagnation in the growth of local demand and, more recently, the pinch of sanctions. The dramatic improvements in productivity suggest that unless demand for coal increases substantially, employment in coal mining will decline, particularly at mines with already low employment levels. One example of such a mine which has already retrenched workers is Ermelo Colliery where productivity is estimated at about 1 130 tonnes per year per worker.

Wages

Chamber coal mines have historically paid coal workers on the same basis as gold-mine workers, although some marginal non-Chamber coal mines in Natal paid wages below these levels. Since the early seventies, as shown in Figure 4, the wage costs of the total workforce (black and white) have declined as a percentage of total sales revenue. Whereas wage costs represented 30–35 per cent of revenues between 1970 and 1975, they have hovered around 25 per cent in recent years – a decline of about a third.

This substantial reduction in relative wage costs – in a period when the rhetoric of mine management claimed that past injustices including the exploitation of black labour were being rectified – was achieved primarily at the expense of white workers, despite the increase in their proportion of the workforce (see Figure 4). The proportion of total white wage costs to sales revenue declined from 21 per cent in 1970 to 9 per cent in 1985, while the corresponding figures for black wage costs were 12 per cent and 14 per cent.

If we look at wage levels, the ratio of white to black miners' wages has been reduced from 22 in 1970 to 9 in 1975 and to 5 in 1985. But change has resulted more from holding back increases in white real wages than from accelerated increases in black real wage levels, at least after the boost in black wage levels in the early 1970s.

Three points emerge from the curves that show average wages of black workers in coal and gold mines (Figure 5). Firstly, coal and gold mine wages have remained very tightly linked, increases in both industries being determined by the Chamber. The skill composition of the coal workforce has increased significantly since the mid-1960s, because of the mechanisation that has occurred. Yet this growing skill differential relative to gold-mine workers has not been reflected in a widening wage gap, the coal workers enjoying a premium of only 10 per cent in 1986.

Secondly, the average real wage of black coal-miners has only risen by about 50 per cent in total from 1975 to 1986, despite productivity improvements of over 100 per cent in this period. A comparison of the real wage curve (Figure 5) with the underground productivity curve (Figure 3) demonstrates that short- and medium-term productivity movements have not been a significant factor in management's wage policy. The curves show that when real wages were substantially increased in the years 1971 to 1975 there were no dramatic improvements in productivity.

Thirdly, unionisation of the black workforce has not substantially im-

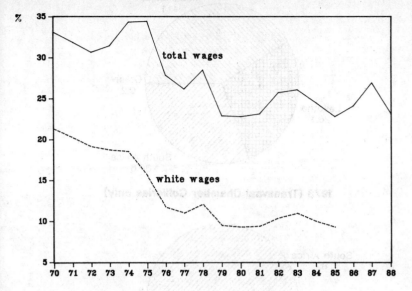

Fig 4. Wages as a percentage of Chamber coal sales revenue

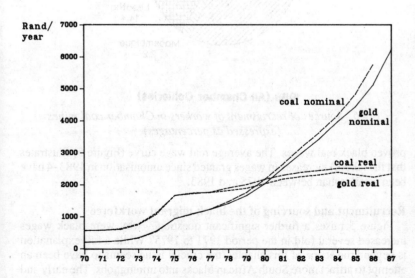

Fig. 5. Real and nominal wages of black workers in gold and coal mines (1980 = index year)

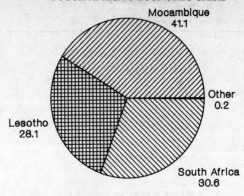

1973 (Transvaal Chamber Collieries only)

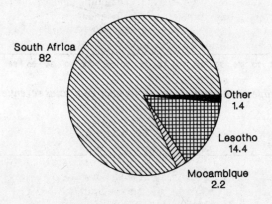

1988 (All Chamber Collieries)

Fig 6. Sources of recruitment of workers in Chamber coal mines (expressed as percentages)

proved black real wages. The average real wage curve (Figure 5) illustrates that the (slight) increases in wages granted since unionisation in 1983–4 have been smaller than between 1975 and 1983.

Recruitment and sourcing of the black migrant workforce

Figure 5 raises a further significant question – why were black wages increased several fold in the period 1971 to 1975? While a full explanation is beyond the scope of this chapter, the crucial issue seems to have been an attempt to attract more South African blacks into mining jobs. The early and mid-1970s were a period in which widespread 'labour shortages' were perceived to exist on the mines. According to the managing director of

Lonrho, 'South Africa has immense reserves of labour able to work on the mines ... but not willing to at the present rates of pay' (Newman, 1972).

Because of these ultra-low wages – lower in real terms in 1970 than in 1911 (Wilson, 1972) – increasing proportions of the mine labour force had to be recruited outside South Africa. In 1973 the proportion of foreign workers recruited by the mine labour organisations (for most gold and certain platinum and coal mines) reached a peak of 80 per cent. While exact figures for the collieries are not available to the author, the proportion of foreign workers was probably not as high as in the gold mines. In the case of the Chamber colliery members in the Transvaal and Free State, 67 per cent of the workforce was drawn from outside the country.[5]

Mine managements were increasingly concerned at their vulnerability to foreign political developments outside of their direct sphere of influence. Speaking at a seminar in March 1974, Harry Oppenheimer expressed his reservations: 'At present wage levels, the gold mining industry obtains a very large part of its labour force from outside the boundaries of the Republic and that is because wages in the gold mining industry have not been competitive with wages in other industries so far as the South African black population is concerned. This is a dangerous state of affairs, both for the industry and the country' (Cited in Anon., 1974b).

Oppenheimer's fears appeared well founded, when just weeks later, a crash of a Wenela aircraft transporting Malawian migrants precipitated the complete withdrawal of all Malawian nationals over the next two years. Though Malawian nationals were insignificant on the coal mines, the importance of Mozambicans raised management concerns as that country moved towards independence after April 1974. The collieries rapidly cut recruitment from this source. From 1976, no Mozambican novices were recruited while the number of experienced miners was cut precipitously from over 10 000 prior to 1975 to less than 2 000 in 1977. Indeed the cut in recruitment of Mozambican nationals working on the collieries was even sharper than on the gold mines.[6]

The real increases in mine wages which continued until 1975 came at a time of growing structural unemployment in the bantustans. The new wage levels rapidly had the desired effect. While there were insufficient mine-workers at the beginning of 1975, the situation quickly changed – Teba's South African recruits (for all mines) nearly doubled from 85 490 in 1974 to 150 825 in 1975, doubling again to 300 630 in 1977 (Malan, 1985). By the beginning of 1978 the Chamber of Mines was describing an 'almost embarrassing flood' of workseekers.[7]

The policy of 'South Africanising' the colliery workforce has also affected workers from Lesotho, the only other foreign country from which a substantial proportion of the colliery workforce has been recruited. From 1977, the first year for which comparative statistics are available for all Chamber collieries, the number of Lesotho nationals has declined from 18 600 to 7 450 (Figure 6) in 1987, a cut of more than half.

Overall, it appears that a pattern is emerging of a more stable workforce drawn increasingly from areas closer to the mines. The proportion of migrants from homelands distant from collieries is also declining. For example, migrant workers from the Transkei have declined from 14 864 in 1977 to 7 142 in 1987, again a reduction by about half. While the compound system persists, at some of the newer open-cast operations – for example, Rietspruit and Duvha (both opened in 1979) – the proportion of migrant workers has been reduced. In the case of Ermelo Colliery, an export operation which started production in 1978, a compound and married quarters were built in the black township adjoining Ermelo, although the mine is some 25 km away.

Labour organisation

The National Union of Mineworkers is the only union to have gained a substantial membership amongst black coal-miners. Membership grew about 10 per cent in the year to June 1987, when NUM had 18 945 members on coal mines. Another union, the Black Allied Mining and Tunnel Workers' Union, had 990 members in 1987. The membership of these two unions represent about 42 per cent of the black workforce of the Chamber collieries.

Table 1 expresses union membership as a proportion of the black labour complement of each mining corporation. Amcoal mines have the highest level of organisation, followed closely by Rand Mines. Since the 'reform' at Gencor involving the appointment of a more liberal top management, NUM membership has increased substantially at TransNatal mines. As a result of a decline in NUM membership at Iscor's Hlobane Colliery, probably due largely to the presence of the Inkatha-linked union Uwusa, the NUM is no longer recognised there.

Table 1. Union membership according to mine corporation

Mining House	June 1986	June 1987		Black labour complement
	No.	No.	%	
Anglo American	9 305	10 969	59.5	18 450
Gencor	2 905	4 712	27.5	17 110
Rand Mines	4 627	3 555	54.2	6 559
GFSA	239	239	9.9	2 413
Lonrho	639	460	13.5	3 403
Iscor	1 429	–	–	–
Total	19 144	19 935	41.6	47 935

Strikes on South African collieries have been of relatively short duration. No strike to date has had more than a small effect on coal deliveries because of the substantial stockpiles maintained both by producers and at the Richards Bay terminal. The latter is approximately 2 mt, the equivalent of two to three weeks' exports, and thus was sufficient to prevent even the three-week NUM strike in 1987 from having any substantial effect. As a result, South Africa has the reputation of being the world's most reliable supplier. Supplies from all other major producers (including Australia) have been interrupted at times by strikes.

MARKETS AND PROFITABILITY IN THE 1980s

The coal industry is dominated by three corporations – Amcoal (Anglo American, 40 mt in 1986), TransNatal (Gencor, 34 mt) and Witbank collieries (Rand Mines, 40 mt). Together with the Sasol collieries, they produce 84 per cent of total output. Other significant producers are Goldfields Coal, Iscor, BP and Shell (through joint ventures), JCI, Lonrho, and Kangra.

Working profits for the major corporations in 1980 varied between R2 (TransNatal) and R4.87 (Rand Mines) per sales tonne. They rose steadily and peaked in 1986 between R7.32 (Rand Mines) and R10.17 (Amcoal), before dropping precipitously in 1987 and 1988 to R5–R5.50 for Amcoal and Rand Mines, and less than R2 per ton for TransNatal. Because of a modest strengthening of export prices to around $30 a ton at the end of 1988, working profits have since risen somewhat.

Only 25–30 per cent of each corporation's total production is earmarked for export. However, the high profit margins give exports great significance. In 1987 the contribution of exports to total working profits of each corporation was as follows: Witbank collieries, 49 per cent; Amcoal, 57 per cent; and TransNatal, 82 per cent. Clearly TransNatal is particularly vulnerable to adverse changes in the world coal market. Indeed in the 1988 financial year the corporation sustained an operating loss of R7.7 million compared to a profit of R112.5 million in the previous year.

The domestic coal market

Domestic consumption is dominated by Eskom, which consumed precisely half (66 mt) of all local coal sales in 1987 (Eskom, 1988). Sasol (40 mt) and Iscor (9,5 mt) together took another three-eighths of the total. By comparison, coal sold for household and merchant purposes only amounted to 4.4 mt, or 3 per cent of total consumption. Indeed, 85–90 per cent of domestic coal is consumed by state corporations and parastatals. Since Sasol itself mines practically all of its own coal and Iscor most of its own, demand prospects are dominated by long-term contracts with Eskom.[8]

In 1986 Eskom purchased practically all of its coal (99.7 per cent) under long-term contracts from tied collieries. The average cost was R14.87 per tonne, slightly less than the (then) controlled price for grade D coal. The profits on coal sold to Eskom are calculated on a cost-plus basis. Pre-tax returns on the older contracts were 15–23 per cent while more recent, larger contracts earn 18–19 per cent per year. On older collieries this works out at between 80 and 220 cents per tonne. On collieries completed more recently, profits range between 355 and 590 cents per tonne. Contracts entered into since 1970 make provision for an escalation of the capital investment for calculating the profit.

One implication of the 'cost-plus' approach is that all mining costs are passed directly on to Eskom. Hence Eskom meets the cost of any wage increases or labour disruptions, and similarly benefits if wages are contained. This probably provides an incentive for coal companies to inflate their costs (and hence their profits). However, mitigating factors are, firstly, that the mining industry, particularly the gold mines, consume 27 per cent of all electricity generated by Eskom and, secondly, the competition for new Eskom contracts.

The older tied collieries were financed by the coal-mining companies whereas recently Eskom has provided up to 50 per cent of the capital required. Since 1983 legislation has permitted Eskom, which can raise finance at a lower cost, to make indefinite loans to corporations establishing a tied colliery.

The export market

The export programme has from the start been subject to strict government regulation on the ground that coal is classified as a strategic commodity. As noted, three phases have already been implemented, the most recent raising the annual level of exports to 44 mt by 1985.

At present the phase IV expansion programme is under way to provide for the next 30 years, although exporters have not been able to obtain orders that allow them to make use of their allocations. Under phase IVa, it is anticipated

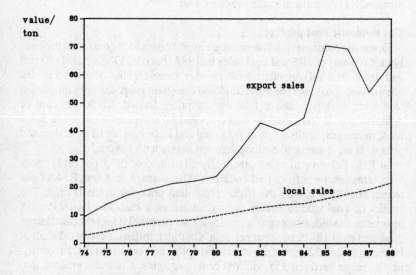

Fig 7. Value per tonne of local and export coal sales

that exports will rise to 67 mt a year, while phase IVb provides for exports of 72.5 mt. In addition, 6.7 mt per year of so-called permit coal (an intermediate grade which it is claimed has no local market) may be exported.

After gold, coal has become South Africa's largest foreign-exchange earner. In 1987 coal exports amounted to about 10 per cent of the total foreign-exchange earnings, comprising nearly half of the value of non-gold mineral exports.

Export sales grew from 2.8 per cent of total production in 1970 to 26.5 per cent in 1985 (see Figure 7). However, their monetary value has grown much more rapidly, and export revenues actually exceeded the value of domestic sales between 1984 and 1987. Clearly, the profitability of exports is far greater than that for domestic sales, and the gap between the two grew between 1970 and 1985. Figure 7 illustrates that the rand value of a tonne of export coal was more than four times the domestic sales value in 1985 – approximately R70 compared with R15.80. In 1970, the equivalent ratio was about three. This change was certainly due in part to the fall in the rand's exchange rate, but the coal price on world markets also rose rapidly during the early 1980s, as Table 2 indicates.

Table 2. F.o.b. prices received for South African steam-coal exports

Year	Rand/tonne	%	US$/tonne	%
1979	19.50		23.15	
1980	21.60	+11	27.75	+20
1981	29.05	+34	33.30	+20
1982	40.75	+40	37.60	+13
1983	37.55	(8)	33.70	(10)
1984	43.20	+15	29.25	(13)
1985	69.00	+60	30.97	+6
1986	68.05	(1)	29.67	(4)
1987	53.91	(21)	27.20	(8)
1988	64.08	19	n.a.	n.a.

South Africa's share of the world steam-coal market increased from less than 4 per cent in 1970 to a peak of 40 per cent in 1979. The expansion of exports from Australia, and the emergence of China and Colombia as major exporters, have trimmed South Africa's share back to about 25 per cent.

The major markets for South African coal exports are given in Table 3. Europe is the major recipient of South African coal. In 1985 Western Europe received 42 per cent and the Mediterranean countries 26 per cent, a total of two-thirds of all local exports. Asia received 30 per cent in 1985, increasing to 36 per cent in 1986 as sanctions began affecting sales. The remaining amounts are divided up between Africa and various other users.

The international competitiveness of South African coal

It was noted above that South Africa is a low-cost coal producer as compared to most other producers. A vital question is the extent that this is attributable to the low wages paid to the black workforce.

Table 3. Major markets for South African coal exports (in million tonnes)

	1981	1982	1983*	1984*	1985*	1986*	1987*
Western Europe	16.2	13.8	11.4	13.8	18.7	12.2	9.9
Mediterranean	5.3	3.8	6.3	9.2	11.5	12.5	13.2
Far East	6.5	8.6	9.8	11.1	13.4	14.3	14.7
United Sates	0.8	0.5	0.7	0.6	0.8	0.9	0.0
Unaccounted for	1.1	0.8	–	0.2	0.5	0.3	0.0
Total	29.9	27.5	28.2	34.9	44.9	40.2	37.8

*Exports from Richards Bay only.

Australia is South Africa's major competitor in the international coal market, selling coal mined in New South Wales (NSW) and Queensland. In NSW, the average wage (including bonuses and overtime) in 1987 was $A872.60 per week, equivalent to a monthly salary of R7 500 ($A1=R2). While miners averaged $A823.50, deputies (equivalent to certified miners in South Africa) were paid $A1 073 a month. These wages are much higher even than those typically paid to white miners and artisans in South African collieries.

In the UK, the average weekly wage (including bonuses and overtime) paid to miners in April 1986 was £214.40, while officials received £248.70. This is equivalent to monthly wages of about R4 030 and R4 670 respectively (£1=R4.37).

Table 4 compares productivity levels of South African Chamber mines to NSW, the UK and the USA. While exact figures for open-cast and underground operations are not available for local mines, the proportion of coal coming from open-cast workings (35 per cent) is less than in NSW (39 per cent) and much less than the USA (59 per cent). While the US mines are clearly the most productive, the productivity of the South African mines approaches that of NSW.

Comparative labour cost figures (Table 5) show that while workers in South Africa (white and black combined) were paid only R3.68 per tonne of coal mined in 1987, United States miners were paid R14.42 per tonne and Australian workers about R25 per tonne! Thus on wage considerations alone, low wages provide South African mines with a competitive advantage of approximately R10 per ton over United States exporters and R21 per ton over NSW mines. While the wages paid to British miners are about half those of NSW miners, the lower productivities obtainable from the much deeper British pits mean that wage costs per ton are considerably higher than in NSW and of course South Africa. With a wage cost alone of the order of R50 per tonne of coal mined, British coal production is evidently vulnerable to imports from South Africa.

Table 4. Production (metric tonnes) per employee per year in selected countries

	SA[1]	NSW[2]	UK[3]	USA[4]
Year	1987	1987–8	1987	1987
Open-cast	n.a.	6 050	–	9 470
Underground	n.a.	2 860	955	4 000
All mines	2 980	3 600	955	5 120
Proportion of total production from open-cast mines	35%	39%	negligible	59%

1. Chamber of Mines members only, saleable coal.
2. Joint Coal Board, 1989a.
3. P. J. Hoddinott, 1987: 71–83 provides a preliminary estimate for UK production as 105 mt by 110 000 employees.
4. Mine Safety and Health administration, 1988: Table 2. Figures for open-cast and underground exclude office and process plant personnel, which are included in calculating the 'all mines' production figures.

Table 5. Wage costs per tonne of coal

	SA[1]	NSW[2]	UK[3]	USA[4]
Average annual wage	10 965	90 000	48 360	79 400
Average annual production per worker	2 980	3 600	983	5 120
Coal sale price per ton	23.07	61.00	163.88	49.82
Wage costs as % of sale cost	16%	41%	30%	29%
Index of wage costs (SA=1)	1	6.8	13.4	3.9
Wage costs R/ton	3.68	25.00	49.20	14.42

1. Chamber of Mine members only, 1987, calculations based on 'at work' figures.
2. JCB, 1989. Pit head sale price = \$A30.50 (June 1987); \$A1=R2. Average wage = \$A872.60 in May 1987.
3. Assume an average wage of £214.40; £ = R4.37. Cost sale price = £1.50/GJ and 25GJ/t, 108.1 mt production (1986).
4. Assume US\$1 = R2.40. Sources: US Bureau of Labour Statistics, *Employment and Earnings Series*; Energy Information Administration. *Annual Outlook for US Coal*. Production worker average weekly earnings in 1987 = \$661.50. Average mine-mouth price (f.o.b. mines) = \$32.78 per short tonne.

South Africa's competitive advantage extends further than favourable geology and cheap labour. The ability of Richards Bay to accommodate very large ships reduces freight charges per tonne. South Africa's competitive position in the Western European and Mediterranean markets, already greatly facilitated by geographical location, improves with increases in shipping rates, since the latter affect its competitors more seriously.

THE CURRENT CRISIS AND FUTURE PROSPECTS

In both the domestic and export markets, difficult market conditions emerged for South African coal mines in the late 1980s. The impact has been made all the worse by the very accumulation, and accompanying expansion of capacity, which underpinned the massive profitability increases during the previous ten years.

The domestic market

The present outlook for growth in Eskom demand is dim. While growth in Eskom's electricity demand averaged 8.7 per cent between 1954 and 1980, in recent years this has dropped markedly to 4.5 per cent per year. The long lead time involved in power station projects, together with poor planning, has meant that Eskom had an excess generating capacity of almost 50 per cent in 1988. In addition, as new, more efficient power stations are commissioned, less coal is required to produce an equivalent amount of electricity.

In consequence, Eskom announced in September 1988 the shutting down or mothballing of 13 power stations. Apart from retrenchments of Eskom power station workers, this decision will result in the closure of several collieries and the retrenchment of an estimated 5 300 miners. Any moves to privatise Eskom are likely to pose a further threat of retrenchments.

Increased coal demand for synfuels hinges on government decisions about financing two private-sector projects through the Central Energy Fund. Firstly, Gencor has a R3bn (1989 values) proposal for mining a reserve of about 70 mt of torbanite, a coal which when heated produces a relatively high oil yield (*FM*, 24 March 1989). Gencor proposes mining torbanite in conjunction with number 5 seam coal, a total of 4 mt annually. Secondly, AECI have proposed a R10.5bn methanol-from-coal synfuel project. About the same size as Sasol 2, the methanol plant would consume about 12.6 mt of high-ash coal per year. Both projects have been in the offing for some time. The government's decision depends on strategic considerations – Hill (1987: 45) states that the government will endeavour to maintain domestic liquid fuel production at approximately 40 per cent of total consumption – as well as on sanctions pressures, the rand price of oil imports, trends in domestic liquid fuel consumption and production from Mossgas. The commissioning of a fourth Sasol plant is not likely before the torbanite and methanol plants. In November 1989, the cabinet refused to approve the latter two plants, on the grounds that the capital expenditure involved was not viable, given prevailing exchange rates and forecast oil prices.

The other major source of local demand, Iscor, depends on growth of its production, and thus in effect on its export capabilities. Finally, there is the small industrial and household market, which has experienced only slight growth over the past decade. With increasing township electrification, it is unlikely that there will be substantial increases in coal demand from this source. In any event, household use is inconsequential as a proportion of overall production. For example, it is estimated that Soweto during the peak winter season consumes some 30 000 tonnes of steam coal per month. Even if this rate were maintained all year round, it would amount to only 0.36 mt per year.

Future coal demand is likely to have an uneven regional impact. In general, the collieries in Natal are less productive than those in the Transvaal because of more difficult mining conditions. To offset this, the sale of Transvaal coal in Natal was prohibited by the price control policy. With the lifting of the controls in April 1987, listed prices for NAC (Natal) coal were 25–36 per cent higher than for equivalent grades of Transvaal coal. Depending on locations of producers and consumers, Transvaal coal may be delivered at competitive prices to Natal consumers. While the TCOA has abided by a government request not to ship coal into Natal, there are no longer any legal restrictions and some smaller non-TCOA collieries are already doing so. In addition, the Department of Trade and Industry has declared the TCOA a monopolistic organisation, and its activities within South Africa ceased in March 1989. If some of the larger Natal producers are forced to close, production will be taken over by the much more efficient Transvaal mines. This is likely to bring about a net decline in employment.

The export market

The first strongly negative factor affecting current export revenues has been the rail tariff hikes on the Witbank–Richards Bay line. In April 1987 SATS increased rates to R19.59 per tonne from Blackhill, the full length of the line. Another increase to R23.68 was imposed from September 1987, making a total increase of 145 per cent on the phase I and II rates of the previous year. SATS justifies these tariffs as necessary to recoup its investment in the upgrading of the line to accommodate the anticipated increase of phase IV exports.

If we assume average costs for extraction and washing at R20 per tonne, and for loading and wharfage at R4 per tonne, Transvaal steam coal could be landed in Rotterdam in 1987 at R55 per tonne. Railage costs, at R23.68, represented the single largest expense: 42 per cent of the total cost. Despite the increase in railage costs, local coal remains competitive with Australian coal and substantially cheaper than Colombian and USA coal in the European market.

As illustrated in Figure 7 and Table 2, both the volume and value per tonne of coal exports have declined since 1985–6. Three factors have affected coal export revenues – the emergence of a buyers' market in international coal, fluctuations in the exchange rate of the rand and, to a lesser extent, the impact of sanctions on volumes and prices for South African coal.

Sanctions

Although at the start of 1987 the prospects of sanctions were presented in alarmist terms, their effects have proved marginal. Sanctions have been imposed on South African coal by Denmark, France and the USA, which purchased a total of 11.3 mt in 1985, approximately 26 per cent of total coal exports. In 1987 producers were able to find markets for most of this coal, primarily by stepping up sales to south-east Asia (South Korea, Taiwan) and the Mediterranean countries. However, South African sellers have evidently been forced by these measures to offer a 'sanctions discount' to prospective purchasers, thus reducing revenues to some extent, even where volumes have remained constant (Gilbertson, 1988: 7). A further effect has been to undermine the perception of South Africa as a reliable supplier.

Table 6. Comparative delivered costs of coal from SA, Australia, Colombia and USA to Rotterdam

	F.o.b. costs inc. debt financing [1]	Delivery to Rotterdam [2]	Delivered ARA cost [3]
South Africa			
Open-cast	23.50	7.49	30.99
Underground	29.00	7.49	36.49
Richards Bay phase IV open-cast	31.00	7.49	38.49
Australia			
New South Wales underground	31.00	12.32	43.32
New South Wales open-cast	30.50	12.32	42.82
Queensland new open-cast	22.00	12.32	34.32
Colombia			
New open-cast	50.00	5.95	55.95
USA			
Large underground	42.00	5.94	47.94
Large open-cast	40.50	5.94	46.44

1. *Source*: Professor Donald W. Barnett, MacQuarrie University, as adjusted by Hill.
2. *Source*: Ocean Shipping consultants.
3. All costs in 1986 dollars.

The employment effects of sanctions have been less severe than the Chamber of Mines suggested was likely. While an initial Chamber estimate of 30 000 jobs lost under a scenario of *total* sanctions seems to be of the right order of magnitude – our estimate, based on data for individual export mines, totalled 24 000 – few workers can be said to have been retrenched directly as the result of sanctions.

A 1989 Chamber publication claims that 'at least 3000 ... retrenchments are directly attributable to the effects of sanctions' (Boers *et al*., 1989: 9). However, this figure requires examination. In 1987 exports declined slightly from 45.5 mt the previous year to 42.4 mt (7 per cent), whereas total sales grew marginally from 172.4 to 172.9 mt. In 1988 exports remained firm at 42.6 mt while total sales rose to 182.4 mt, an 11 per cent increase on 1987.

Employment on South African coal mines dropped by about 16 000 jobs between 1985 and 1989. Since overall sales have increased, it cannot be argued that sanctions have cost jobs. Where workers have indeed been retrenched, this has largely been due to rationalisation of operations and improved productivity. Firstly, there have been shifts of production from less efficient to more productive mines, and secondly, cutbacks in Eskom demand.

Ironically, South African coal producers have performed exceedingly well in comparison with producers in Australia where, of course, sanctions are not an issue. In New South Wales alone, because of the over-supply in the world coal trade, fourteen mines were forced to close between January 1986 and

June 1988 with the loss of 3 400 jobs (Joint Coal Board, 1989: 5). Raw coal production was cut by 12 mt.

The world coal market

Three principal factors determine the world price of coal – the price of oil, world coal demand and the level of competition amongst coal producers.

Oil is the major substitute for coal, and in many applications the two can be interchanged. The spot oil price (dollars per barrel) declined steadily after 1980 and dropped dramatically in 1986. The effect has been to force the price of coal down also. The oil price remains at low levels because supply has continually outstripped demand, particularly as Iraq and especially Iran have tried to boost export revenues to pay for their war. It is anticipated that the low price will not be sustained beyond the medium term.

The most significant alternative energy sources are hydro-electric and nuclear power. World hydro-electric power utilisation is thought to be approaching its full level of exploitation. While installed nuclear capacity increased substantially in the 1960s and 1970s, disasters and near-disasters such as Chernobyl and Three Mile Island, as well as the unresolved problem of waste disposal, have seen the freezing of much nuclear development. Nuclear power is unlikely to be an important contributor to new electricity-generating capacity in the foreseeable future.

In Europe, South Africa's major competitors are Poland, the United States and Australia. However, South Africa has the advantage over the US in terms of costs of production and over Australia in terms of costs of transportation. In the Pacific Rim market, the major competitors are again the US and Australia. Here Australia has a transport cost advantage.

Exports from Australia more than doubled from 42 mt to 92 mt between 1980 and 1986. Australia depends heavily on demand from Japan. The latter uses imports from South Africa to undercut Australian prices. USA and Canadian prices are seen as too high to maintain their share of world exports, while Poland's exports have declined in recent years.

Two further considerations are important in determining world coal supply and demand. Firstly, on the supply side, coal exports provide many developing countries with an avenue to earn foreign exchange, even though the state may have to subsidise production. Production is likely to continue, provided returns are not too far below acceptable levels.

Major potential new competitors which fall into this category are China, Colombia, Venezuela and Indonesia. China mines coal on a massive scale – 469 state-run mines, 69 000 worker cooperatives, and local authority and privately owned mines employing 4 million workers. Costs are reputedly similar to those of South Africa. In 1986 China mined 894 mt of coal although only 10 mt was exported. A new port with an export capacity of 50 mt is under construction (McQuaid, 1988: 183). With substantial Japanese capital being invested in Chinese mines, China may well become a major exporter, especially to the Pacific Rim market where it will enjoy a considerable transportation advantage. However, to date, China has not been able to meet contractual delivery commitments.

Secondly, on the demand side, countries such as West Germany, Britain and France have in effect subsidised their coal industries to avoid the unemployment consequent on pit closures. There is mounting pressure to do away with these subsidies for both economic and political reasons. In Germany it is estimated that if the subsidies were removed, only 25 per cent of the 90 mt mined each year would remain economically viable, requiring imports of some 67 mt per year. Similarly, in Britain privatisation of the Central Energy Generating Board (CEGB) and British Coal could result in British demand for imports rising to well over 20 mt per year. Indeed, British imports during the 1984 miners' strike increased from 4.5 mt to 12 mt, and have remained at this level since. That South African coal owners are interested in the export potential of CEGB privatisation is evidenced by the visit to Britain of a delegation drawn from the Chamber's Collieries Committee in April 1988.

CONCLUSION

The prospects for South African coal are intimately dependent upon developments in the international energy markets. The current crisis in the South African coal-mining industry is characterised by much lower profit rates than those prevailing until 1986, and by reductions in employment levels. Prospects for major increases in local coal consumption are limited, especially following the shelving of the proposed liquid fuel-from-coal project. Anticipated increases in consumption will probably not generate new employment as they will be achieved through productivity increases.

Industry analysts estimate that the 1988 capital cost for a new export colliery was R100–R140 per annual tonne, whereas new capacity at an existing mine would require R90 per annual tonne. At a discount factor of 18 per cent and a mine life of 25 years, capital costs alone would be R16.50–R26 per tonne of saleable coal. If one adds this capital cost to production and transport prices, it is clear that with the relatively low price of exports (Table 2), a new colliery would not provide a sufficient return to attract capital investment.

The export market remains fraught with pitfalls and possibilities. Export prices at Richards Bay have risen significantly to $30 per tonne f.o.b. in the first quarter of 1989, compared to $19–$23 in early 1987. For reasons of geology and labour control, South Africa is able to produce coal that is amongst the cheapest in the world. But in an oversupplied market, it is difficult to dispose of coal tainted by its origins in apartheid South Africa. A rise in oil prices or a privatisation assault against British and German miners could increase world demand and raise export coal prices. This could lift profits substantially for South African mines, improve employment prospects to a small degree and restore some of the bargaining position of mineworkers. In short, the very success of the South African coal-mining industry in promoting exports in the 1970s has made it ever more vulnerable to the yo-yo booms and busts of the increasingly competitive world energy economy.

POSTSCRIPT

The invasion of Kuwait by Iraq late in 1990 has important implications for South African coal exports. The crude-oil price rise from $18 a barrel in 1989 to above $30 increases the attractiveness of coal as an energy source. If higher oil prices are sustained, there is likely to be a substantial increase in the world steam-coal trade. With a political settlement on the horizon and declining sanctions pressures, South African coal exporters will be able to compete for a share of this expanding market on a more equal footing.

Prior to the Gulf crisis, the United States Energy Information Administration predicted that steam-coal trade would rise from 179 mt in 1988 to 500 mt in the year 2000. If oil prices above $28 a barrel are maintained, an increase in coal trade of this order is likely. A two- to three-fold increase in South African exports is feasible, assuming that South African coal-owners are able to maintain their share of world steam-coal trade (25 per cent in 1988). While this will substantially improve the overall profitability of the industry, the scope for increased employment is limited. Rationalisation and productivity improvements are likely to offset the potential for job creation that increased export demand will create.

7
Manufacturing development and the economic crisis: a reversion to primary production?

ANTHONY BLACK

The recession of the 1980s was the worst in the post-war period while the brief upturn of 1987–8 was stalled by the re-emergence of balance of payments problems. That an economic crisis exists is now commonly accepted although orthodox commentators may prefer to use terms such as 'structural slow-down' or 'stagnation' to refer to South Africa's dismal economic performance since the mid-1970s.

Problems in the manufacturing sector are central to an understanding of South Africa's economic crisis. Not only is manufacturing responsible for major contributions in terms of output and employment,[1] but it provided a key expansionary stimulus for the economy during much of the post-war period. In a developing economy with a rapidly growing population, industrialisation is the *sine qua non* of economic development. But in the 1980s manufacturing employment has been declining in both absolute terms and relative to total non-agricultural employment. Nor is slowing (or negative) growth a recent phenomenon. As Table 1 shows, employment growth rates in manufacturing have been declining since the early 1970s and contrast sharply with the boom years of the 1960s.

During the 'apartheid boom' of the 1960s, wages of black workers rose little but the economy grew rapidly throughout the decade and new employment opportunities were generated. In the 1980s, however, the economy appeared unable to provide even the inequitable growth of previous decades. Moreover, stagnating aggregate employment levels understate the impact of the economic slow-down. For, paradoxically, manufacturing employment has grown in peripheral areas, particularly the bantustans where huge financial incentives and wages as low as one-third of metropolitan levels have attracted a growing share of investment in manufacturing industry. The overall picture, then, has been of negligible growth in the industrial workforce, a growing proportion of which is, moreover, employed at extremely low wages in the periphery. What is clear is that recent economic and especially political factors have exacerbated the current recession. But as this chapter tries to demonstrate, the slow-down has been of a much longer duration, dating back to the mid-1970s and is attributable more to structural problems that began emerging in the earlier period of rapid growth. In particular, it is symptomatic of the failure of South Africa's post-war growth model.

Table 1. Gross domestic product and manufacturing growth, 1946–85

	Average rate of growth per annum		
	Manufacturing output (%)	Manufacturing employment (%)	Total GDP (%)
1946–1950	9.1	6.6	4.7
1950–1955	7.5	3.0	4.8
1955–1960	4.5	0.9	4.0
1960–1965	9.9	6.8	6.0
1965–1970	7.4	3.2	5.4
1970–1975	6.0	4.1	4.0
1975–1980	4.1	1.5	3.4
1980–1985	−1.2	−1.0	1.1

Source: Based on data from *SARB Quarterly Bulletin* Supplement (September 1981) and *Quarterly Bulletin* (June 1986), *Union Statistics for Fifty Years* and *SA Statistics* (1968, 1978 and 1982).

The first part of the chapter outlines the forces that produced rapid growth in the early post-war phase and then proceeds to analyse the structural weaknesses of the manufacturing sector, which began emerging during that period. The motor industry is used as a case study to illustrate some of the structural problems of South Africa's manufacturing sector. In the final section of the chapter it is argued that the combination of weaknesses in the manufacturing sector and current policy shifts may be leading to a situation where South Africa's industrialisation process is being 'forced back' along the lines of what has happened in Argentina and also in Chile during its experience of extreme economic liberalisation under the tutelage of the 'Chicago boys'. The primary sector is thus reasserting itself at least in relative terms as the previous growth model collapses. Primary exports, especially from the mining sector, may become even more dominant, and as a corollary of this process, manufacturing, which from the 1920s to the 1970s expanded its share of GDP considerably, may be entering a period of relative decline.

MANUFACTURING DEVELOPMENT: THE ISSUES

The broad issues central to an analysis of South Africa's manufacturing development are shared by many other middle-income or semi-peripheral countries. These concern the question of inward- and outward-oriented development paths, and more generally, the role of the state in the industrialisation process.

In the Third World, industrial development has been seen as virtually synonymous with economic development – the economic strength of the advanced countries clearly hinged on their highly productive industrial might. Thus the influential UN Economic Commission for Latin America (ECLA) argued that Latin American and other Third World countries, as exporters of primary products and importers of manufactures, were disadvantaged by the long-term tendency for the terms of trade to turn against them (United Nations, 1950). These countries therefore needed, the ECLA argued, to develop their own manufacturing capacity, which would take place through inward-oriented policies.

The nurturing of infant industry through tariffs and the development of manufacturing aimed at the internal market have for these reasons been important characteristics of industrial development in most Third World countries. Such policies were promoted by developmental states of nationalist, populist or socialist orientation which were keen to reduce dependence on imported manufactures and to develop and diversify their economies. They also had the support of nascent capitalist interests anxious to secure a niche in the peripheral economy, perhaps in partnership with foreign capital.

The *dirigiste* pursuit of industrialisation has come under attack from exponents of the neo-classical revival in economic theory (eg. Lal, 1984) and also in the policy prescriptions of international agencies such as the IMF and the World Bank. In particular, neo-classical economists have predictably attacked inward-oriented and 'statist' policies, arguing for the opening up of economies and reduced interventionism (Krueger, 1985). To this end, they have cited the rapidly growing newly industrialised countries (NICs) as examples of the benefits of export-led growth and the free market.

While it has been clearly shown that the NICs such as South Korea, Taiwan and Singapore have been highly interventionist in the realm of economic policy (Luedde-Neurath, 1984; Wade, 1984), the case for outward-oriented policies is less easy to refute. Undoubtedly, protectionism in Western countries renders the generalised promotion of manufactured exports problematic (Kaplinsky, 1984) but the comparative performance of the fast-growing Asian NICs and the stagnating inward-oriented economies[2] remains striking and of great relevance to the South African case. While the south-east Asian NICs are characterised by authoritarian political systems, all sections of society have benefited at least to some extent from their phenomenal economic growth. They have also proved fairly resilient to downturns in the world cycle and to increasing protectionism in the advanced countries.

Latin America presents an altogether different experience – there, economic stagnation and growing indebtedness are the products of the downturn in the world cycle and the exhaustion of the import-substitution model. Restructuring in the face of these adverse circumstances has taken a number of forms. On the one hand, Brazil and Mexico have sought to maintain high rates of accumulation and to expand manufactured exports. This has been accompanied by their running up massive foreign debts.

Opening of the economy, a growing reliance on primary exports, and wage repression has been the course followed by Argentina and Chile in the 1970s and early 1980s. The economic performance of Argentina during this phase was particularly dire. Among its economic strengths it could count self-sufficiency in energy, a substantial food surplus and a low import coefficient (less than 10 per cent). Nevertheless, it managed to treble its external debt in the five year period up to 1983, while per capita income fell by 10 per cent (Fitzgerald, 1983). The response to this series of events has recently been the virtual dismantling of its industrial system.

EARLY INDUSTRIAL GROWTH IN SOUTH AFRICA

A number of factors contributed to rapid manufacturing development in the early post-war period in South Africa. Firstly, the expansion of mining provided not only a growing market for a wide range of manufactures but also the foundations for the growth of national capital. In the early post-war period, rapid development in the international economy and buoyant commodity prices favoured South Africa's very open primary exporting economy. Using the vast revenues from this source, the mining houses were able to diversify into manufacturing on a large scale, establishing enterprises such as AECI and Highveld Steel.

A second factor was the particular form state intervention took. From the 1920s the state was committed to a systematic policy of industrialisation, and used as its main policy instrument tariff protection to encourage import substitution. The state also played a direct role through the establishment and expansion of parastatal corporations such as Iscor, Eskom and Sasol, and made indirect interventions through involvement by the state-owned Industrial Development Corporation in joint ventures with the private sector.

Economic development in South Africa has also relied on substantial inflows of foreign capital, and notwithstanding recent disinvestment by major foreign firms, South Africa has been the major destination of foreign investment in Africa. According to one estimate, 28 per cent of manufacturing employment in South Africa is in foreign-controlled firms (Rogerson, 1982: 201).[3] The bulk of this investment took place during the 1960s, with the main attraction being a substantial and rapidly growing domestic market and relatively low wages. These inflows of direct investment reflected the high profitability and rapid growth of South Africa's economy in the 1960s, just as the decline in foreign investment in the 1970s and the outflow of capital in the 1980s reflected the onset of slowing growth.

As a result of protectionist policies, the first stage of import replacement was well advanced by the late 1950s. From 1946 to 1956 manufactured imports as a percentage of total domestic production declined from 63 per cent to 42 per cent, there being a marked reduction in imports of non-durable consumer goods (Marais, 1981). Manufacturing output growth reached unprecedented levels in the 1960s, spurred by increased local production of durable consumer goods.

By 1970 South Africa's manufacturing sector had also become substantially diversified. Sectors that could loosely be described as 'heavy' industries increased their share of total net manufacturing output from 44.1 per cent in 1946 to 57.3 per cent in 1970. Most striking was the fall in relative share of output of the food, textiles and clothing industries, which declined from 38 per cent to 27.3 per cent over this period. This was counterbalanced by substantial growth in the output shares of more advanced sectors such as chemicals, transport equipment and machinery. Indeed, the share of the machinery (including electrical) sector increased from an insignificant 2.2 per cent in 1946 to 10.7 per cent in 1970 and 15.0 per cent by 1982 (McCarthy, 1988: 11).

Manufacturing was also growing at this time in relation to total output. Its share of GDP rose from less than 13 per cent in 1946 to over 23 per cent by 1970. Thus in a period when manufacturing was growing rapidly and the process of import substitution was being accompanied by considerable diversification into more advanced sectors it was possible to argue in the mid-sixties that 'given the built-in propelling force of industrialisation on the one hand, and the positive national approach to modern economic development on the other, it appears that economic growth will not constitute a major problem in South Africa during *at least* the next twenty years' (emphasis in original) (T. A. du Plessis (1965) cited in Scheepers, 1982: 22).

THE ROOTS OF STAGNATION

Having briefly outlined the major forces of economic development that were responsible for rapid growth in South Africa in the 1950s and 1960s, we turn now to the more negative aspects of this growth process, the implications of which are now so readily apparent. The emphasis of my argument falls on the failure of the growth model which had underpinned economic development and particularly manufacturing development until the 1970s. This authoritarian import-substitution model was premissed on a rapidly growing but narrowly based (mainly white) market for consumer durables. Accordingly manufacturing development was spurred by domestic demand, the expansion of which was a function, firstly, of growth in the internal market and, secondly, of the replacement of imported goods with locally manufactured products.

The manufacturing sector has a number of structural weaknesses which seriously constrain growth. Firstly, the sector remains very dependent on imports. Any further reduction in the degree of import intensity will be difficult and expensive to achieve. A second problem has been extremely weak export performance. As a result of these two factors, the sector is a net importer, and expansion depends therefore on foreign-exchange 'subsidies' generated through primary exports or capital inflows or both. A third structural feature relates to the pattern of demand arising out of South Africa's highly skewed income distribution. On the one hand, this has limited the development of a broad mass market while on the other it led to the early expansion of a substantial market for consumer durables such as motor vehicles. This has increased the pressure on the balance of payments because such sectors are particularly import-intensive. What is more, state industrial policy has failed to provide a coherent framework for resolving these problems.

Import dependence

A striking feature of South Africa's industrial development is that in spite of the policies of import substitution, overall import intensity (imports/GDP) has not changed since the 1920s and imports still accounted for as much as 24 per cent of GDP in the period 1980–4 (see Table 2). By comparison, the more advanced Latin American countries – Brazil, Argentina and Mexico – have import coefficients of around 10 per cent.

Table 2. Import co-efficients (imports/GDP) for selected five year periods

1920–24	0.25
1935–39	0.23
1950–54	0.27
1965–69	0.20
1980–84	0.24

Source: McCarthy, 1988:13.

What has happened in South Africa is that the profile of imported goods has changed considerably. The importation of consumer goods has been substantially reduced. For example, in 1985 local production comprised 93 per cent of total domestic demand for clothing and 89 per cent and 92 per cent, respectively, in fabricated metal products (except machinery) and food (Kahn, 1987: 248). This has been achieved at the expense of increased imports of equipment and materials. South African industry remains highly import-intensive with regard to its inputs, particularly of producer goods. The domestic capital goods sector, always small, has actually been declining in recent years (Kaplan, 1987). In 1986 the major categories of imported manufactured goods were machinery and transport equipment excluding motor vehicles (R8 420 million), chemicals (R3 590 million) and motor vehicles (R2 180 million). There is also a high degree of dependence on foreign technology. While a 1983 survey indicated that South African manufacturing was becoming less reliant on foreign technology, the extent of dependence remains considerable and no less than 84 per cent of the new technology embodied in machinery is imported (Black, 1985: 155).

Thus any further import replacement would have to be directed mainly at the capital, intermediate goods and high-technology sectors. This could be achieved with state support and tariff protection but would be extremely expensive unless sufficient economies of scale, probably requiring significant penetration of foreign markets, could be achieved.

The failure of exports

Not only has South Africa's manufacturing sector remained highly import-intensive but it has also manifestly been unable to generate significant export growth. Thus, although the real value of fully manufactured commodity exports[4] increased at 9.5 per cent annually after 1970, they still accounted for only 10 per cent of total exports by 1982. Growth in the export of manufactured goods has occurred mainly in the basic metals sector, which comprised 49.4 per cent of total manufactured exports in 1982 (up from 35.4 per cent of manufactured exports in 1970). In the more advanced branches of manufacturing, export growth has (with the notable exception of armaments) not kept pace with the growth of non-gold exports. This is particularly notable in the case of machinery and transport equipment where export volumes fell by an average of 0.7 per cent per annum between 1964 and 1984 (McCarthy, 1987).

For the manufacturing sector as a whole, the high costs and inefficiencies generated by South Africa's particular form of import-substituting industrialisation have worked against any competitiveness on international markets. One example here would be the additional costs imposed on the clothing industry through having to use domestically produced (and protected) textiles. Tariff protection has been partly responsible for the establishment of high-cost plants, the proliferation of motor-vehicle producers being the prime example here.

Potential exporters have been unable to rely on long-term government policies regarding export incentives and other forms of support. During the past decade, uncertainties about exchange-rate fluctuations and sanctions have also led to a situation where in spite of the weak rand there has been little new investment in export-oriented production.

A number of additional factors have borne an adverse effect on specific industries. For example, the motor industry is particularly affected by comprehensive licence agreements with source companies that not only limit exports of locally manufactured vehicles and components but also necessitate frequent introductions of new models. Recent massive capital flight and continuing pressures for disinvestment place the state in a particularly weak position in negotiations with transnational corporations (TNCs). It will be extremely difficult, therefore, for the state to develop policies of 'assertive pragmatism'[5] successfully adopted by the Asian NICs in their dealings with TNCs. Not only have these countries been selective about the type of investment made but they also have been able to exert considerable leverage as regards technology transfer, minimum export levels, etc.

Given the import intensity of South African manufacturing and the failure of production for the export market, it is not surprising that the sector is a net importer. By the early 1970s 'a persistent and fundamental structural imbalance on the current account' had become apparent (Reynders, 1975: 129). It was in response to this issue that the Reynders Commission into Export Trade was set up in 1970. In its 1972 report, it noted the limitations of the import-substitution model, and advocated greater emphasis on exports. However, little concrete action was taken.

High capital intensity

A number of factors have contributed to high capital intensity in the manufacturing sector. Levy (1981) has argued that state and foreign investment has been concentrated in capital-intensive sectors and that these were deliberately fostered as a strategy to improve the situation of white workers, who were heavily concentrated in the capital-intensive branches of industry. Restrictions on urbanisation, and measures such as the Physical Planning Act (1967) which limited industrial expansion in metropolitan areas, encouraged capital-intensification. Moreover, legislation such as the Group Areas Act, which restricted black participation in commerce and industry in 'white' areas, has also dealt a severe blow to small-scale, labour-intensive industry.

Mechanised plant has been artificially cheapened through subsidies in the form of capital investment allowances and, until 1981, through artificially low interest rates. Over-valued exchange rates and the fact that import tariffs are not usually applied to capital equipment have also contributed to relative factor prices that favour increasingly capital-intensive production techniques. Moreover, substantial real wage increases and the upsurge in union organisation over the past decade in South Africa have certainly led to a degree of mechanisation, but this should not be overestimated. Survey results indicate that the rising cost and shortage of skilled labour was of greater importance than the rising cost and shortage of unskilled labour in channelling investment towards fixed assets (Black, 1985). For this, state policies have been mainly responsible, especially if one considers inadequate education and inadequate technical training for blacks and job reservation, as well as closed-shop agreements limiting the advancement of black apprentices.

The market profile of final demand is also important in determining the choice of technique in manufacturing (Stewart, 1983). A more equal distribution of income would tend to increase demand for low-income goods such as food, clothing, basic housing and simple manufactured articles, which are labour-intensively produced. Instead, South Africa's very unequal income distribution has tended to limit the development of the domestic market for these basic commodities.

THE MOTOR INDUSTRY: A CASE STUDY

The motor industry has been central to overall manufacturing development: industry sources suggest that about one-eighth of all economic activity in South Africa is linked to it (B. McCarthy, in *Financial Mail*, Supplement, 28 November 1986). In consequence, the motor industry contributes significantly to the trends already identified in the manufacturing sector as a whole and usefully illustrates them as well. By 1960 South Africa was the largest producer of motor cars of any developing country, producing over three times total Asian (excluding Japanese) production (see Table 3). While the industry has recently declined in relative importance, South Africa is still a major developing-country producer.

Table 3. South African car production in comparative perspective

	(1000 units)			Local content (%)
	1960	1970	1980	
S. Africa	87.4	195.0	277.0	50
Nigeria	–	7.1	151.0	30
Brazil	62.2	343.7	977.7	98
Argentina	30.3	163.4	204.4	95
Mexico	24.8	136.7	303.0	60
Venezuela	6.5	48.0	94.0	40
Chile	–	20.7	29.0	40
Colombia	–	7.7	43.0	20
India	19.1	37.4	30.5	100
S. Korea	–	14.5	57.2	90
Philippines	2.9	7.6	26.6	70
Taiwan	0.4	n.a.	132.0	60
Malaysia	–	7.5	81.0	50
Iran	2.5	31.8	80.0	50
Indonesia	2.0	2.0	41.0	40
Thailand	–	6.6	25.0	35
Total (Developing Countries)	238.1	1 029.7	2 552.4	
World Total (Million)	13.0	22.8	28.6	

Source: Adapted from Jones and Womack (1985: 396–7).
Note: Production levels in some countries have changed significantly since 1980. South Korea now produces close to a million units per annum. South Africa's output has declined since 1980.

The motor industry has always been highly cyclical but overall growth was rapid in the post-war period, reaching a peak in 1981. Since then, the sector has experienced what industry sources themselves refer to as a depression. While 1988 showed a significant improvement over the previous year, car sales were still only 76 per cent of levels attained in 1981, and long-term forecasts have been substantially downgraded.

As is to be expected, the depression in the motor industry has been considerably deeper than for manufacturing generally. For example, capacity utilisation in motor-vehicle assembly, parts and accessories was 62.3 per cent in 1986 compared to 78.5 per cent for the manufacturing sector as a whole (BTI, 1988: 16–17). The industry has suffered massive employment losses, with the total workforce in the automotive and related industries declining by 19.3 per cent from 1980 to 1986, compared to a decline of 9.7 per cent for manufacturing as a whole. The labour force in the assembly plants themselves has been worst hit, and blue-collar employment in seven major plants was reduced by a third between 1981 and 1986 (BTI, 1988: 110).

The pattern of demand

The pattern of demand can have important implications for the import intensity of the industrial structure. Viewed historically, production for the local market has been primarily oriented to meeting the growing demand from a fairly narrow middle and upper-income group. This feature should of course be linked to South Africa's highly unequal income distribution.

I have argued elsewhere that unequal income distribution has constrained the development of a more broadly based demand which could provide a major stimulus to local manufacturing (Black and Stanwix, 1987; also Freund, 1986). The counterpart of this process has been the 'overdevelopment' of sub-sectors such as the motor industry that supply upper-income groups. At fairly low levels of average per capita income, demand for a luxury consumer durable such as a new motor vehicle is clearly limited to a small section of the population. But increases in income inequality produce a greater demand for cars, while conversely a more progressive distribution of income would tend to limit the demand for cars.

Jenkins (1987) cites a study for Mexico by Lustig, which argued that if an increase in income was wholly concentrated in the top income groups in urban areas, the growth in demand for cars would be over a hundred times higher than if income growth were concentrated in the lowest income groups. Similarly, in a study of the Brazilian economy, Morley and Smith (1973) have shown that a more regressive distribution of income, by which the income share of the top 10 per cent of families rose at the expense of the bottom 40 per cent, increased the rate of growth of demand for transport equipment by almost one-third, compared to a situation in which all groups maintained their income share. Such a regressive shift would double the rate of growth of transport equipment production that would have been achieved by a redistribution away from all other groups towards the bottom 40 per cent. This trend, furthermore, would be considerably reinforced if one was considering demand for motor vehicles rather than more broadly defined transport equipment.

South Africa's unequal income distribution thus spurred the premature development of a substantial motor industry in the 1960s, and per capita car-ownership levels have been substantially higher than for most other countries with similar income levels. Given the high degree of import intensity of the motor industry, its rapid development has therefore tended to reinforce rather than alleviate high levels of import dependence.

Import intensity and local content

The motor industry illustrates the difficulties of reducing import intensity even in a sector where there has been substantial state intervention to promote local content. With the exception of machinery, the motor vehicle and transport equipment sector is more import-intensive than any other manufacturing sector (Kahn, 1987: 248). Import payments by the motor and component industries accounted for over 9 per cent of total merchandise imports in 1986.

Regulations designed to promote local content in motor vehicle production have been introduced in a series of phases. In 1958, before the implementation of the local content programme, only 17 per cent (by mass) of components used in local assembly were domestically produced. The first phase of the local content scheme was introduced in 1961, and by 1969 under phase II of the programme local content levels had reached approximately 50 per cent on a mass basis (BTI, 1988: 4). Phase V (now ending) required a minimum local content of 66 per cent by mass. Heavy penalties were imposed on vehicles produced with local content of less than 66 per cent, and rebates on excise duties were available to encourage local content levels above the minimum.

Using vehicle mass as the basis for measuring local content has encouraged domestic sourcing of components with high mass and relatively low value. Body pressings, for example, are almost totally produced locally. Local production of these components accounts for 19.2 per cent of total mass but only 8.6 per cent of value (BTI, 1988: 20). On the other hand, many components with a high value–mass ratio are imported. The result is that 66 per cent local content on a mass basis is equivalent to little more than 40 per cent by value. This is low by the standards of many other developing countries, particularly in view of the substantial relative size of the South African motor industry (see Table 3).

The difficulties currently faced by the motor industry, as well as by South African manufacturing more generally, were neatly captured by the Board of Trade and Industries' estimate that, if motor vehicle sales improved, foreign-exchange usage for phase V manufacture (including tooling) could reach R6 000 million by 1990, compared to R2 288 million in 1986 (BTI, 1988). In other words, the improving market would lead directly to a rapid widening of the foreign-exchange gap (net usage) and so make it necessary to restrict sales and force the industry back into a slump. The improved export incentives now available to the industry are unlikely to change this outcome.

The import costs incurred for the expensive tooling required to produce certain components locally have been further increased by the weaker rand in the 1980s, as well as by new production technologies, materials and more sophisticated products. The amortisation rate for a low-volume manufacturer such as South Africa has become prohibitively high. Average tooling costs for new models at 66 per cent local content (by mass) have risen from between R3m and R8m in 1977 to between R40m and R100m in 1987, and according to manufacturers' estimates could exceed R300m by 1990 (BTI, 1988: 26).

The recognition that the mass-based local content programme has been a failure has brought about a recent shift in state policy. The central objective is now to reduce net foreign-exchange usage in the sector. A new local content programme (phase VI) based on value rather than mass came into operation in 1989. Manufacturers will have to increase average local content value in their vehicles to at least 55 per cent by the beginning of 1990. This is expected to rise to 75 per cent by 1997. The important criterion will be foreign-exchange usage which is defined as 'all payments in foreign currency for components, tooling and royalties less all foreign exchange earnings in the form of exports and royalties received' (BTI, 1988: 94). Under this system, targets for foreign-exchange usage would be set based on the value of vehicles sold by the manufacturer. These targets could be reached by means of local content or exports or both. In addition, local sourcing of tooling would also be taken into consideration. Thus, manufacturers with no exports will be tied to a minimum of 55 per cent local content by value and to the planned increases in local content levels. Export earnings will, however, allow these companies to increase imports.

Increasing local content requirements is certain to raise car prices. As Table 4 illustrates, premiums are incurred on unit costs as local content rises above 50 per cent on a mass basis. These are likely to become substantial as the minimum level of local content increases from the 1990 level of 55 per cent (by value) to 75 per cent by 1997.

Table 4. Impact of increases in local content on vehicle unit cost

Level of local content	1–20%	20–40%	40–50%	50–55%	55–60%	60–66%	66–70%
Medium cars (931–1 250 kg) Weighted by unit sales	1.716	.684	15	(96)	(138)	(237)	(265)

Source: Adapted from Board of Trade and Industries, 1988: 19.
Note: Bracketed figures indicate unit cost premiums incurred as local content levels rise. Unbracketed figures indicate unit cost savings at lower local content levels.

The objective of the new programme is to cut the motor industry's foreign exchange usage by at least 50 per cent. The effectiveness of this new scheme remains, however, to be seen. One important obstacle is clearly the rapidly rising cost of the new tooling required to increase local content levels. Higher levels of local content require substantially increased tooling investments, and these costs together with low volumes constitute a significant constraint on increasing local content.

Constraints on exports are likely to be a further limiting factor. Here, too, the motor industry is typical of the manufacturing sector as a whole. Exports by motor manufacturers are expected to reach R240 million by 1990 (see Table 5) on the basis of existing export incentives. This is less than 1 per cent of total non-gold exports (at 1988 levels). The rapid growth in 1986–7 is mainly attributable to the introduction of incentives which took effect in 1985.

South Africa's motor industry has been established on the basis of the domestic market, and a really significant expansion of exports would require that the domestic industry achieve a measure of integration into the broader global system of automotive production. Rapid world-wide changes in technology and the advent of new forms of production organisation have great implications for the manner in which developing countries are integrated into this global system and hence for the future prospects of automotive exports from semi-industrialised countries such as South Africa (Jones and Womack, 1985; Kaplinsky, 1988).

Table 5. Actual and projected value of exports by Phase V manufacturers (R millions)

Category	Registered exports							Projected exports		
	1981	1982	1983	1984	1985	1986	1987	1988	1989	1990
1. Components	4	2	2	2	10	77	167	191	199	207
2. Vehicles	1	36	37	35	19	28	27	26	30	33
Total	5	38	39	37	29	105	194	217	220	240

Source: Board of Trade and Industries, 1988: 43.

The transformation now sweeping through the motor industry is linked to the development of state-of-the-art microelectronics and automated flexible manufacturing systems. The significance for our discussion of these technological improvements lies in their flexibility, and the implications this has for optimal economies of scale and the location of plants supplying components to final assemblers. Flexible automation and robotics allow for the production of different models without the expense of complete retooling required by the previous system of dedicated production lines. Under the latter production-line system, minimum efficient scale was estimated to be in excess of 200 000 units per year for one model, and a single producer needed to produce of the order of 2 million units per year to remain internationally competitive and profitable (UNIDO, 1986: 50). While minimum efficient production levels might well be lower depending on the extent of local content and local manufacture, limited domestic markets have led to substantial inefficiencies in developing countries because of extremely low levels of output per firm.

The new technology may open up possibilities for smaller firms and for efficient production in smaller markets, developments which are of particular significance in developing countries (Jones and Womack, 1985). On the other hand, however, the reduced level of wage costs as a proportion of the total limits the attraction of sourcing components from low-wage countries. High levels of automation and heavy initial investments, together with a production system based on minimal inventories, mean that component supply is increasingly being sited at the main final assembly site. This may leave little scope for developing countries to supply components, except minor mechanical components. The adoption of the 'just in time' (JIT) system of production organisation places South African exporters at a considerable disadvantage because of their remote location from major international markets. While some developing countries, notably South Korea and Brazil, have achieved significant export success in this area, there is little prospect that South Africa will emulate this without substantial state support.

RESHAPING OF INVESTMENT PATTERNS

Capital intensification and the productivity of capital

In spite of minimal real wage increases and negative rates of growth in the capital stock, the rate of capital intensification has been rising rapidly, accompanied by a sharp fall in the productivity of capital. To some extent, this reflects falling rates of capacity utilisation and large-scale lay-offs.

The rate of growth of capital stock per worker has increased significantly since the early 1960s (see Table 6). Rapid increases in the amount of capital per worker have not, however, produced substantial growth in labour productivity.[6] This is indicative of a rising incremental capital–output ratio (ICOR), whereby more and more capital is required to produce an additional unit of output. These patterns have been much more pronounced during the 1970s, and especially the 1980s. Indeed, average capital output (the inverse of the capital–output ratio) in manufacturing fell by nearly 40 per cent between 1970 and 1986 (McCarthy, 1988: 7).

Table 6. Average annual change of capital stock in manufacturing industry

	Manufacturing fixed capital stock (per cent per annum)	Manufacturing capital stock per worker (per cent per annum)
1960–65	9.0	2.6
1965–70	6.9	3.6
1970–75	8.6	4.3
1975–80	5.8	4.1
1980–85	4.5	5.9

Source: Adapted from Bell and Padayachee (1984), Thompson (1987), *SA Statistics*.
Note: Figures denoting growth rates of capital stock are in real terms.

The collapse of investment

Perhaps the most disturbing feature of the current crisis in the manufacturing sector is the collapse of productive investment. During 1986, real gross domestic fixed investment in manufacturing was only 40 per cent of the 1980 figure and at its lowest level for 16 years. In a revealing statement, the Minister of Constitutional Development and Planning appealed to big corporations to 'face the risks', claiming that 'the fact that large local investors seem to remain unwilling to take a forward view beyond 6 months is causing a self-inflicted form of internal disinvestment' (*Daily News*, 23 March 1986).

The profitability of foreign investment in South Africa, compared with other countries, declined in the 1970s. McGrath and Jenkins (1985: 25) reported that throughout the 1960s, US foreign investment in South Africa had been substantially more profitable than elsewhere, but after 1971 the trend was reversed. Inflows of foreign direct investment consequently decreased after the early 1970s, with adverse balance of payments results that had to be offset by an increase in foreign borrowings by the public sector. Large-scale disinvestment especially by US firms has, of course, been a feature of the 1980s. At the same time, internal stagnation led to an increase in direct foreign investment by South African corporations as domestic firms sought to place their funds elsewhere. Kaplan (1983) reported, for example, that during 1982 Anglo American was the largest single foreign investor in the USA. All these responses are indicative of the limited scope perceived to exist for productive investment domestically.

Spatial shifts

To some extent also, a spatial reorganisation in industrial production is taking place, as labour-intensive processes are relocated to peripheral, cheap-labour areas. The proportion of the labour force employed in decentralised or peripheral areas rose from 12.3 per cent in 1972 to 19.3 per cent in 1984 (Wellings and Black, 1986: 12).

The global manifestation of this process is the so-called 'new international division of labour': profound structural changes have forced Western firms to reorganise production internationally. This has involved substantial relocation of productive facilities, especially in labour-intensive light industry, to Third World production sites mainly in Asia (Fröbel *et al.*, 1980). Particularly in respect of the clothing industry, imports from developing countries have made significant inroads into the domestic markets of the advanced countries. In 1968, such imports accounted for only 1.9 per cent of total clothing sales in the advanced countries, but they had increased to 11.6 per cent by 1978 (Jenkins, 1984: 42). Restructuring in the world garment industry has therefore tended to take the form of relocation rather than innovation and new technology.[7]

Such changes tend to occur when industry is under severe competitive pressure or affected by recession as is currently the case in South Africa. The South African clothing industry, in particular, is under considerable pressure. Recessionary conditions, rising imports from the Far East and, lately, competition from decentralised areas are producing a 'shake-out' in the industry. On account of this scenario, as well as very generous decentralisation benefits, new investment in the clothing sector is increasingly taking place in growth point locations. As the executive director of the National Clothing Federation points out: 'if you look at the world trend it is evident the industry has tended to go to low cost labour areas that might be comparable to the homelands ... Taiwan, Sri Lanka, the Philippines' (*Financial Mail*, 7 July 1985).

Manufacturing firms have thus responded to the economic crisis by reshaping investment patterns. This has involved greater emphasis on more mechanised production techniques, on investment in the form of take-overs and sweeping rationalisation programmes, often associated with closures of older (more labour-intensive) plants, and increased investment in decentralised production sites, particularly by firms in labour-intensive subsectors. Another response has simply been a refusal to invest in productive assets. Gross domestic fixed investment has fallen to very low levels, and this has been accompanied by increasing foreign investment by South African corporations, a virtual halt in new direct foreign investment and, of course, the much publicised disinvestment by major foreign corporations. In sum, the limits of the authoritarian import-substitution model have led to much slower growth and a virtual collapse in productive investment within the manufacturing sector.

New directions in state policy

As noted above, there has been concern within, but little action by, the state for at least two decades about the limitations of the import-substitution approach to industrialisation, as implemented in South Africa. The increasing severity of the problems of the manufacturing sector during the early 1980s has demanded a more substantial shift of direction in policy. Yet while this is recognised by policy-makers, there exists only limited agreement amongst them on the question of the future orientation of secondary industry.

The familiar story of constraints on further import replacement was reiterated by the Kleu Study Group on Industrial Development Strategy (1983). The subsequent White Paper (Republic of South Africa, 1985) set out a series of broad policy guidelines and objectives which emphasised the improvement of productivity and international competitiveness of the manufacturing sector. Considerable weight was placed on the need to reorientate South African industrial strategy from import-substitution to export promotion. Export incentives are now being revamped and the state is considering methods of restructuring industries to make them more competitive.[8]

In these tentative moves towards the promotion of manufactured exports, policy-makers are clearly influenced both by the new theoretical orthodoxy of outwardly oriented development and by the strikingly rapid growth in the south-east Asian NICs which has been fuelled by manufactured exports. There are, of course, important objections to generalised export promotion of manufactured goods by developing countries, the most important of these perhaps being the world-wide lurch towards protectionism. For South Africa, such difficulties are exacerbated by the inefficiencies and high costs engendered by import-substituting industrialisation. Furthermore, because South African manufacturing is so highly geared to the domestic market,[9] the increase in the export of manufactures would have to be dramatic in order to have a significant impact on local industrial output. Interestingly, the fall in the exchange rate of the rand, while leading to greater manufactured exports, has not brought about significant investment in export-oriented production, with the possible exception of the clothing industry in the bantustans.

These limitations on both import substitution and export promotion have been recognised by a number of influential state advisors, and one group has proposed a strategy of 'inward industrialisation' as a solution to the stagnation of the manufacturing sector. Lombard (1985) sees inward industrialisation as drawing its impetus from 'orderly urbanisation' which would expand the domestic market, as a result of the rapid rise in expenditure by low-income urbanised groups. This would promote labour-intensive development and the rapid growth of small firms and the informal sector. In a study of the composition of final demand, Dreyer and Brand (1986) argue, for example, that increased demand for basic goods will have positive spin-offs for the economy because such products are labour-intensively produced.

While a strong case can be made that redistribution will allow scope for expansion which is both more labour-intensive and less import-intensive than the current structure, these formulations tend to abstract from the powerful class interests which reinforce existing patterns of inequality. Both Lombard (1985) and Dreyer and Brand (1986) imply that measures such as 'orderly urbanisation' would enable significant redistribution to occur. This is by no means necessarily the case. While the abolition of influx control has been undoubtedly a prerequisite for the improvement of living conditions for low-income groups, it will not necessarily be accompanied by increased spending on housing and social services. Nor would it automatically follow that incomes will rise. While many people, who were effectively trapped in rural areas, may find greater opportunities in, for example, the informal sector of the urban peripheries, the existence of a growing labour reservoir in urban areas may well put downward pressure on formal sector wages. Significant redistributive measures are clearly not on the agenda of South Africa's ruling group and even milder measures (such as increased spending on urban social services and rural development) are constrained by the state's fiscal crisis and the indifference of a political constituency with very little interest in such action.

CONCLUSION: THE REASSERTION OF PRIMARY PRODUCTION

With the manufacturing sector unable to expand, the primary sector of the economy is reasserting itself. Agriculture's contribution to GDP declined many years ago to its currently modest proportions. The contribution of mining fell steadily until the 1970s but has since shown signs of renewed importance. However, because this process is still in its early stages, it is in investment terms that it is most noticeable. In the decade up to 1975 fixed investment in mining was on average little more than 30 per cent of investment levels in manufacturing. In 1986–8 fixed investment in mining was roughly equivalent to that in the manufacturing sector. Although the last few years have been exceptional, there is clear evidence of a reversal in the long-term trend according to which manufacturing grew in importance relative to mining until the early 1970s (Figure 1).

In spite of difficulties in the coal industry, major new investments in gold

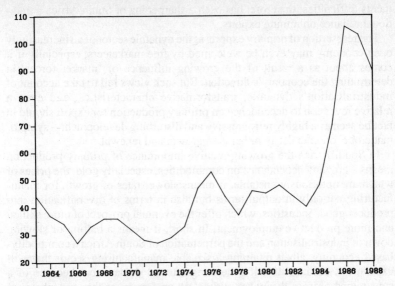

Fig. 1. Mining investment as a percentage of manufacturing investment

and platinum mining will ensure that for the next few years at least investment levels in mining will remain high. Three recently completed or current projects at Vaal Reefs, Western Deep Levels and the new H. J. Joel mine alone amount to nearly R3 billion. For the sake of comparison, fixed investment in manufacturing in 1986 amounted to less than R5 billion in current terms. Notwithstanding the recent weakness in the gold price, the South African gold-mining industry is apparently poised for a third wave of expansion, which could boost production from the present level of 610 tonnes per annum to 670 tonnes per annum by 1995 (*Business Day*, 5 October 1987). Platinum mining is going through one of its most rapid growth phases since the industry was established in the 1920s.

For these reasons South Africa's dependence on primary exports and particularly on gold has been increasing recently and is set to rise further. This trend reflects the high rand gold price which, in turn, is partly a function of a domestic currency weakened by crisis conditions in the broader political economy. The former Governor of the Reserve Bank repeatedly stated his desire to keep the rand at a fairly low level, with one of the prime objectives being to maintain the profitability of the gold-mining industry. Such a course has been regarded as essential to restore external equilibrium and in particular to allow for the repayment of foreign debt.

A weak domestic currency has benefited exporters and led to major new mining developments, but there has been very little new investment in export-based manufacturing. The response to continuing balance of payments difficulties, therefore, has been a sharpening of South Africa's traditional reliance on mining exports.

The reassertion of primary exports as the dynamic sector of an increasingly open economy may even be welcomed by free marketeers, especially if it comes about as a result of the growing influence of 'market forces' in determining the economy's direction. But such views fail to take account of industrialisation's dynamic, transformative characteristics, and even a relative reversion to dependence on primary production for export should in fact be seen as a highly retrogressive and disturbing development – symptomatic of economic decay rather than growth and renewal.

In South Africa the growing relative importance of primary production means a growing dependence on commodities, especially gold, the prices of which are notoriously unstable. It means slower rates of growth for manufacturing not only in output terms but also in terms of diversification into producer goods industries, which offer the eventual prospect of more skilled and more productive employment. In short, it means a halting or slowing down of industrialisation and the perpetuation of South Africa's commodity-based economy, albeit retaining a sizeable manufacturing sector that will expand not because of self-generating internal demand but because of a constrained process dependent ultimately on foreign-exchange 'subsidies' earned by primary exports.

8
The South African capital goods sector and the economic crisis

DAVID KAPLAN

THE SIGNIFICANCE OF THE CAPITAL GOODS SECTOR

The term 'capital goods' refers in this chapter to machinery and equipment used in production or transportation. The significance of the capital goods sector in the process of development has long been widely acknowledged. One of the distinguishing features of a non-developed country is the reliance on imported capital goods, and any expansion of the economy, leading to an increased demand for (imported) capital goods, tends therefore to run up against a foreign exchange constraint. Thus, the developing economy can be contrasted with the developed economy – the latter internally generating the means for its own expansion and the former acquiring such means only externally. This contrast is central to the notion of dependency and is, for example, at the heart of Amin's distinction between the integrated, autocentric centre and the disarticulated, dependent periphery (1974: 286–99).

However, apart from considerations of foreign exchange, economists are presently paying increasing attention to the impact of the capital goods sector upon processes of technical change. This emphasis has come about principally as a result of the emergence of new technologies, especially microelectronics. Many economists now accept that the capital goods sector is central to any process of technical change since the latter invariably requires the introduction of new or modified capital goods. And improvements in the capital goods sector will be diffused to users throughout the economy. Some economists have laid emphasis on capital-saving innovations that may be of particular importance to developing countries (Rosenberg, 1976).

Moreover, each country has unique features which are constantly in flux and which affect users (for example, market structure, prices and availabilities of factors of production, climate and legal regulations), and for this reason a local capital goods sector will produce items that are better suited to local conditions than imported varieties. There is abundant evidence that the capital goods sectors of Latin America and Asia have produced machinery and equipment that are well suited to their local environment and for export to other Third World countries (Fransman, 1986: 24).

Even where the technology is imported or 'transferred' from abroad, the local producer still has to apply the technology. Assimilation of technology – entailing such activities as adaptation, support, maintenance, repair, updating and reconditioning – will be rendered much more effective in the context of a local capital goods sector which has developed technological capabilities[1] and is therefore capable of providing such services. The development of an indigenous capital goods sector as the site for enhancing indigenous technological capabilities may be essential if developing countries are even simply to remain efficient users of overseas technology in the long term (for example, in telecomms; see Hobday, 1986a: 370)

In addition, the existence of a local capital goods sector has other advantages. For example, it is frequently an important site for the training of skilled workers and the further spread of skills throughout the economy. In other words, the development of the capital goods sector, especially in the Third World context, gives rise to positive externalities.

In South Africa, as we shall see, the capital goods sector has not expanded rapidly. This is particularly evident when a comparison is made with some of the other countries that are at similar stages of development, especially the Asian NICs (above all, South Korea and Taiwan) and Brazil. Very broadly speaking, the capital goods sector in South Africa was comparable to, and in many cases more advanced than, that prevailing in these countries up until at least the early 1970s. Since that date, the expansion of the capital goods sectors in the NICs has far exceeded that in South Africa. Moreover, the divergences in growth rates have become even more accentuated with the contraction of the capital goods sector in South Africa since 1982.

The slow rate of growth of the South African capital goods sector has had adverse repercussions on the balance of payments, on the development of skills, and the generation and diffusion of more productive technologies through the wider economy. While it is easy to measure the impact upon the balance of payments, the impact on the diffusion of skills and productive technologies is impossible to quantify. But this will be one important determinant of the rate of productivity growth for the economy as a whole.

This chapter has three major objectives. The first is, simply, to chart the development of the South African capital goods sector since the early 1970s. The second is to examine the factors that have impeded the development of the capital goods sector, with particular attention paid to the contemporary economic crisis. The third is to sketch, albeit highly tentatively, what would be required to enable a more rapid development of the capital goods sector in the future.

The structure is as follows. The first section examines the macro-statistics, that is, the development of the capital goods sector *in toto*. However, for a number of reasons, and especially because the capital goods sector is a catch-all category, comprising in reality a number of industries with differing production characteristics, the macro-picture is not very informative. Sections 2 and 3 therefore consist of two case studies – namely, those of machine tools and telecommunications equipment manufacture. Both products can be categorised as complex capital goods since their production requires a significant degree of technological complexity.[2] These two industries were selected because they both form a 'leading edge' in the diffusion of new technologies. The capabilities of these industries to design and produce original equipment are assessed. Finally, section 4 concludes by stressing the impediments to the further development of the South African capital goods sector and the implications for the longer-term development prospects of the South African economy.

THE CAPITAL GOODS SECTOR SINCE 1970

A number of indices of the development of the capital goods sector over the period 1970–85 are presented below. Broadly speaking, these are measures of output and of import penetration and export success. It is difficult to distinguish long-term trends from short-term fluctuations. However, most indices suggest that the expansion of the capital goods sector has not been rapid. This slow rate of expansion was partially disguised by the short 'boom' which occurred between 1979 and 1982, but since then the sector has declined considerably.

Output measures

In terms of the gross output of local production, over the 15-year period 1970–85 there was considerable growth in real terms in electrical machinery (171 per cent), but much slower growth in machinery and equipment (45 per cent) and transport equipment (excluding motor vehicles) (23 per cent). Overall, the percentage increase was 78 per cent – or a compound growth rate of 3.9 per cent per annum. The capital goods sector expanded in the first half of the 1970s, but then declined so that 1975 output levels were only surpassed in 1980. There was a period of rapid but very short-lived growth between 1980 and 1982, but in the three-year period 1982–5, gross output in the capital goods sector declined by some 13 per cent.

In terms of net output – a more significant measure of development,[3] the trends are similar. However, the net output growth rates for two of the sub-sectors were marginally lower than the gross output growth rates – electrical machinery and equipment (170 per cent), machinery and equipment (38 per cent) and for transport equipment (excluding motor vehicles) (2 per cent) very much lower. The overall increase was somewhat lower than for gross output – 67 per cent – or a compound growth rate of 3.5 per cent. In South Africa, between 1970 and 1985, the value added in machinery and transport equipment, as a percentage of total industry, declined from 17 to 16 per cent. By contrast, in all the NICs, value added in machinery and transport equipment, as a percentage of total industry, in the same period expanded very significantly and now substantially exceeds the South African figure (IBRD, 1988: Table 8).

Table 1. Imports – machinery and transport equipment (excluding motor vehicles) 1972–1985

	1972	1973	1974	1975	1976	1977	1978	1979	1980	1981	1982	1983	1984	1985
1 Imports of K goods	983	1.123	1.531	2.231	2.520	2.110	2.644	2.766	4.273	5.926	6.173	5.201	6.627	7.017
2 Total imports	2.813	3.275	4.909	5.577	5.859	5.118	6.253	9.904	14.381	18.511	18.391	16.288	21.804	22.698
3 Local production	1.266	1.578	1.877	2.443	2.746	2.771	2.744	3.282	4.692	5.907	6.087	7.012	7.026	8.985
4 Local sales (1+3−Exports)	2.142	2.588	3.262	4.514	5.099	4.664	5.174	5.761	8.671	11.477	12.860	11.907	13.285	15.399
5 K good imports as % of total imports	35	34	31	40	43	41	33	28	30	32	34	32	30	31
6 K good imports as % of local sales	46	43	47	49	49	45	51	48	49	52	48	44	50	46

Sources: Import data from the *Monthly Abstract of Foreign Trade Statistics* (chs 84–9 excl. ch 87). Local production – *Statistical News Release* (Census of Manufacturing 1985) 20/08/87; Central Statistical Service and *Quarterly Bulletin of Statistics*, Value of Sales.

Table 2. Exports – machinery and transport equipment (excluding motor vehicles) 1972–1985

		1972	1973	1974	1975	1976	1977	1978	1979	1980	1981	1982	1983	1984	1985
1	Imports of K goods	107	113	146	160	167	217	214	287	294	356	400	306	368	664
2	Total exports	3.129	4.134	5.571	5.948	6.540	8.212	11.312	14.811	19.880	18.129	19.292	20.620	21.804	22.698
3	Merchandise exports	1.977	2.364	3.006	3.408	4.194	5.417	7.449	8.809	9.739	9.791	10.666	10.691	14.537	20.845
4	K goods exports as % of merchandise exports	3.4	2.7	2.6	2.7	2.5	2.6	1.9	1.9	15	2.0	2.1	1.5	1.4	1.8
5	K good exports as % of merchandise exports	5.4	4.8	4.9	4.7	4.0	4.0	2.9	3.3	3.0	3.6	3.7	2.9	2.5	3.2
6	K good exports as % of local production	8.4	7.2	7.8	6.5	6.1	7.8	7.8	8.7	6.3	6.0	5.6	4.4	5.2	7.4

Sources: Import data from *Monthly Abstract of Foreign Trade Statistics* (chs 84–9 excl. ch 87).

Table 3. Balance of trade on capital goods 1972–1985

		1972	1973	1974	1975	1976	1977	1978	1979	1980	1981	1982	1983	1984	1985
1	Imports of K goods	983	1.123	1.531	2.231	2.520	2.110	2.644	2.706	4.273	5.926	6.173	5.201	6.627	7.017
2	Exports of K goods	107	113	146	160	167	217	214	287	294	356	400	306	368	664
3	Balance: Exports – Imports	–876	–1.010	–1.385	–2.071	–2.353	–1.893	–2.430	–2.419	–3.979	–5.570	–5.773	–4.895	–6.259	–6.354
4	'3' deflated by 1980 CPI	–2.147	–2.260	–2.776	–3.659	–3.741	–2.704	–3.131	–2.752	–3.979	–4.835	–4.370	–3.299	–3.777	–3.299

Sources: *Monthly Abstract of Foreign Trade Statistics*; *Quarterly Bulletin of Statistics*, Value of Sales.

Imports and exports

Capital goods as a percentage of total imports tended to rise in the mid-1970s and then fall back to approximately one-third of all imports (see Table 1). Overall, there is no clear trend evident that capital goods have become a less significant proportion of the total import bill.[4] If one looks at the share of the local market captured by imports, again it is difficult to discern any overall trend. Over the entire period, imports of capital goods have captured between 46 and 52 per cent of the local market, and the percentage is exactly the same in 1970 as in 1985. However, as we shall see, the true significance of imports is considerably understated by these figures.

While the import data reveal no clear trend, the export data give a distinctly negative picture (see Table 2). Capital goods exports as a percentage of total exports show a pronounced decline from 3.4 per cent in 1972 to 1.8 per cent in 1985. There is also a clear tendency for capital goods exports to decline as a percentage of merchandise exports. The percentage of local production exported is very low and while there is no clear trend, overall there is a tendency for this magnitude to decrease.[5]

A comparison of the gross figures for imports and exports reveals the net foreign exchange requirement in respect of capital goods, that is, the extent to which capital goods exports have been able to compensate for capital goods imports (see Table 3). In real terms, the excess of imports over exports rose steadily until 1976, then tended to decline until 1980. It then rose sharply, tending to fall again after 1982. This largely reflects the cyclical fluctuations in the economy. When the economy is expansive, imports especially tend to rise and, to a lesser extent, exports to fall, while when the economy contracts, exports tend to rise and imports especially tend to fall. The effect on the balance of payments is evident. Overall, in real terms, with imports rising more rapidly than exports, the adverse balance of trade in respect of capital goods has shown a pronounced tendency to increase.

By comparison with the NICs, the local production of capital goods in South Africa, at a macro-level, is much less developed. Thus, for example, the domestic supply ratios for Brazil (the domestic supply ratio measures the proportion of local purchases of capital goods that are supplied through local manufacture and is the inverse of column 6 on Table 1) is above 75 per cent whereas in South Africa it has fluctuated around the 50 per cent mark. As a consequence, the importation of capital goods accounts for a far larger percentage of total imports in the case of South Africa. Moreover, South African exports of machinery and equipment are far smaller than such exports from the NICs, both in overall terms and as a percentage of capital goods imports.

The macro-indices presented above suggest that there have been substantial impediments to the development of the capital goods sector and that these impediments have become more pronounced over time. However, there are some serious deficiencies in the data. For example, all the South African import data are f.o.b. and for this reason the final market value for imports and thus, for example, the percentage of the local market accounted for by imports, will be much higher when other 'charges' are taken account of and imports calculated c.i.f. Moreover, the capital goods sector is an agglomeration of many different products and industries. For example, the production of a simple wood-working tool and an advanced machine tool would both fall into this statistical category. For these reasons, the macro-data conceal the very significant differences within the sector. The true picture can only be drawn in a series of concrete case studies rather than simply extrapolated from the macro-data. Moreover, it is only through such case studies that the factors inhibiting the development of the sector can be precisely identified.

Table 4. Imports of machinery and equipment as a percentage of total imports

	1979	1980	1981	1982	1983	1984
Brazil	19.2	15.7	15.6	14.6	12.0	11.5
Korea	23.9	16.4	16.0	18.1	19.9	20.5
South Africa	33.2	26.3	27.8	28.0	—	—

Source: The data are SITC 7, *United Nations Yearbook of International Trade Statistics, 1986*. Data for South Africa for 1983 and 1984 are not given. The data for South Africa in the UN *Yearbook for International Statistics* do not correspond exactly to the data released by the Department of Statistics, hence the difference with the figures in Table 1.

Table 5. Exports of machinery and equipment overall and as a percentage of imports of machinery and equipment ($m)

	1981	1982	1983	1984	1985	1986
Brazil						
Exports	3,392	4,214	3,461	3,036	—	—
Ex/imports %	76%	116%	126%	129%	—	—
S.Korea						
Exports	4,839	6,152	7,980	10,453	11,378	11,658
Ex/imports %	80%	102%	105%	153%	107%	108%
South Africa						
Exports	495	456	378	339	904	521
Ex/imports %	5.6%	6.2%	5.8%	5.5%	9.5%	11.1%

Source: Exports are SITC 7, *United Nations Yearbook of International Trade Statistics, 1985* and *1986*. The percentages for South Africa are derived from *South African Monthly Trade Statistics*.

THE MACHINE-TOOL INDUSTRY

The state of the industry

Machine tools cut and shape precision metal parts. Since virtually all manufacturing utilises machine tools or equipment produced with machine tools, the production of better and cheaper machine tools will lower capital

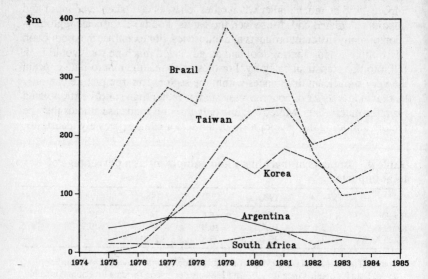

Fig. 1. Machine tool production: South Africa and selected countries

costs and stimulate productivity throughout industry.[6]

The production of machine tools in South Africa began during the Second World War, in order to facilitate armament production. However, this ceased at the end of the war. Production began again on a small scale in 1952 but grew steadily, so that by the mid-1970s there were some 30 independent producers.[7] By the early 1970s, the industry was expanding and was comparable in terms of volume of output and product range with those of, for example, South Korea and Taiwan. However, since the mid-1970s, output growth has been far slower in South Africa than in any of the NICs, particularly the Asian NICs (see Figure 1).

Since 1982, the industry has tended to decline dramatically. (Although the data suggest that the decline began only after 1983, this is because of the way in which the statistics are collected, resulting in a time lag.) Export performance has been very poor while local production has satisfied a diminishing share of the local market (see Table 6). This is in stark contrast with the situation of the machine-tool industry in the Asian NICs and Brazil, where growth has been very rapid.

The poor performance of the South African machine-tool industry, in quantitative terms, has been accompanied by a poor performance in terms of more qualitative indicators. Since the mid-1970s, there has been very little product development in the South African industry. The industry's products are overwhelmingly at a low level of technological sophistication. In particular, the industry has only very recently and on a very small scale begun the production of numerically controlled (NC) machine tools. The latter have captured a large share of the world market from conventional machine tools and offer major increases in productivity.

Table 6. South African machine-tool industry: production and trade, 1975–85 (US$m current prices)

	(1) Prodn.	(2) Export	(3) Import	(4) Total market (P+I+E)	(5) Prodn. as % of total
1975	15.5	2.4	88.4	101.5	15
1976	15.0	4.9	88.7	98.8	15
1977	13.8	4.6	46.0	55.2	25
1978	14.9	4.0	80.5	91.4	16
1979	21.3	4.6	140.5	157.1	14
1980	29.6	5.1	205.7	230.2	13
1982	36.8	5.0	250.0	281.8	13
1983	23.4	4.4	169.1	188.1	12
1984	10.4	0.7	132.4	142.1	7
1985*	5.9	0.5	75.2	79.6	7

Source: *American Machinist*, various February issues. (Compiled from annual data submitted by the South African Machine Tool Manufacturers' Association. This data is in turn compiled from the returns submitted by members of SAMTMA.)
Notes: * Estimates.

Moreover, while no precise data are available, it is clear that the local content of the South African machine-tool industry is low. By comparison with, for example, South Korea and Taiwan, a large number of important items still have to be imported. Local content appears to be increasing only slowly.

Indigenous technological capability in the machine-tool industry is very limited. The firms employ only a few people whose primary function is design or adaptation. There are no formal R and D departments. Firms rely overwhelmingly on foreign principals or licensers. This is expensive in terms of foreign exchange and serves to inhibit the development of local design capabilities. The firms are all small in scale and none of the large South African conglomerates have significant investments here.

In summary, the stagnation of the machine-tool industry since the mid-1970s and its actual decline since 1983 have had an adverse impact on the balance of payments. In consequence, the industry has not acted as an effective site for the generation and diffusion of new technologies.

Accounting for the industry's limited development

Two widely accepted explanations are advanced for the constrained development of the South African capital goods industry. First, it is argued that limited research and development is undertaken locally. This in turn is seen to be a consequence of the dominance of foreign subsidiaries and the reliance of local companies on licensing agreements with foreign firms. Thus, technical know-how has to be purchased abroad and is said to be expensive and inappropriate to South African conditions. Technological dependency is often seen as almost total[8] and is therefore viewed as one of the principal restraining factors on development.

Second, there is the argument that emphasises the limited size of the domestic market. Given economies of scale, the production of most capital goods locally, in the absence of significant levels of protection through tariffs or import licensing, is simply not competitive, even in the local market (Kleu Report, 1983: 71). Similar arguments are often advanced by many producers in the machine-tool industry. They particularly stress the constraining effect of the small domestic market size, which is said to create much higher costs of production.

Both arguments have some cogency but neither is adequate to explain the constrained development of the South African machine-tool industry. As regards technological dependency, the dramatic growth of the machine-tool industries in the Asian NICs has occurred in conjunction with, and arguably through the effective assimilation of, imported foreign know-how (Fransman, 1984b: 12). Moreover, the dominance of foreign producers in the machine-tool industry, in Singapore for example, is frequently more pronounced than is the case in South Africa. In any case, if local producers have chosen to remain dependent on foreign firms for technology rather than develop their own in-house technological capability, this itself needs explanation.

As regards the second argument, the size of the local market compares favourably with a number of other countries whose machine-tool industries have been far more dynamic. For example, in the decade between 1975 and 1984, the market in South Africa was nearly 30 per cent larger than that of Taiwan, 9 per cent larger than Australia's and comparable with those of some major developed countries producing machine tools, such as Sweden. In contrast with other countries, South African machine-tool producers are currently only able to satisfy a very small proportion of the local market demand. What requires explanation is why South African domestic procurement ratios are so low when compared with other countries. Since this is true even by comparison with countries whose domestic market is far smaller than South Africa – Argentina for example – it cannot be explained simply by reference to the size of the domestic market.

The explanation is more complex and involves the interplay of a number of factors. International developments within the machine-tool industry have to be integrated with developments occurring within the South African economy as a whole. A brief synopsis of the key factors is presented below. (For a fuller discussion of these issues, see Kaplan, 1989a.)

Firstly, in all the Asian NICs, exports have been central to development. Especially in the early stages, the development of local regional markets was important (Fransman, 1982, 1984; UNIDO, 1986: 15). Exporting in and of itself provides producers with knowledge of customer requirements and developments internationally – so-called 'learning by exporting' (for example in South Korea, see Westphal *et al.*, 1984: 296). In the South African case, as we have seen, exporting has been distinctly limited and the South Africa machine-tool industry exports a significantly lower percentage of its output than any of the NICs. This poor export performance has both reflected and reinforced the limited capacity of the industry to design or adapt differentiated products for the market. There are a number of reasons for the limited export of South African machine tools. Exchange-rate fluctuations and lack of governmental support are of some importance. But what is of greater significance is that the industry's principal market is confined to southern Africa. This is not simply a function of proximity and of similar market and environmental conditions. Many local producers are either subsidiaries of, or manufacture under licence from, overseas corporations and are restricted, in terms of these arrangements, to exporting to the southern African region alone. The southern African market has been growing only very slowly and South African companies have been increasingly encountering 'political obstacles' as neighbouring countries attempt to seek alternative sources of supply. Moreover, this market is technologically undynamic and therefore offers very little scope for 'learning by exporting'.

On the other hand, penetration of the more sophisticated markets in the industrialised countries depends in large part upon local product development, which in turn depends on local technological capabilities. This is the second explanatory factor. As we have seen, local technological capabilities, especially in regard to product design and development, are very weak. The relationship between these two factors is best illustrated by considering the shift from conventional to NC machine tools.

In order to undertake the development and production of NC machine tools, a major expansion in technological capability is required. Developing these capabilities is expensive, the returns are uncertain, and the payback period is often lengthy. In both Korea and Taiwan, government-financed research institutes helped producers design their first NC machine-tool products. Other forms of assistance such as cheap credit have also been available. In South Africa, by contrast, no such state aid has been forthcoming. In the present recessionary conditions, the self-financing of such investments can simply not be justified by the individual firm. The machine-tool producers are all small-scale, and the absence of large conglomerates means that there are few external sources of funding for the development of technological capabilities. While the machine-tool industries of the NICs are rising up the 'learning curve', the South African firms, especially in respect of the more sophisticated products, are falling ever further behind.

Thirdly, although the local industry enjoys higher nominal tariffs than is the case in the Asian NICs, these have been rendered of little significance. This is in part because of special trade agreements, for example with Taiwan;

in part because second-hand machinery is exempted; and in part because tariff protection is confined to those machine-tool products extensively produced locally (which have come to constitute an ever smaller share of the market). Unlike many of the NICs, where tariff protection has been extended precisely in order to encourage the local production of new products, the South African government has been reluctant to grant protection in advance of production.

Fourthly, the economic recession and the fluctuations in the business cycle, particularly since 1980, have severely impeded the development of the machine-tool industry. In the period 1980–2, the South African economy, unlike the economies of its principal trading partners, entered an upswing and the exchange value of the rand was very high. Locally, demand was strong and rand prices were cheap. Moreover, on the international market, there was a precipitous decline in orders for machine tools as the recession bit. Internationally, machine-tool producers found themselves with tremendous excess capacity. At the same time, in order to meet the threat of Japanese competition, European producers began to cut their prices, and provoked a major price war.

As a result of all of these developments, machine tools flooded into South Africa. Tariff protection was rendered almost totally ineffective and importations were massive – for example, in 1981, South Africa imported twice as many NC lathes as South Korea, Taiwan and Brazil put together. When the downturn began, around March 1982, importers were left with massive unsold stocks, which have depressed the market ever since. The recessionary conditions prevailing since 1982, combined with the existence of unsold stocks of imported machine tools, have served to ensure that demand for locally produced machine tools has remained very restricted. Thus, in the upswing, the high value of the rand and the substantial price discounts offered rendered tariffs largely ineffective, while in the downswing the depressed level of demand generally and the extensive import overhang rendered tariffs mostly irrelevant.

Finally, two other considerations have also served to constrain the development of the South African machine-tool industry: firstly, the high cost of certain locally produced inputs such as quality steels and electronic components: and secondly, the high price and especially the shortage of appropriately skilled labour. As the industry becomes more design-intensive, this latter constraint is of ever-greater significance.

In the Asian NICs, machine-tool producers have been expanding by utilising a strategy that, like the Japanese before them, emphasises cost leadership. This entails broadening the product range, into NC machine tools for example, but competing in the lower end of the range on the basis of price and simplicity of operation rather than performance (for a discussion, see Chudnovsky and Nagao, 1984: 206–24). Firms following this strategy need to enhance substantially their R and D spending, acquire new, highly skilled manpower and, with the associated increase in the minimum scale of production, engage increasingly in exports. This is a long-term strategy, and the increases in cost will not initially be associated with concomitant increases in turnover, especially in view of the highly competitive international market. Hence the need for some 'support' and the provision of risk capital from outside the industry. Such a strategy is being pursued by the machine-tool producers in the Asian NICs – South Korea and Taiwan, in particular. In these countries, the state has financed R and D programmes as well as education policies which have greatly increased the supply of highly skilled labour. In addition, the risk capital necessary to finance this cost-leadership strategy has been made available.

But in some other countries, such as South Africa and Argentina, an alternative strategy has been adopted by machine-tool producers (for the case of Argentina, see Chudnovsky and Nagao, 1984: 216–18). This strategy is inward-looking and focused on production for the local market. Technological know-how is acquired via licensers rather than more unpredictable in-house development. Cost leadership is not attempted, but instead extensive tariff protection is sought. Protection, it is argued, is necessary in order to ensure local production. Faced with fluctuating exchange rates, high local input costs and shortages of skilled labour, producers have concentrated on the local market. Since the local market is restricted, South African firms have tended to diversify into a wider range of products than would firms of a similar size in South Korea or Taiwan. Product diversification entails fragmentation of such planning skills and technological capabilities as exist within firms. As a result, where design does occur, design cycles tend to be more lengthy and the emphasis is on minor modifications to externally generated designs. Given the situation in which they find themselves, machine-tool producers in South Africa and in Argentina have elected to 'play it safe'. Indeed, the absence of significant risk capital has made any other strategy untenable.

An analysis of the development of the machine-tool industries in other industrialising countries suggests that the state will need to play a far more active role, if the industry is to expand in South Africa. In this regard, it is worth noting the conclusions of a major study of the world machine-tool industry which recommended 'an explicit policy of direct subsidies for: (a) the training of engineers, technicians and other specialists; (b) the organisation of firms towards specialisation or diversification; (c) product and process research; (d) marketing; (e) assistance in overcoming financial crisis at a time of economic depression. Alternatively, the establishment of institutes for design improvement, standardisation of quality control, marketing and provision of capital at low interest rates may be effective' (UNIDO, 1984: 118).

However, in the absence of a sustained revival of investment in productive assets, and, to a lesser extent, of an expansion of exports to regional markets, governmental financial aid alone would not suffice to ensure the long-term development of the South African machine-tool industry. In the meanwhile, the stagnation of the local machine-tool industry adversely affects the development of the capital goods sector and secondary industry as a whole. Indeed, the stagnation of the machine-tool industry is both a cause and a symptom of the current economic crisis. By contrast, the industrial expansion of the NICs, some of which, at least in the realm of exports of industrial products, are South Africa's principal competitors, is being facilitated by, and is in turn facilitating, the dramatic development of the domestic machine-tool industries in these countries.

TELECOMMUNICATIONS IN SOUTH AFRICA

The design, manufacture and operation of telecomms is a very large and rapidly growing industry. If present rates of growth continue, by the year 2000 telecomms will have surpassed all other major industrial sectors, contributing around 7 per cent of GDP world-wide (Arthur D. Little study quoted in Hobday, 1986: 27; Mansell, 1988: 243). But apart from being of major importance in its own right, a well-developed communications infrastructure is a vital prerequisite for economic development. In particular, telecomms constitute the 'cutting edge' of all microelectronic-based information technology. As a result, most OECD governments give large-scale support to their telecomms-equipment manufacturing industries and more than 60 developing countries have established governmental departments specifically designed to formulate national policy with respect to telecomms (Hobday, 1985: 2).

There are two broad perspectives to adopt in evaluating a country's telecomms sector. On the demand side, one must consider the efficiency, capacity and actual utilisation of the telecomms network; on the supply side, the efficiency and capability of the local industry, in the production and the design of telecomms equipment.

The SA telecomms network

In terms of the Post Office Act of 1958, the South African Post Office (SAPO) has exclusive rights to the provision of telecomms services in South Africa. SAPO orders constitute some 75 per cent of the domestic telecomms market.

In the mid-1970s, with the advent of digital replacing the old electro-mechanical technology, telecomms began to change fundamentally. For a variety of reasons,[9] SAPO was one of the first telecomms administrations in the world to decide unequivocally in favour of fully digital telecomms equipment. That decision was to have an ambiguous impact upon the local telecomm equipment manufacturing industry but, in terms of the network itself, was to give South Africa a head-start in the operation of a modern, sophisticated telecomms service. Thus, the first advanced Siemens EWSD exchange was installed in South Africa – some two years before West Germany itself – making this country the first in the southern hemisphere to have a digital exchange. Presently, over 20 per cent of the subscriber ports are connected to digital exchanges – which is higher than for many European countries. By 1990, over 80 per cent of transmission circuits will be digital. The system is expected to be fully digital by 2010. This is similar to the time-frame for many industrialised countries.

The expansion of the digital network has allowed for the modernisation of the service and an expansion in the range of services offered. There is extensive use of fibre-optics (South Africa was also the first country in the southern hemisphere to have a working fibre-optic line), very modern and extensive telex and packet switching systems, as well as an extensive system of leased-line data services over digital transmission lines. Moreover, much of the digital equipment installed is compatible with major anticipated technological developments such as ISDN (the integration of all transmission services down a single wire), which has recently begun to operate locally.

Apart from being technologically sophisticated and offering a wide range of services, the cost of telephone services is low by international standards. As at 1 January 1985, in terms of prevailing exchange rates, South African prices were fourteen lowest out of 74 countries, with a price (the price is a composite of installation fees and call charges for local and long-distance) of 368DM compared to an average of 609DM. In terms of purchasing parities, South Africa was eighth lowest of 34 countries. In terms of the average number of working hours required to pay the cost of telephone services, South Africa was nineteenth lowest of 40 countries. It required 67 working hours to pay the cost of telephone services in South Africa, whereas the world average was 108.

The actual rate of expansion of the network is somewhat less impressive. Growth has been steady, but much less spectacular than in the NICs. The annual average rate of expansion has been about 6.6 per cent since the early 1970s but it has been decelerating since 1982. Presently there are just over 5 million telephones installed in South Africa. In terms of overall telephone penetration,[10] South Africa falls somewhere between the developed industrialised countries (which typically have penetration ratios of 2–5 times that of South Africa) and the developing world. In 1985, South Africa had 118 telephones per 1 000 population. By comparison, the telephone penetration ratio of Taiwan was 228, South Korea 183 and Brazil 84.[11]

Moreover, there is a very significant maldistribution of telephones and telecomms services generally, firstly by racial group and secondly as between metropolitan and rural areas. (Since a far higher number and proportion of blacks live in the rural areas, these two inequalities somewhat overlap.) This can be best illustrated by looking at the distribution of telephones. Over three-quarters of all telephones are in the metropolitan areas and over 72 per cent of the telephones in metropolitan areas are in white hands. Coloureds have 12 per cent and Indians 7 per cent, whereas Africans have only 9 per cent.[12] This gives a very high penetration ratio for the white metropolitan community, comparable to any of the advanced industrialised countries, and a penetration ratio for Africans equivalent to those found in the middle range of developing countries. Correlated with this, the wide range of sophisticated services is overwhelmingly made use of by the metropolitan business sector. But the white and, to a much lesser extent, Asian and coloured markets have reached, or are reaching, saturation point. The key growth segments for telephone services are therefore African domestic consumers, especially in the metropolitan areas and, for the growing range of sophisticated data services, larger and more sophisticated business enterprises, primarily in the metropolitan areas.

Telecomms has always made an operating profit and has subsidised the loss-making postal service. Since the early 1970s, telecomms revenue more than kept pace with telecomms costs, and thus net profit tended to rise. However, costs began to rise more steeply than revenue after 1983/4 and net profit declined as a proportion of operating expenditure. With declining net returns from telecomms, SAPO has had to turn increasingly to loans to finance capital expenditure. While SAPO's debt situation improved steadily between 1973/4 and 1980/1, subsequently there was a massive increase in loans contracted, and SAPO's debt expressed in real terms has more than doubled. The massive increase in SAPO debt has been in foreign loans – in 1975/6 foreign loans were 20 per cent of the total, by 1985/6 they were 90 per cent of the total. Following the debt standstill, foreign loans as a percentage of total loans have declined (Annual Report of the Postmaster General, 1987/88: Table 10).

The high foreign-loan commitment, the real burden of which is increased by the prevailing low value of the rand, and the lack of further roll-over foreign lending, will put considerable pressure on the continued expansion of the telecomms network. Alternatively, the cost of the service will have to rise or loan finance could possibly be raised by the privatisation of telecomms services.

The SA telecomms equipment manufacturing industry

By contrast with machine tools, the manufacture of telecomm equipment is highly concentrated; four firms account for over two-thirds of local production and South Africa's largest conglomerates have significant interests in these companies. This is a highly profitable industry and the large conglomerates are very keen to expand their stakes here. There is talk in the industry that there will, at some stage, be a restructuring such that the industry will be divided between two of the large conglomerates – probably Barlow Rand and Anglo American.

As with most of the other sub-sectors of the electronics industry, telecomms imports are increasing more rapidly than local production. Moreover, exports of telecomms and of electronic products generally are very minor, and for the electronics industry as a whole the adverse balance of trade was over R1.5 billion in 1985. Electronic products represented 4 per cent of merchandise imports in 1980, 9 per cent in 1984 and will exceed 16 per cent by 1990.

The South African telecomms market is currently the fourteenth largest in the world. In 1984, imports of telecomms equipment and parts represented about 30 per cent of the total domestic market. However, when allowances are made for imports being valued f.o.b., and the fact that many imports are direct inputs into local production, it is likely that about 44 per cent of South Africa's telecomms equipment needs are imported (TIR Market File – South Africa, 1987: 2). For the electronics industry as a whole, the proportion of the local market satisfied by local production is estimated to be much lower – between one-quarter and one-third.[13]

Table 7. Corporate shareholdings of the major SA telecomms companies

Firm	Local shareholding	Foreign shareholding
Altech	20% Anglo American	12.5% CIT Alcatel
Siemens	16% Sankorp 16% General Mining 16% IDC	52% Siemens (AG)
Plessey SA	26% Sanlam	74% Plessey (UK)
Telephone Manufacturers	50% Plessey (SA) 25% Reunert	25% GEC (UK)
Reunert	80% Barlow Rand 6% African Finance Corp	

Source: Annual financial reports.

The telecomms equipment industry, therefore, is the most developed part of the local electronics industry, constituting almost half of total domestic electronics production. The reason for this is that SAPO has, since 1958, exercised its overwhelming predominance in the market to encourage local production. This is effected through SAPO entering into long-term agreements (previously ten and currently fifteen years) to purchase specified items produced by local manufacturers. SAPO monitors the companies' costs and, in terms of a complex set of formulas which distinguishes between established and new items, determines the price and hence the profit allowed. In addition, any company profits above a certain percentage return on assets employed have to be shared between the company and SAPO. Moreover, SAPO has entered into what it terms 'arm's length agreements' with suppliers of electronic components, in order to encourage the local components industry. These agreements run for four years and stipulate that suppliers of telecomms equipment to SAPO must utilise components supplied by approved local contractors.

In broad terms, until about the late 1970s, this policy worked well. Although composed solely of subsidiaries of foreign multinationals, the local telecomms equipment industry was growing, local content was generally high and rising, and the price differential between the local and the equivalent imported product was small. By comparison with many of the NICs, the

South African telecomms industry was advanced – in Brazil, for example, there was no attempt on the part of the state to encourage the industry until the mid-1970s. But this 'success' rode on the back of the old electro-mechanical technology, where the rate of change in both product and process was very slow.

When SAPO made the decision to 'go digital', there was an immediate problem of how this would affect the local telecomms equipment manufacturing industry. To summarise a very complex and involved story (see Kaplan, 1989b), two overseas manufacturers of digital exchange equipment – namely Siemens and the French Compagnie Generale d'Electricité (CGE) – were recognised as suppliers to SAPO. Local production of the Siemens EWSD exchange was effected through the local Siemens subsidiary and a local firm, Telephone Manufacturers of South Africa (owned by British-based GEC and Plessey), and of the CGE E10 exchange through the licensing of a local firm, Altech.

Siemens received two-thirds of the critical local switching market and in return they were prevailed upon to transfer core technology.[14] This involved the technology required for the local assembly of digital exchanges, ensuring an ever-greater degree of local content. But, in addition, since an increasing proportion of value added is embodied in integrated circuits (IC), the establishment of a local IC foundry was seen by SAPO as essential in order to ensure a reasonable level of local content. Siemens possessed this technology and they were prepared to license it; as a result South African Microelectronic Systems (SAMES) was set up and began manufacturing ICs in 1980.

Despite this technology transfer and the early lead taken by SAPO in going digital, the local telecomms industry has not developed as expected. It is now widely recognised that the industry is 'at the crossroads' and that new policies are needed if the local industry is to be economically viable. These policies are presently being debated within the industry and by various state bodies.

To summarise very briefly, the following 'indices' are suggestive of the need to rethink policies designed to support the continued existence of a local telecomms equipment manufacturing industry.

(1) As mentioned above, local production is growing more slowly than imports – there is a growing dependency on imports.

(2) A very low proportion of local telecomms output, presently about 1.5 per cent, is exported. Exports represent only 3 per cent of the value of imports – which has an adverse effect on the balance of payments.

(3) A large price disparity exists as between local and international prices. This disparity has increased with the introduction of digital telecomms products. While a number of qualifications must be made – no precise data are available; the disparity is much greater in certain areas than in others; dynamic rather than static price comparisons are more valid; price comparisons are very sensitive to changes in the rate of exchange – the price disparity probably exceeds 25 per cent for telecomms equipment and is substantially more in electronic components.

(4) Local content in the telecomm equipment industry is low. Local content has fallen significantly with the introduction of digital telecomms products. Again, a number of similar qualifications need to be made as in the assessment of price disparities. The Board of Trade and Industries (BTI) calculated the average local content at 53 per cent. If a more satisfactory definition of local content is employed, the true figure is well below this. In terms of local inputs utilised, calculations show that the percentage of local inputs hardly exceeds 40 per cent (see Kaplan, 1988: 79).

(5) Although the overall trend is ambiguous, the evidence suggests that local content is rising very slowly. Moreover, the direction of technological change may result in local content falling even further.

(6) Local technological capability – measured by the degree to which the complete product can be designed and produced locally – is low and was certainly reduced with the advent of digital technology. This is particularly evident in the limited design capabilities of local producers who rely almost exclusively on foreign franchises and licences. Typically, R and D spending on the part of local telecomm equipment producers is of the order of 4 per cent of turnover, which is well below the average for the industry world-wide. As a result, there are often long delays before new products can be introduced, few unique products are produced for the local market, and the industry is unable to produce differentiated products with which to enter export markets. Moreover, the lack of local design capability also adversely affects local content. Licensing agreements often stipulate that use must be made of proprietary items. In any case, designs generated abroad tend to incorporate inputs produced abroad for which there may be no available local equivalent.

(7) The limited resources devoted to R and D in the case of the IC foundry have meant that there has been no process enhancement. As a result, with the rapid technological advance in this area, what was initially a 'sunrise' technology has become a 'sunset' technology. The costs of upgrading would be very high indeed. With ever more design changes being incorporated on the chip, the lack of an up-to-date facility here is especially significant.

(8) At the same time, those companies which produce telecomms equipment under long-term agreements with SAPO, and which account for the bulk of all telecomms production – namely, STC and Teltech (recently merged) in the Altech group, Siemens, Plessey and Telephone Manufacturers of South Africa [15] – are all highly profitable. Needless to add, these companies contemplate any changes in governmental policy towards the industry and especially the long-term agreements, with great unease. They have a quasi-monopoly both because of the long duration of the agreements and because each company tends to have a monopoly or near-monopoly in the production of particular items for SAPO. In addition, the telecomm equipment manufacturers also dominate the local components industry. Moreover, all these companies have 'connections' via equity holdings to the largest local conglomerates (see Table 7). The barriers to entry for smaller firms, which are

often technologically dynamic and have elsewhere played an important role in the dynamism of the industry, become difficult to surmount.

The future of the telecomms industry in South Africa

The debate over possible solutions to the present impasse facing the local telecomm equipment and electronic components industries centres on the proposals made in a recent Board of Trade and Industries' report concerned with the entire electronics industry (BTI, 1987). The BTI recommended a number of changes in governmental policy with respect to the electronics industry, of which the following are most significant:

(1) there should be state subsidisation of R and D expenditures undertaken by companies;

(2) the operation of the long-term agreements should be limited by possibly making them of shorter duration, but especially by SAPO purchasing more products on open tender;

(3) the entire electronics industry should become export-oriented so that international market niches in which South African producers might compete can be identified. Once identified, the necessary support in terms of technological and marketing support should be given to enable firms to 'exploit' these niche markets.

With respect to the first proposal, the state support envisaged was very modest – of the order of R50m per annum. In any case, it seems that this has already been rejected by the Cabinet. The role of the state in the NICs stands in marked contrast. In South Korea, Taiwan, Brazil and India, for example, major state support has been forthcoming for telecomms R and D. Moreover, this is centrally effected through state-financed and state-directed central research institutes rather than by subsidising R and D within the firms themselves. Once again, as with the machine-tool industry, the South African government is not prepared to finance the R and D expenditures essential for the adoption of digital product and process technologies nor to envisage expanding the economic role of the state necessary to render such expenditures effective. Without a substantial increase in local R and D and the consequent enhancement of local technological capabilities, as the BTI itself recognised, the aim of 'exploiting' international niche markets is a mere chimera. Since the characteristics of niche markets are constantly changing, design capabilities have to be constantly renewed, and therefore the comparative advantage of firms attempting to 'exploit' such niche markets must rest in their design capability rather than in their manufacturing strength. While such a policy has worked, for example in the case of South Korea or Taiwan, there are a number of factors on which the export successes of these countries are built that are not replicated in the case of South Africa.[16] In any case, as we have seen, there are at present severe 'political' obstacles to exporting from South Africa, especially to the adjacent regional markets.

Limiting the operations of the long-term agreements and encouraging more open tendering will, in the absence of any effective policies to enhance the local telecomms equipment industry, simply lead to the expansion of imports and the further undermining of the local industry. Similarly, were SAPO to privatise or deregulate its telecomms operations as many forces are stridently calling for,[17] this would almost certainly undermine the SAPO system of long-term agreements upon which the existence of the local telecomm and, to a large extent, electronic components industries is based. Nevertheless, calls for privatisation and deregulation of SAPO's telecomms operations seem, at this point, more likely to succeed than any expansion of the state's economic role, which effective support for the telecomms industry would necessarily entail.

Apart from the armaments and (possibly) automobile industries, the South African state has done more to promote local production and technological capabilities in the telecomms industry since the Second World War, than in any other single industry. Nevertheless, as elsewhere, the South African state has always been constrained to operate via the private sector – in the case of telecomms, through SAPO's system of long-term agreements – and has kept its own direct involvement to a minimum. In South Africa, the private sector is empowered to make all the key decisions concerning production, the identification of product priorities, and the allocation of resources to R and D, albeit subject to various state incentives. In the NICs, on the other hand, state involvement, particularly in the development of the higher technology industries such as telecomms and machine tools in the capital goods sector, has been both far more pervasive and direct.

Whatever its efficacy in the past, this limited role for the state will not suffice to ensure the development of certain key industries such as telecomms, where the rate of technological change is rapid and competing producers can rely on extensive state support. Apart from the very viability of the industry, its capacity to produce products appropriate to the domestic market and environment, as well as for export, will be severely curtailed.

Conversely, enhancement of local technological capabilities through state support could be 'twinned' with the development of products appropriate to the local market – especially the 'African' market where, as we have outlined, telephone penetration is very low. Such products would also have export potential, especially in the Third World and particularly in adjacent regional markets. Realising this export potential would, to some degree, depend on a settlement of the current political impasse, while generating demand for these telecomms products locally would similarly depend on expanded SAPO spending on telecomms equipment for this market.

In sum, in addition to the continued expansion of the telecomms network (which itself will require an increase in SAPO's capacity to finance additional capital expenditures and thus eventually a 'solution' to the current economic crisis), what ultimately is required is a reordering of economic priorities. This does not mean that the advanced digital telecomm services demanded by business should be abandoned. These services are associated with major advances in productivity and provide as we have outlined, the necessary infrastructure for implementing microelectronic-based information technology. But here, close to the technology frontier, South African producers are, by and large, at a competitive disadvantage and such products will, to a considerable extent, have to be imported; where they are made locally this would be based largely on foreign designs and licences. By contrast, in the mass market where demand is largely unsatisfied and (in terms of the required technology) relatively straightforward, the potential does exist for the local design, development and production of telecomm equipment. Such a policy of 'walking on two technological legs' is being attempted in other developing countries, such as India and Brazil. In these countries, state support has been indispensable in the design and development of cheap and robust telecomm equipment suitable for the low traffic volume of many consumers.

THE FUTURE OF THE SOUTH AFRICAN CAPITAL GOODS SECTOR

The machine-tool and telecomms equipment industries were chosen for study in this chapter as both important in themselves and as indicative of the problems facing the local capital goods sector more generally. In the case of the machine-tool industry, state aid has been completely absent and the impact of the current economic recession has consequently been very severe. In the case of the telecomms equipment industry, the state has played a fundamental role in establishing local production and ensuring the transfer of overseas core technologies. The local telecomms industry has been more effectively insulated from the direct effects of the recession as SAPO's capital expenditures have been generally expanding. However, the continuing recession has also begun to set limits to the expansion of the local telecomms network while the adoption of digital technology has rendered existent policies of state aid much less effective.

There are two keys to the further expansion of the South African capital goods sector. Firstly, as is the case in the NICs and indeed many industrialised countries, the state has to adopt an expanded role, especially with the aim of enhancing local technological capabilities as this sector becomes more knowledge- and design-intensive. This would entail not only direct state investments in R and D, and increased expenditures for the training of engineers and technicians, but also the monitoring and control of technology contracts so as to limit undesirable features. Secondly, there has to be a reordering of social and economic priorities which would focus on the encouragement, by the state, of production to serve the needs of the local mass markets where the requisite technology is less sophisticated. To reiterate, this would not mean abandoning advanced digital-based telecomms services for example, but rather a reordering of priorities. Accordingly, local production and hence state expenditures would focus principally on the design, development and manufacture of less sophisticated products – for example, a simple digital exchange for the urban 'African' market – products that would be identified as appropriate for the domestic market and would also have an export potential. The bulk of state resources devoted to R and D would be devoted to fostering the local design and development of those products identified as appropriate.

It is here that what is being proposed intersects, in part, with calls being made for policies of 'inward industrialisation'. The expansion of local production of less sophisticated but nevertheless complex capital goods (in terms of our definition) based upon local demand can result in significant import savings. A more dynamic capital goods industry and one with greater technological capabilities, especially product design and development capacity, itself facilitates development, in particular by producing more products appropriate for the local environment. In this way, a significant virtuous circle is established. However, all this depends upon the deepening of local technological and especially design capabilities of the capital goods sector. Herein lies the key to satisfying local demand through higher local output with a higher local content, thus saving on imports.

But, in addition, without the capacity to produce distinctive products, the South African machine-tool and telecomms equipment industries, and the more research- and design-intensive industries within the capital goods sector, will not succeed in capturing export markets. Of course, the deepening of local technological capabilities hardly guarantees success on the export market, but it is an essential prerequisite. The advantages of developing local technological and design capabilities are, at least so far as capital goods are concerned, not confined to facilitating 'inward industrialisation'.

Such capabilities are the *sine qua non* for facilitating exports. In the current debate over the desired direction of economic policy in South Africa, 'inward industrialisation policies' are often counterposed to 'export-oriented policies'. Our analysis of the machine-tool and telecomms equipment industries suggests that this may be a false opposition. Broadly speaking, where local production is significant but local content is low and exports are insignificant, the twin objectives must be to raise local content and at the same time promote exports. As regards complex capital goods with a high design content, the key to achieving both objectives may lie in the expansion of local technological and especially design capabilities.

A change in social and economic priorities, an expanded role for the state, and a solution to the contemporary economic crisis – all the most essential ingredients for the expansion of the local capital goods sector – are ultimately dependent upon a 'new political dispensation'. Without this, the South African capital goods sector seems doomed to a long period of stagnation which will, over the long term, have significant adverse effects upon secondary industry and the economy as a whole. At the same time, South Africa's principal competitors and the NICs are experiencing major developments in this area. The gap is growing and so are the 'costs of catch-up'.

9
The accumulation crisis in agriculture[1]

MIKE DE KLERK

For the greater part of the 1980s, agriculture in South Africa has been in crisis. Arguably this crisis is likely to be the most prolonged, and the accompanying restructuring the most fundamental, since the Marketing Act of 1937 and the Land Acts of 1913 and 1936 laid down the foundations of the present structure. For a sector that is still the second largest employer of black labour, that is the source of the most essential of wage goods, and that provides the backbone of the rural economy, the potential importance of this crisis not only for agriculture but for the structure, level and distribution of activity in the economy as a whole needs no stressing.

This chapter explores the symptoms and nature of the crisis; the causes, immediate and underlying; and the implications of the crisis for agriculture itself and for the economy at large.

The focus is on commercial agriculture – referred to simply as 'agriculture' (unless otherwise indicated).

SYMPTOMS AND NATURE OF CRISIS

The ability to accumulate capital depends on a number of factors. In the first instance, it is a function of profitability. But this is not all that matters; to be able to 'stay in business' and continue to accumulate, firms need both to remain 'liquid', that is, able to meet their short-term obligations, and, ultimately, to remain 'solvent', that is, able to cover all their liabilities to outside parties on the sale of their assets.

Failure to make profits can be sustained, in the short run, by a reduction in the owners' capital (or 'equity' as it is sometimes called) and/or by an increase in borrowed capital, that is, debt. Consistent failure to make profits will generally result in insolvency and the winding up of a firm. Failure to remain sufficiently liquid can also be sustained, for a period, by increasing borrowings, but this makes a firm vulnerable to being wound up at any moment at the discretion of its creditors, whether or not it is insolvent.

The degree to which capital is being accumulated by a firm is measured by the growth or decline in the value of its capital assets, while the degree to which its owners are accumulating capital is gauged by the change in the 'net worth' of the firm, that is, total assets minus total debt. Both measures need to be adjusted to eliminate the effects of inflation if a true or 'real' indication of the rate of accumulation is to be obtained.

Though Marx's notion of profit may differ from conventional accounting measures, it is in terms of the latter that one is usually obliged to assess the process of accumulation of capitalist firms – the data needed to do it by any other method are not often available. Likewise, the ideal in assessing the performance of an entire economic sector, such as agriculture, would be to break it down by sub-sector and region, but the data are seldom available.

Table 1 sets out the essential information for a conventional analysis of accumulation in agriculture as a whole in South Africa between 1970 and 1988. Where available, additional fragmentary findings are referred to, to give the overall picture some regional and sub-sectoral flavour.

Profitability

The profitability of farming can be assessed in various ways, the most basic of which is 'net farming profit'. The data in Table 1 (col. 6) show a rising trend through the 1970s until 1981, followed by a decline in 1982 and 1983, and then a gradual recovery until the 1981 peak was passed in 1986.

However, this measure does not take into account the resources applied to achieve these returns. A more comprehensive measure which reflects the latter is

$$\text{net return on assets (investment)} = \frac{\text{net farming income}}{\text{value of capital assets}}$$

Line 1 in Table 2 records the net return on assets between 1970 and 1988. The pattern is very similar to that of net farming profit: from a 5.5 per cent starting point in 1970, the return climbed to 11.4 per cent in 1981, fell to a low of 5.3 per cent in 1983, and then rose steadily to a new peak of 12.1 per cent in 1988.

The valuation of capital – on which the calculation of the net return on assets depends – is no simple task. Analysis of the estimates of the stock of agricultural capital made by the RSA Department of Agricultural Economics and Marketing, shows that the basis of the calculation was revised in a number of ways between 1978 and 1982. Each of these revisions had the effect of understating the value of capital assets in the later period relative to the earlier period, though it is hard to say which part of the series is the more reliable. Appendix A elaborates. In respect of the net return on assets, this suggests that the data in Table 2 (line 1) are relative overestimates for the 1980s. The nominal 'peak' in 1988 may therefore be no higher than the rates of return that were achieved in the late 1970s, while the 'trough' reached in 1983 is likely to have been a good deal lower than the rates of return in the 1970s.

Table 1. Income and expenditure, assets and liabilities of agriculture: South Africa, 1970–1988

Year	1 gross farming income	2 expend. on intermed. goods, services	3 expend. on salaries, wages, rent, depreciation	4 net farming income 1−(2+3)	5 interest payments	6 net farming profit 4−5	7 value of capital assets	8 short-term debt	9 total debt
1970	1 265	418	337	510	74	436	9 202	395	1 402
1975	2 833	906	458	1 469	134	1 335	16 974	702	2 004
1980	5 882	2 213	887	2 782	323	2 459	28 579	1 668	3 839
1981	7 104	2 658	1 075	3 371	545	2 826	29 574	2 184	4 839
1982	7 496	3 192	1 352	2 952	785	2 167	33 053	2 967	5 786
1983	7 122	3 410	1 791	1 921	1 074	847	36 259	4 034	7 409
1984	8 533	3 575	2 012	2 946	1 344	1 602	38 709	5 203	9 495
1985	9 270	4 144	1 661	3 465	1 698	1 767	42 067	6 069	11 118
1986	11 513	4 775	2 185	4 553	1 585	2 968	45 879	6 517	12 431
1987	13 696	5 242	2 456	5 998	1 650	4 348	49 783	6 980	13 286
1988	15 366	6 028	2 728	6 610	1 700	4 910	54 463	6 889	13 582

Sources: RSA Department of Agricultural Economics and Marketing, 1989: 104; personal communications with the Department, April 1989 and November 1989.

Note: Short-term debt (col. 8) defined as loans from commercial banks and agricultural cooperatives.

Table 2. Agricultural financial indicators, 1970–1988

	Indicator	Table 1 col. ref	1970	1975	1980	1981	1982	1983	1984	1985	1986	1987	1988
1	net nominal return on assets (%)	4/7	5.5	8.7	9.7	11.4	8.9	5.3	7.6	8.2	9.9	12.0	12.1
2	net real return on assets (%)	note 1	0.2	-4.8	-3.4	-3.8	-5.8	-7.0	-9.1	-8.0	-8.7	-4.1	-0.4
3	net nominal return on owners' equity (%)	6/7–9	5.6	7.4	9.9	11.4	7.9	2.9	5.4	5.7	8.8	11.9	12.0
4	net real return on owners' equity (%)	note 2	0.3	-6.1	-3.2	-3.8	-6.8	-9.4	-11.3	-10.5	-9.8	-4.2	-0.5
5	net worth (Rm)	7–9	7 800	14 970	24 740	24 735	27 267	28 850	29 214	30 949	33 448	36 497	40 881
6	C.P.I.	note 3	36.1	56.6	100.0	115.2	132.1	148.4	165.7	192.6	228.5	265.3	298.5
7	inflation rate (%)	note 4	5.3	13.5	13.1	15.2	14.7	12.3	16.7	16.2	18.6	16.1	12.5
8	real net worth (Rm)	note 5	21 814	22 985	24 740	20 604	20 752	19 826	18 962	19 108	19 927	20 219	20 133
9	real capital stock (Rm)	note 6	25 698	26 526	28 579	24 805	25 132	24 829	24 692	24 881	25 367	25 227	24 683
10	debt burden (%)	9/7	15.2	11.8	13.4	16.3	17.5	20.4	24.5	26.4	27.1	26.7	24.9
11	short term debt (%)	8/9	28.2	35.0	43.4	45.1	51.3	54.4	54.8	54.6	52.4	52.5	50.7
12	average nominal interest rate (%)	5/9	5.3	6.7	8.4	11.3	13.6	14.5	14.2	15.3	12.8	12.4	12.5
13	average real interest rate (%)	note 7	0.0	-6.8	-4.7	-3.9	-1.1	-2.2	-2.5	-0.9	-5.8	-3.7	0.0
14	net farm profit: short-term debt	6/8	1.10:1	1.90:1	1.47:1	1.29:1	0.73:1	0.21:1	0.31:1	0.29:1	0.46:1	0.62:1	0.71:1
15	capital–output ratio	7/1	7.27:1	5.99:1	4.86:1	4.16:1	4.41:1	5.09:1	4.54:1	4.54:1	3.98:1	3.63:1	3.54:1

Sources and Notes:
1. Net nominal rate of return on assets *less* rate of inflation.
2. Net nominal rate of return on owners' equity *less* rate of inflation.
3. Calculated from: SA Reserve Bank, 1980: S65; 1981: S79; 1988: S80; Central Statistical Services, 1988a: 5; 1989:6.
4. Year-on-year change of Consumer Price Index.
5. Real capital stock (line 9) *less* real debt. Real capital stock and real debt calculated at 1980 prices. Real debt adjusted by Consumer Price Index. For real capital stock, see Appendix A.
6. See Appendix A.
7. Average real interest rate *less* rate of inflation.

Perhaps the most refined indicator of profitability is:

$$\text{net return on owners' equity} = \frac{\text{net farming profit}}{\text{capital assets} - \text{total debt}}$$

(See Table 2, line 3.) This equation most accurately reflects the return to farmers on the capital they have invested. Though both the trend and the level are very similar to that of net return on assets up to 1981, the low to which net return on owners' equity fell in 1983 was considerably lower (2.7 per cent as against 5.3 per cent), and the recovery thereafter was slower. The immediate reasons for this divergence – the increased reliance on borrowed capital and the rise in the interest rate payable on borrowings – are discussed below. The qualifications made in respect of the estimates of the rate of return on assets apply equally to the rate of return on owners' equity, and the conclusions that follow are similar.

At first glance, these rates of return, though clearly low in 'bad' years such as 1983, do not seem to represent an unsatisfactory record from the point of view of farmers and indirect investors, even taking into account the relative overstatement in the 1980s. However, they conceal at least two important sets of indicators, namely, variations by sub-sector and region, and returns net of inflation. Information by sub-sector and region is not readily available, but from the negative net farming income received by field-crop farmers in the Transvaal and Orange Free State and by meat-producers in the Transvaal in 1983, it can be seen that important sub-sectors and regions have, at times, actually experienced negative nominal rates of return while the return to agriculture as a whole has still been positive (SA Agricultural Union, 1984: 51).

The rates of return in lines 1 and 3 of Table 2 are expressed in nominal terms. When calculated in real terms – net of inflation – with the exception of 1970, they are all found to be negative, the more so when the estimates are adjusted for the relative undervaluation of capital in the 1980s. Lines 2 and 4 set out the details. Were net real return on owners' equity the sole criterion, the average farmer would have been well advised to sell up and reinvest in another sector. In practice, for most farmers, hidden returns in the form of salaries received out of current income, lifestyle and the nominal appreciation of capital assets accompanying inflation were sufficient attraction to retain their investment in agriculture.

Perhaps the most important impact of these trends has been on the real stock of capital in agriculture. The series in Table 2 (line 9) shows a significant rise (of about 11 per cent) during the 1970s before a sharp drop in 1981, followed by narrow fluctuation during the rest of the 1980s. However, the analysis in Appendix A suggests that much of the rise in the 1970s should be discounted if all of the data are to be compared on a consistent basis. It therefore appears that the total real capital stock in agriculture has remained substantially the same for most of the past two decades.

Moreover, to maintain this level of the capital stock, farmers in the 1980s have had to rely increasingly on borrowed capital. Line 10 of Table 2 shows the extent of the increase in the burden of farmers' debt: whereas in 1975, debt constituted 11.8 per cent of the value of farming assets, by 1986 this had risen to 27.1 per cent. Though part of this increase must be discounted because of the relative undervaluation of capital in the 1980s, the escalating dependence on loans from outside parties is clear. Farmers have become either unwilling or unable to retain the proportion of the capital stock they owned in previous years.

There is some evidence that farmers have become less willing to maintain the relative level of their involvement in agriculture. Data collected by the South African Agricultural Union (SAAU) in 1984 showed that the proportion of farmers' assets held in non-agricultural investments increased from 8.8 per cent in 1970 to 13.9 per cent, valued at R6 441m, in 1983 (1984: 34). The average annual growth rate of investments outside agriculture during this period was 16.7 per cent as against 12.5 per cent for investments in farming. Though this could have been caused merely by a more rapid rate of appreciation of non-farming investments, the SAAU concluded that it was largely the result of an increase in the number of part-time farmers, 45 per cent of whose assets were located outside agriculture in 1983 (1984: 34–37). Some active diversification has therefore taken place.

But other factors have also played a role. Fiske (1988, personal communication) points out that credit has, until recently, been readily available to farmers at lower than market rates, while the returns to be had on investments in agriculture have generally been below those obtainable elsewhere. It has therefore paid farmers to borrow cheaply and invest in beach-cottages, town-houses, insurance policies, share portfolios and so on – not directly with borrowed funds, but with internal funds which could otherwise have been used for farming purposes.

A second more powerful argument suggesting that farmers may have become less willing to retain as large a share of the agricultural capital stock, could be mounted on the basis of the surprisingly steep decline in the capital–output ratio, which measures the value of capital required to produce each rand's worth of output. From 7.27:1 in 1970, the capital–output ratio (calculated at the current values recorded by the Department of Agricultural Economics and Marketing) fell to 4.16:1 in 1981, and then rose somewhat before finishing at 3.54:1 in 1988 – less than half the ratio at the beginning of the 1970s (see Table 2, line 15). This would indicate a declining need for capital in farming. Together with the poor rates of return on agricultural capital described above, it would therefore appear to have been rational for farmers to transfer capital from agricultural to non-agricultural investments.

In fact, it is probable that the capital–output ratio did not fall as rapidly as the figures in Table 2 indicate. There are two main strands to the explanation. The first is the relative undervaluation of capital, starting in 1978 and growing in disproportion until 1982, when the procedure for valuation resumed some stability. (Appendix A gives details.) This accounts to some degree for the considerable drop in the capital–output ratio in the late 1970s. Thereafter, the greater part of fluctuations in the ratio can be explained by corresponding fluctuations in weather patterns.

The second strand concerns the rapid increase in the application of intermediate inputs. Real expenditure on fertilisers, seeds, pesticides, herbicides, fuel, etc. rose by about 50 per cent between 1970 and 1982, when the drought began. Real output, between 1970 and 1981, grew a little more than commensurately. During the 1980s, the relationship between the two has remained close, though modified by the drought.

Together, the two explanations suggest that the substantial fall in the capital–output ratio is more apparent than real – and can be accounted for, on the one hand, by statistical aberration, and on the other, by the omission of growing inputs of working capital. If so, the argument that farmers became less willing to invest in agriculture because of the declining requirement for capital should not be overemphasised.

Conversely, there is a wealth of evidence to suggest that growing financial pressures sapped the ability of most farmers to fund capital needs from internal sources, making it difficult to avoid the accumulation of debt. The evidence is provided by an analysis of debt and liquidity trends in agriculture.

Liquidity

The particularly rapid build-up of debt in 1983 and 1984 focuses attention on liquidity as a measure of financial health. Liquidity is the capacity to repay short-term debt at short notice. The most common measures of liquidity are:

$$\text{'current ratio' or } \frac{\text{current assets}}{\text{current liabilities}} \text{ and,}$$

$$\text{'liquidity ratio' or } \frac{\text{cash, marketable securities, receivables}}{\text{current liabilities}}$$

The rule of thumb in general use (Helfert, 1967: 59) and adopted by the SAAU (1984: 79) is that these ratios should not be below 2:1 and 1:1 respectively. Of the two, the SAAU considers the liquidity ratio – popularly referred to as the 'acid test' – to reflect the true position of farmers more accurately (1984: 80).

Data that would make the annual calculation of these ratios possible are hard to come by. The best approximation of the trend, though not of the level, of the liquidity ratio is:

$$\frac{\text{net farming profit}}{\text{short–term debt}} \quad \text{(See Table 2, line 14.)}$$

This is, in practice, the liquidity ratio, omitting the non-farming liquid assets (cash and marketable securities) in the numerator, since by far the largest 'receivable' is net farming profit. Table 2 shows a marked deterioration in this ratio between 1975 and 1980, which becomes quite dramatic thereafter, reaching a low point of 0.21:1 in 1983.

The financial survey conducted by the SAAU in 1983 allows the accurate calculation of both the liquidity ratio and the current ratio for that year. While the average current ratio for farmers as a whole stood at 1.70:1 – significantly less than the acceptable level – the liquidity ratio was on average only 0.32:1 – less than one-third of what is satisfactory (SAAU, 1984: 80). Just one region, the Eastern Cape, was able to show a liquidity ratio above the safe minimum. Others, including the key Transvaal and Orange Free State regions, were as low as 0.17:1. This indicates an acute degree of illiquidity. Farmers with strong cash and marketable security reserves might have been able to meet their short-term obligations by drawing on these reserves. But by 1983, few farmers had such reserves. (To some extent, this has been encouraged, at least in the major grain sectors, by the ready availability of Land Bank finance for short-term needs through cooperatives.) Consequently, most farmers were obliged to borrow further to honour their short-term debts. In most instances, this borrowing took the form of the 'consolidation' of existing short-term debt, that is, an extension, normally by a year or more, of the period for repayment of such debt. This effectively converts short-term debt to medium- or long-term. By 1986, more than a quarter of the total debt of crop farmers in the summer rainfall region consisted of 'hard-core', normally short-term, debts of this nature (Potgieter, 1987: 5).

Perhaps the greatest problem created by this form of financing is that it increases the burden of interest payments, making it more difficult for farmers to generate sufficient income to cover their short-term costs and repay loans. It is for this reason that the threshold of insolvency is reckoned to be much lower than the actual level of debt which signifies insolvency (see below).

Acute illiquidity in the period from 1982 onward is thus a substantial – though by no means the only – cause of the rapid movement of many farmers towards insolvency. Furthermore, the combination of increased debt and high interest rates has helped turn illiquidity into a chronic problem, prevalent even in years of relatively good returns (see results for 1986–88 in Table 2). It will take not one but several years of high net income to re-establish an acceptable level of liquidity.

Solvency

In terms of the ultimate debt criterion – solvency – the overall position of the agricultural sector is still sound. The rule of thumb for financial health in this respect is that total debt should not exceed half of the value of total assets (SAAU 1984: 72). As line 10 of Table 2 shows, the average debt burden which has been rising continuously since 1975, still stood at about 27 per cent in 1986 and 1987, although some part of this rise should be discounted because of the relative undervaluation of capital in the 1980s. Nevertheless, as far back as 1983, when the average debt burden was considerably lower (18.9 per cent), many farmers were at or beyond the critical level. The SAAU records that the average debt burden of the 15 200 farmers most seriously in debt in that year was exactly 50 per cent (1984: 78). If one assumes a normal distribution, this would have placed at least 11 per cent of farmers in immediate danger of insolvency. And, as is pointed out below, the situation appears to have deteriorated substantially since then.

However, solvency on its own is no guarantee of financial stability. As the level of debt grows, it becomes increasingly difficult for farmers to cover their interest payments and repay loan capital. Beyond a certain point – usually much lower than the critical level for solvency – it is reckoned to become effectively impossible to farm without a progressive increase in debt. Of course, what is critical at any moment depends not only on the burden of debt but also on the rate of interest, expected crop yields, input and output prices, asset structure and so on. In the circumstances prevailing in 1983, the SAAU calculated the critical debt burden for several of the largest sub-sectors as: 16.7 per cent for summer crops; 34.1 per cent for winter crops; 10.2 per cent for red meat; 17.2 per cent for milk; and 14.2 per cent for wool (1984: 56). Though this level rises by about half if non-farming income is included, the broad standards adopted by the Union are that farmers in the summer-crop and meat sub-sectors with a debt burden in excess of 20 per cent should be regarded as being financially unsound, and that for all other producers the critical burden should be 30 per cent (1984: 58).

Against these criteria, no fewer than 15 200 farmers – 22.4 per cent of the total – were assessed to be critical in 1983, concentrated chiefly in the Transvaal and Orange Free State, particularly younger farmers. The sub-sectors worst affected were summer crops, where 52 per cent were beyond the critical level, followed at a distance by winter crops (22.6 per cent). By the end of 1984, these estimates were expected to have grown to 22 700 farmers (33 per cent of the total), 65 per cent (summer crops) and 38 per cent (winter crops) (1984: 58–66, 86).

Since then, the position would appear to have worsened: while the average interest rate has changed little, the debt burden has grown significantly (see Table 2, line 10), and it will be shown below that input prices have grown faster than output prices. Only total output has improved. Estimates of the number of maize farmers who would not survive, given conditions prevailing in 1987, put the figure at around 6 000, or more than half of those involved (*Farmer's Weekly*, 8 May 1987: 75; Potgieter, 1987: 5).

Confirmation of these trends is to be found in court records. Though relatively few farms that change hands under financial duress are actually sold on sequestration, the number of farmers sequestrated for insolvency has risen sharply: whereas between 1980 and 1984 the average number of agricultural sequestrations per year was 75, between 1985 and 1987 the average jumped to 232. In 1987, it was 313 (Central Statistical Service, 1986: 15.23; 1988b: 10.67), and recent reports suggest that the rate has not receded (*Maize News*, September 1988: 7).

The number of sequestrations would have been far greater had it not been for extensive state aid. Quite apart from the 'normal' forms of financial assistance, state aid designed specifically to alleviate the extraordinary financial pressures of the 1980s has included subsidies on:

—the consolidation of debt (R344m between 1981 and 1987);
—crop production loans (R470m between 1981 and 1987);
—interest on consolidated debt and production loans (R90m between 1981 and 1987 with a further 'interest subsidy equivalent to 10 per cent of the Land Bank's interest rate on cash credit loans to agricultural cooperatives in respect of carry-over debts' approved for 1988–9);
—stock feed loans;
—input costs for farmers in drought-stricken areas (R120m 'paid to creditors of farmers to help clear production debts incurred in the 1987–88 season');
—the conversion of sub-marginal crop-lands to planted pasture (R280m budgeted for 1987/8–1991/2); and
—export losses for summer grains, chiefly maize (up to R200m per annum available from 1988).

In addition, the state stands as guarantor of consolidated debts to the value of R900m. Direct state aid to farmers in its various forms – but excluding the indirect effects of tariff protection, import control, etc. – amounted to more than R2.7 billion between 1981 and 1987. About 25 000 of the 59 000 farmers on the land during this period were beneficiaries – an average of more than R1m per recipient. The National Maize Producers' Organisation (NAMPO) estimated that 'at least 40 per cent of South Africa's grain producers would be forced into liquidation ... if State aid to farmers was summarily withdrawn' (*FW*, 11 September 1987: 83–4; 5 February 1988: 75; 15 April 1988: 75–6; 5 August 1988: 76).

Of the various indicators discussed in this section, arguably the single most comprehensive is the burden of debt, or the ratio of total debt to total assets (see Table 2, line 10), since this reflects not simply the year-to-year fluctuations in liquidity and return on investment, but the cumulative results of these fluctuations over an extended period. Perhaps more important, its inverse provides a rough idea of the capacity of the agricultural sector to accumulate capital. The higher the debt burden, the lower the capacity to accumulate, both because of the increased interest and loan capital repayment drain on net farming income, and because banks and other creditors as effective part-owners of farms are unlikely to want to 'plough back' profits into farms. Even mitigated by the relative undervaluation of capital in the 1980s, the steady rise in the burden of debt over the last decade indicates a progressive weakening of the capacity of agricultural capital to accumulate. Unqualified by undervaluation, the present debt burden of 25 per cent would indicate that the average farmer is still close to the threshold of sliding into insolvency – that is, total cessation of the capacity to accumulate.

The analysis of the causes of the crisis in the following section focuses on the processes that have brought the burden of debt to its current high level, and examines the likelihood of these processes – and hence of the crisis in agriculture – persisting.

CAUSES, IMMEDIATE AND UNDERLYING

The causes of the crisis can be grouped into three broad categories: drought; monetary policy, or more specifically, the structure and movement of interest rates; and the deterioration of agriculture's terms of trade with industry. It is important to identify not only the degree to which each cause has been responsible for the crisis and the mechanisms by which this has occurred, but also the proximity of each to the cyclical or structural end of the spectrum.

Drought

Nearest the cyclical end is the prolonged drought of 1982–5 in the summer rainfall region. Rainfall has improved since 1986 and is expected to be more favourable in the 1990s (Tyson and Dyer, 1983: 6; *FW*, 21 November 1986: 19–21), but the financial legacy of the drought is likely to be felt for some years yet.

The most immediate effect of drought is on farming income and hence on liquidity: for arable farmers, crop failure reduces liquidity in the current year, while for pastoral farmers the effect is usually delayed for a year or so by the slaughter or sale of stock. Either way, adverse weather conditions call for cash to build up current assets. For farmers without cash reserves, this means additional debt. The onset of drought, which was at its most severe in 1983 and 1984, accounts for a substantial part of the steep rise of agricultural debt in those years. However, the State President's Economic Advisory Council has estimated that only 22 per cent of the increase in farming debt between 1980 and 1985 can be directly ascribed to drought (Economic Advisory Council, 1986: 105).

Interest rates

Interest rates are most often expressed in 'nominal' terms, that is, at current prices, or the rate quoted by the institution concerned. An alternative, which has particular significance in economic analysis, is to express them in 'real' terms, that is, net of inflation: the real rate of interest is therefore calculated by deducting the current rate of inflation from the (current) 'nominal' rate of interest.

Trends in nominal interest rates

Line 13 of Table 2 shows a steady rise in the average effective nominal interest rate, net of state subsidies, paid by the farming sector. Though the most rapid increase occurred between 1980 and 1982, prior to the drought, nominal interest rates have remained on a high plateau, at or above the 1982 level, since then. The period of historically high nominal rates therefore coincides with the prolonged drought and its financial aftermath. If one bears in mind that the drought made it necessary to 'consolidate' much of the sector's short-term debt, the effect of high interest rates was to compound the growth of farming debts at a particularly rapid rate. The State President's Economic Advisory Council attributes 31 per cent of the increase in the agricultural debt burden between 1980 and 1985 to interest rate movements (1986: 105).

Until fairly recently, one would have had little hesitation in placing high interest rates, with drought, at the cyclical end of the spectrum. But with high rates of inflation and a tight balance of payments constraint expected to become more permanent features of the economy in the foreseeable future, high nominal interest rates may be more structural than cyclical. The decision by the state to phase out various policies which reduced the rate of interest payable by farmers to the Land Bank and cooperatives will add to this, although the practical effects of this decision have not yet been felt. And the present high level of direct interest subsidies paid to farmers by the state is unlikely to be maintained indefinitely. High nominal interest rates have therefore contributed materially to the growth of farming debts, at least in the 1980s, and seem likely to do little to ease this burden in the foreseeable future.

Trends in real interest rates

However, there is a further, less obvious, but more fundamental mechanism by which interest rates have influenced the debt structure and the capacity of agriculture to accumulate capital. The rise in nominal interest rates was accompanied in most years by a still more rapid rise in the rate of inflation (see Table 2, line 7), which meant that, in real terms, the rate of interest payable by farmers was negative. As line 13 of Table 2 shows, only in 1983 did the real rate of interest, net of state subsidies, rise to a positive value. For most of the past two decades, many farmers have therefore felt it sensible to increase, rather than reduce, borrowing. This has been encouraged further by the ready availability of credit from banks and cooperatives; by the basis on which income tax for farmers has been calculated (see below); and by the relatively low cost of credit available to farmers (see above).

Capital investment in agriculture can be divided into three main categories: in descending order of overall magnitude, they are land and fixed improvements; livestock; and machinery and implements. In respect of the first and third of these, the effect of persistently negative real interest rates on debt and the process of capital accumulation can be clearly discerned.

The borrowing encouraged by very low positive or negative real interest rates has pushed land prices up, well beyond a level commensurate with the productive capacity of land in most regions. One of the main determinants of the price of land is the value of the expected stream of net income from that land discounted at a certain rate of interest. The lower the rate of discount, the higher the value of the income stream and the price of the land. Persistently low real interest rates have led most farmers to use an equally low discount rate, and hence to value land at an inordinately high price – inordinate, that is, relative to the real profitability of production on that tract of land.

In other words, most of the profitability of farming, at least over the last decade and a half, has come from an appreciation of the capital value of land, brought about not so much by physical improvements to the land as by increases in the price that farmers have been prepared to pay for land (of a constant productive capacity). Put still more simply, it is speculation in land rather than the fundamental profitability of agricultural production that has been the main source of profit in farming. To a large extent this has been brought about by very low real interest rates since the start of the 1970s (Janse van Rensburg, 1984).

In the present context, two consequences are worth noting. First, the level of debt is higher than it would otherwise have been. And second, much of the nominal capital accumulation that has occurred is of a precarious nature. With higher real interest rates, intense financial pressures on farming and a rising number of sequestrations, land prices and nominal capital values must be vulnerable to significant falls. Paradoxically, what is probably shielding farmers most at present is the very degree of their indebtedness. As substantial part-owners of farms, banks are wary of precipitating a slide in land values by accelerating the pace of legal action against insolvent farmers. They are, to a degree, 'locked in'. Indeed, the threat of substantial capital losses, and the range of disruptive effects that these could have, have held up the entire process of reconstituting the accumulation process in agriculture.

Low real interest rates have also encouraged the purchase of machinery and implements. Broadly speaking, mechanisation on farms seems to have been labour-complementing prior to 1970. Few analysts have questioned the productiveness of capital investment of this nature. Post-1970, it seems by and large to have been labour-substituting (Fenyes *et al.*, 1988: 189), and there is greater doubt about its productiveness. Though the indications are not all uniform, it seems more than probable that there has been a degree of over-mechanisation. The Marais Commission drew attention to this tendency in 1970 (RSA Commission of Enquiry into Agriculture, 1970: 165). In addition, there are numerous local studies of over-mechanisation (Fenyes *et al.*, 1988: 190). The SAAU's survey of farm finances in 1983 showed that those farmers most deeply in debt had invested twice as large a proportion of their capital in machinery and implements as those least in debt (1984: 30), although there is some ambiguity in this: the stock of machinery and implements has remained more or less constant in real terms since 1982.

If one assembles the evidence, it would appear that, though there is little direct connection between negative real interest rates and the rapid rise of agricultural debt in the 1980s, such low real rates have brought about a higher level of borrowing than would otherwise have occurred. More important, they have helped induce relatively unstable and unproductive forms of investment which, along with changes in the terms of trade (see below), have eroded the fundamental profitability of agricultural production and, with it, the sector's capacity to generate a surplus for accumulation. It is reasonable to conclude that this, as much as any other consideration, is why the state has begun to shift away from policies which reduce the cost of investment in

Table 3. The terms of trade and agricultural goods, 1970–1987

Year	Rand price indices (1975=100)							Other indicators	
	1 domestic agric. inputs	2 domestic agric. outputs	3 maize exports	4 wool exports	5 1/2	6 1/3	7 1/4	8 manuf. goods/ crude foods (US prices)	9 rand/ US $
1970	56.20	54.90	46.40	33.20	1.02	1.21	1.69	1.57	0.72
1971	59.10	56.90	47.50	39.30	1.04	1.24	1.50	1.50	0.72
1972	63.40	64.80	47.70	112.90	0.98	1.33	0.56	1.53	0.77
1973	70.30	81.80	75.30	109.80	0.86	0.93	0.64	1.08	0.69
1974	82.30	92.00	103.10	74.40	0.89	0.80	1.11	0.86	0.68
1975	100.00	100.00	100.00	100.00	1.00	1.00	1.00	1.00	0.74
1976	115.60	109.50	109.00	117.90	1.06	1.06	0.98	1.18	0.87
1977	130.20	119.60	89.60	120.00	1.09	1.45	1.09	1.44	0.87
1978	146.70	127.10	93.60	133.50	1.15	1.57	1.10	1.44	0.87
1979	176.90	151.30	103.20	149.80	1.17	1.71	1.18	1.44	0.84
1980	207.40	179.30	108.90	149.40	1.16	1.90	1.39	1.41	0.78
1981	230.30	201.00	129.40	190.60	1.15	1.78	1.21	1.50	0.88
1982	270.50	225.80	127.40	180.20	1.20	2.12	1.50	1.83	1.09
1983	308.30	255.00	167.60	208.60	1.21	1.84	1.48	1.76	1.11
1984	330.20	284.50	224.20	298.40	1.16	1.47	1.11	1.74	1.48
1985	392.90	310.40	275.90	334.60	1.27	1.42	1.17	1.99	2.23
1986	467.20	341.10	223.70	353.00	1.37	2.09	1.32	(2.28)	2.28
1987	520.40	380.10	161.60	614.50	1.37	3.22	0.85	n.a.	2.04
1988	575.50	441.30	219.20	767.20	1.30	2.63	0.75	n.a.	2.27

Sources: Columns 1 and 2: RSA Department of Agricultural Economics and Marketing, 1988: 90, 97.
Column 3: calculated from International Monetary Fund, 1977: 44, 46; 1978: 35, 37; 1981: 51, 53; 1984: 71, 73; 1988: 81, 83; and from SA Reserve Bank, 1980: S80. Assumed: international maizeprice = current wholesale price, Chicago.
Column 4: SA Wool Board, 1986/7: 21; 1987/8: 10; personal communication with SA Wool Board, November 1989.
Column 8: International Monetary Fund, 1986: 201. Assumed: US export unit values for manufactured goods and crude foods.
Column 9: SA Reserve Bank, 1980: S65; 1981: S79; 1988: S80; 1989: S80.

agriculture, and can be expected to pursue this line – short-term measures notwithstanding – in the foreseeable future.

Finally, a question arises as to why capital was so cheaply and readily available for relatively unproductive forms of investment. The various forms of direct interest subsidy – which, it must be remembered, are very recent – and indirect subsidy through favoured treatment by the Land Bank and cooperatives have already been discussed. A change in tax legislation in 1977 had the effect of making it still cheaper to borrow for some purposes: farmers were granted permission to write off the full cost of machinery and implements against taxable income in the year of purchase (compared to a three-year period for all other businesses). The lure of short-term tax savings must have outweighed the burden of longer-term debt repayment for more than a few undiscerning farmers – before the weather changed in 1982. Following the Margo Commission's recommendation, this provision is now to be scrapped.

Perhaps most important is the basis on which banks have granted credit. Solvency, not liquidity, has been the main criterion. In other words, loans have been granted fairly freely against the security of a farmer's net assets, rather than against his capacity to fund interest charges and capital repayments out of current income. Especially with nominal land values rising rapidly, many farmers have been allowed to borrow beyond this capacity (Potgieter, 1987: 9–10). As banks have now become partly the prisoners of their own policies, this too is starting to change. But whatever the changes, the financial damage of past policies seems likely to remain with the agricultural sector for many years to come.

Terms of trade

Trends in the terms of trade

The most enduring cause of the deterioration in farm finances has been the gradual but fairly consistent adverse movement in agriculture's terms of trade, that is, in the rate at which agricultural goods exchange for those of other sectors, primarily manufacturing – the very recent improvement notwithstanding. The Economic Advisory Council's calculations also suggest that it was the most significant single cause of the increase in farming debt between 1980 and 1985, accounting for as much as 47 per cent of the rise (1986: 105).

There are several ways in which this rate of exchange manifests itself. The most immediate is the domestic terms of trade, or the ratio of farm input prices to farm output prices in South Africa. In keeping with international trade trends (see Table 3, column 8), this ratio improved significantly from farmers' point of view at about the time of the first oil crisis in 1973. Since then it has deteriorated almost unbrokenly. If the terms of trade were at parity, or 1:1, in 1975, by 1986 they would have reached a ratio of 1.37:1 (see Table 3, column 5). In other words, if the average South African farmer had had to exchange 1 000 bags of maize for, say, a tractor in 1975, by 1986 he would have had to part with an additional 370 bags.

Agricultural output can, of course, also be sold abroad. No composite index of the ratio of domestic input to export output prices is published, but rough calculations for two of the country's most important agricultural exports, maize and wool, show similar trends until very recently. For wool, the ratio fell from 1:1 in 1975 to 1.32:1 in 1986 – close to the domestic average – before improving dramatically in 1987 and 1988. For maize the drop was considerably greater, from 1:1 in 1975 to 3.22:1 in 1987, in spite of the large boost to the rand price of farm exports provided by the depreciation of the rand (see Table 3, columns 5, 6, 9). By 1987, maize farmers would therefore have had to export more than three times as many bags to pay for a tractor as they would have in 1975. So, regardless of whether farm output has been sold domestically or abroad, the terms of trade have moved substantially against South African farmers for the last decade and a half.

Determinants of the terms of trade

Farmers attempted to offset the negative effects of this trend on profits by simply producing more. This was made possible by the extended period of favourable climatic conditions lasting until 1981, and was encouraged by the basis on which the prices of several of the most important agricultural products – notably maize and wheat – were determined. For many years now, the prices of these commodities have been fixed annually on what would appear to be essentially a 'cost-plus' basis by the marketing boards concerned – although the precise method of calculation is never disclosed. This has had two important consequences: first, it has allowed the domestic producers' price to escalate more in line with the rate of inflation in South Africa than with supply and demand conditions locally and internationally. And secondly, by placing the emphasis on a 'fair return' for farmers whatever their input costs and by guaranteeing a fixed price whatever the size of the crop, it has created incentives for farmers to produce more, rather than more efficiently.

At the same time that agricultural producer prices determined by marketing boards have risen rapidly in South Africa, they have tended, after the boom that accompanied the first oil crisis, to stagnate or fall on international markets, both absolutely and relative to the price of manufactured goods. Only recently have they turned upward materially. The cost-advantage of South African agricultural exports has therefore been eroded, and in most cases eliminated. A survey in 1983 showed that of tradable agricultural goods that South Africa produced in significant quantities, only the various categories of fruit, wine, wool, mohair, karakul pelts and ostrich feathers could be exported at a profit (Stadler *et al.*, 1983: 14–23). Though the subsequent sharp fall in the rand offered temporary relief to exporters of some other farm products, the rate of inflation in South Africa soon counteracted that, and the position at present is probably much as it was in 1983. It is about ten years since South Africa was last able to export maize at a profit.

The losses on exports engendered by this process and exacerbated by rising output and export volumes have slowed the rise of domestic producer prices, but not that of input prices. Particularly in the 1980s, this has been an important cause of the widening gap in the terms of trade.

The most important factor in the sustained rise in input prices has undoubtedly been the general rate of inflation in South Africa. The establishment and protection of import-substitution industries has added to this: it was estimated in 1982 that the cost of intermediate inputs (cattle feed, fertiliser, fuel, etc.) was 6.9 per cent higher than it would have been under tariff-free international trade (Stadler *et al.*, 1983: 6). The effect on the costs of capital inputs, whose life extends over several years, is more difficult to assess, but the same study estimated that over the ten year 'phasing-in period' of Atlantis Diesel Engines from 1982, the 'tractor bill' would be 16 per cent higher than it would otherwise have been (Stadler *et al.*, 1983: 12–13).

It is rather surprising then that, on balance, agriculture has probably gained rather than lost from industrial protection. The reason is the growing degree of protection afforded to agricultural outputs. Without tariffs and import controls but including transport costs, Stadler and colleagues estimated that consumers would have paid R916m less for the same volume of agricultural output than they actually did in 1982, as against an additional cost to farmers of R221m from the protection of intermediate inputs (plus a small additional amount on capital inputs). This increased the value added by agriculture by 19.4 per cent. Most significantly, Stadler and colleagues point out that the degree of net protection of agriculture has grown as the competitiveness of South African farmers has declined (1983: 24–5). Without state intervention to protect industries, the adverse movement in agriculture's terms of trade would therefore have been still greater.

The overall impact of state intervention on the terms of trade can be summed up as follows: the manner in which key output prices have been fixed and the net protection of outputs have brought about a more rapid increase in domestic producer prices than would otherwise have occurred. Intentionally or not, this has helped slow the deterioration in agriculture's terms of trade. But the export losses induced in the process have, at least for the present, placed a ceiling on the capacity of these policies to limit the rate of deterioration.

The most fundamental determinants of the terms of trade in the long term are the relative rates of change of supply and demand for different categories of goods. In relation to manufactured goods, the international supply of agricultural goods has expanded more rapidly than demand over the last decade and a half, leading to a fall in the relative price of the latter. With some qualifications, similar trends can be identified in South Africa. The details are worth examining briefly because of their implications for future trends.

The global supply of agricultural commodities has expanded comparatively rapidly in the 1970s and 1980s. The two most important reasons are technological advance and the increased emphasis on agricultural self-sufficiency in many countries (Schuh, 1986). The former has been a significant enabling factor for the latter, particularly in less industrialised countries, but other, primarily economic mechanisms have also been employed to achieve self-sufficiency. These include the protection of domestic agriculture by tariffs and import controls, and the use of price incentives, notably export subsidies, to stimulate production behind these barriers. The result has been that countries which for many years were substantial importers of agricultural commodities, chiefly grains, have now become either marginal importers or exporters. India, China and the European Community are the most important examples (Groenewald, 1987a: 200–1; 1987b: 226–8). And of the present major importers, only Japan appears certain to remain a major importer (Van der Vyver, 1988: 303–4).

On the demand side, several developments have acted to slow the growth of consumption. The population growth rate is falling in most parts of the world except Africa. The demand for agricultural goods is, in general, income-inelastic: that is, as incomes rise, the proportion spent on food tends to fall (even though total spending on food may increase). And the industrial demand for agricultural raw materials appears to be weakening (Groenewald, 1987b: 226).

The combined effect of these forces has been to lower the relative price of farm outputs, as the rough index in Table 3 (column 8) shows. It is chiefly the depreciation of the rand that has kept the rand price of exported maize increasing (Table 3, column 3) and, to lesser degree, the same is true of exported wool.

Though commodity markets are notoriously unstable, most indicators suggest a continuation in the long run of the decline in the terms of trade of agricultural goods on world markets in the medium-to-long term. There is no shortage of technological capability; population growth rates are unlikely to rise; the 'hierarchy of needs' which makes the demand for food income-inelastic is unlikely to change; and countries which have gained or regained agricultural self-sufficiency will, for the most part, be reluctant to lose this capacity, especially if they have balance of payments problems. Environmental calamities excepted, only a significant movement towards freer international trade in agricultural commodities, involving less competitive subsidisation of farm production, appears capable of reversing the trend.

Within South Africa, the supply–demand relationship is more ambiguous and is complicated by foreign trade. A very rough approximation for the growth of domestic demand for agricultural commodities can be obtained by multiplying the rate of growth of real national income by the income-elasticity of demand for food (since no estimate of the latter for agricultural commodities as a whole is available, and food, in any case, makes up the bulk of agricultural output). The resulting average annual growth rate of demand for the period 1970–85 varies between 0.79 per cent and 1.76 per cent, depending on one's choice of elasticity estimate (Groenewald, 1987b: 231; SA Reserve Bank, September 1978: 572; March 1980: 575;

March 1985: 578; June 1988: 583). The latter figure is the more likely, given the high population growth rate. Over the same period, the physical index of food production grew at an average annual rate of 2.19 per cent (RSA Department of Agriculture, 1989: 82). These are only the crudest of indicators, but the results do lend support to the argument that supply has tended to grow faster than demand.

A conclusion to this effect should be qualified, since the market for domestically produced agricultural goods is not closed. While agricultural imports make up only a small proportion of domestic consumption, exports form a very significant component of demand. Much the most important export commodities in a typical year are maize, fruit, sugar and wool. As noted earlier, maize has been exported at a loss for the last ten or more years, so domestic supply cannot have been increased for the export market. A similar conclusion must be arrived at for sugar, whose export price has been well below the domestic price for an extended period. In the case of fruit and wool, domestic supply responds largely to international demand, only a small portion of output being marketed locally. So for these commodities, it cannot be argued that a domestic over-supply situation has arisen, although, as with all exports, this could change radically if trade sanctions were effectively imposed.

With some qualifications, therefore, it would appear that the strong growth of agricultural supply relative to demand has been an important underlying cause of the steady shift of the terms of trade against agriculture in South Africa for the past fifteen years. With the outlook for exports generally less than favourable and with limited scope for import substitution, a sustained improvement in agriculture's terms of trade is unlikely in the medium term, not only internationally but also within the domestic economy, despite the recent counter-trend.

This finding puts the adverse trend in the terms of trade nearer the structural end of the spectrum of causes of the build-up of farming debt.

To sum up the causes of the current crisis, it appears that of the three issues, only one – drought – is likely to be of short duration. The higher current incomes from improvements in the weather and the short-run terms of trade will certainly help to stabilise the burden of debt, and there are indications that this is already occurring. But there is reason to believe that the unfavourable trends of the other two – interest rates and the terms of trade – will resume or persist for some years, which makes a rapid reduction of the burden of debt unlikely. The capacity of agriculture to accumulate will therefore probably remain at low levels. Indeed, the prolonged relative decline in agricultural commodity prices is a strong signal to shift resources out of agriculture. Some far-sighted farmers have been willing to do this on their own. But for most less-efficient producers, the process – which now appears well-established in South Africa – has been and will continue to be both involuntary and painful.

Ultimately, much of the cause of the present crisis must be ascribed to state policy – in its many forms – to keep white farmers on the land. It is beyond the scope of this chapter to investigate the motives for this policy, but party-political and 'security' interests come immediately to mind. And, needless to say, farmers themselves have been a most vociferous lobby. For longer than the period of this study, the level of investment in agriculture has exceeded what is justifiable on purely economic grounds. But it has taken the developments of the 1970s and 1980s to raise the costs – both economic and political – to a level at which the policy has become unsustainable. Seen in these terms, the crisis through which agriculture is now passing should be regarded as a healthy development, where 'crisis' should be identified more closely with 'restructuring' than with 'catastrophe' – though large parts of the farming community might not perceive it in this light.

IMPLICATIONS FOR AGRICULTURE AND THE WIDER ECONOMY
A wide range of implications follows from the present crisis.

Agricultural production

Perhaps the most far-reaching of these is that the rate of growth of domestic farm output can be expected to slow down for much of the 1990s. Several decelerating influences are at work. First, there is the strong price signal, reflected in the declining terms of trade, to shift resources out of agriculture. Farmers are much more aware of this signal now than they were at the end of the 1970s. More important, state policy is now taking the long decline in the terms of trade into account. Moves to discourage inefficient forms of investment were discussed above, but the most significant shift is in the basis on which the major grain prices are determined. The first indications came in the early 1980s, when maize producer price increases started to become less sensitive to input price increases. However, the fixed single-price policy was maintained, which, as was pointed out, created a perfectly elastic demand curve and still encouraged maximum output at the ruling price. Recently, however, the Wheat Board (*FW*, 15 April 1988: 75) and the Maize Board (*FW*, 9 September 1988: 68) have made it known that the producer price in the immediate future at least will vary inversely with the size of the crop, with the highest prices – still well below the present price in the case of maize – being paid only for deliveries that leave little or no surplus for export. Though the price-elasticity of the Boards' demand curves is greater than unity, so that the total revenue received by farmers collectively will still increase with the size of the crop, the incentive to try to beat the decline in the terms of trade by expanding output is now very much less than before.

The policy for the foreseeable future would appear to be to try to limit output to a little above the level of domestic consumption. For maize, this will mean a drop of about 20 per cent on the average year's production between 1977 and 1987. For wheat, it will mean holding output levels steady at the eleven-year average, which may not be easy with many maize farmers wanting to move into wheat production.

The main alternative for marginal maize farmers is to plant pastures, and the state has offered financial assistance to those undertaking this change. But there is little scope for import substitution in most animal product sub-sectors. In the red meat market, only about 6 per cent of local consumption has been imported annually over the last eleven or so years. On the other hand, studies suggest that the domestic demand for most sorts of meat – and hence for yellow maize as a stockfeed – would expand fairly rapidly if prices were reduced (Groenewald, 1987b: 232–3).

The market for wool and mohair is effectively unlimited since most of South Africa's output is exported. But the transition would be costly for most farmers, and it is not clear whether wool or mohair production could be undertaken profitably on planted pastures. There is also the additional uncertainty of trade sanctions hanging over exported commodities. Indeed, those sub-sectors which rely heavily on exports – wool, mohair, almost all types of fruits and fruit products and, to a lesser degree, sugar, hides and skins – would be in a far worse position than the maize industry if trade sanctions were made effective. The only comfort for farmers, in the absence of a substantial depreciation of the rand or an unexpectedly sustained strengthening of international agricultural commodity prices, is that the state is most unlikely to risk the country's capacity for self-sufficiency by lowering the protection they presently receive.

With certain qualifications, then, the growth of agricultural production seems likely, at most, to be slow over the medium term. This, in turn, has a number of consequences – for employment, wages and the degree of industrial concentration in agriculture and dependent sub-sectors; and for the gross domestic product, the balance of payments and population distribution in the wider economy.

Agricultural employment, wages and industrial concentration

Employment in agriculture has been on an erratic downward trend since the late 1960s. Reliable and comparable data are hard to come by, but farm employment has fallen from about 1.6 million in 1968 to about 1.3 million at present (RSA Department of Agricultural Economics and Marketing, 1989: 4). Technological change and the growth in the average size of farming units have probably been the main causes of the decline. It was argued above that little further substitution of capital for labour can be expected in farming in the next few years. On the other hand, there are also indications that any fall in the real wages of farmworkers that may occur during this period is unlikely to lead to a significant substitution of labour for capital (Van Zyl, 1986: 69). But the average size of farming units will almost certainly continue to rise, given the financial pressures that agriculture is currently experiencing. Many smaller farms are likely to be incorporated into larger units in an attempt to generate economies of scale. More often than not, this involves the retrenchment of workers on the smaller farm. While it was seasonal workers that bore the brunt of mechanisation, it is permanent workers that are most directly affected by the process of farm consolidation (De Klerk, 1985: 14–15).

Another negative influence on farm employment is the switch from arable to pastoral farming in marginal arable areas. Pastoral farming generally employs fewer workers per hectare. All of these factors, along with the slow-down anticipated in production, will weaken the demand for agricultural labour.

For this reason, most workers who remain on farms are unlikely to see their real wages rising noticeably. Those with skills which are readily marketable in urban areas, such as truck drivers, may be an exception (De Klerk, 1985: 20), and attempts to pre-empt the growth of unions may improve the wages and working and living conditions of others. But, in general, one would expect a weak trend in real wages. This, together with a similar trend in on-farm employment, will increase the degree to which farm residents already rely on off-farm income (De Klerk, 1984: 47–8; Seleoane, 1984). If anti-squatter legislation is used to drive members of extended families off farms, the result will be a marked fall in income for those who are allowed to stay, irrespective of wage rates. It is also worth noting that salaries and wages now form a comparatively small part of total costs – about 13 per cent in 1987 (RSA Central Statistical Service, 1988c: 2) – so that lower real wage rates will probably not reduce output prices significantly or have any material effect on the competitiveness of exports.

Developments in agriculture will have an impact on industries directly dependent on farming – input manufacturers and output processors – and on the many small-town activities that exist primarily to service the farming community. The agricultural input industry is likely to remain in the doldrums as its market contracts or grows only very slowly. Attendant developments would be a transfer of capital out of the industry, increasing concentration of ownership, and weak employment and wage trends – some of which have already started to occur, in particular in fertiliser and farm machinery manufacturing. Disinvestment by foreign input suppliers has a sound economic basis. In some instances, this has led to a rather unusual form of vertical integration – the purchase of input manufacturing firms by agricultural cooperatives. Whether farmers will be any more successful in controlling rising input costs thereby remains to be seen.

In general, with so many members in financial ill health and heavily indebted to cooperatives, and with the advantage of subsidised funding by the Land Bank diminishing, one would expect the cooperative movement to be on the retreat. Some weaker groups have already merged with stronger ones, and others have formed partnerships with firms in the private sector.

Agricultural output processors which are orientated towards export, such as fruit canners, are in a vulnerable position. Employment and wages in such firms will depend largely on the imposition and effectiveness of trade sanctions. Those who produce primarily for the domestic market are comparatively secure, but, like most of agriculture, are limited by the rate of growth of this market.

The widest group affected by developments in agriculture are the many employers and workers in the secondary and tertiary sectors in small towns throughout the platteland whose livelihood depends on demand from the farming community. Industrial decentralisation and mining will make growth possible in some instances, but in most others the outlook for employment is less than favourable. In fact, as the comments on changes in population distribution below suggest, there are indications of an increase in the rate of unemployment in rural towns, not only amongst urban workers but also amongst those formerly employed on farms (Wilson and Ramphele, 1989: 88–9). In the Cape, where the majority of farmworkers are coloured, this is a well-established phenomenon, but in the northern provinces, the recent relaxation of influx control may be starting to increase the number of black workseekers in this category too.

Paradoxically, the relatively high risks and low profits of agricultural production may also open up some opportunities which have long-term significance. In the sugar industry, for instance, the exit of many medium and large-scale white commercial producers has led milling companies to encourage the growth of the already substantial numbers of small, part-time cane producers, who are mostly black.

Not only input prices, but also retail food prices have moved well ahead of the producer prices received by farmers, especially in the 1980s, while producers' share of consumer prices has tended to fall – particularly in the case of sugar (RSA Department of Agricultural Economics and Marketing, 1989: 99). This is the essential reason for a rather surprising general trend: the increasing concentration of farm ownership evident since the 1950s has, in most sub-sectors, not been accompanied by a similar increase in the penetration of industrial capital into farm-operating, ultimately because industrial capital has found it more remunerative to take its profit at other points in the agricultural production chain. As was pointed out above, the most profitable aspect of owning a farm has generally not been in operating it, but in the long-term appreciation of the value of the land – though even this is far from assured in the foreseeable future. Since most firms need to show an acceptable annual trading profit, capital appreciation of this nature is not generally a sufficient attraction, despite being less taxable when it is finally taken. With the number of sequestrations rising, banks may temporarily become the owners of more farms, but if farm ownership was unattractive to industrial capital in the past, it will, over the next few years, become even more so. The age of monopoly capitalism in South African agriculture is not on the doorstep.

The growth of part-time farming is not confined only to small-scale sugar producers. In other sectors, such as grain and livestock production, full-time farmers are becoming or being replaced by part-timers (SAAU, 1984: 35). Though not as efficient as full-time operating (Nel *et al.*, 1987: 25), part-time farming diversifies assets and income sources, thereby reducing risk, and increases liquidity. About 15 per cent of farmers were part-timers in 1983 (SAAU, 1984: 33), and in current circumstances this percentage is likely to increase. One noteworthy consequence of this trend is the increase in the number of black farm managers, or their equivalents, that must be occurring, though few such positions would involve the acquisition of essential financial skills.

In this context, one other probable consequence of the financial pressures on farmers can be identified, namely an increase in the occurrence of covert black tenancy on marginal and sub-marginal commercial farmland – as well as the need perceived by financially stronger farmers and the state for more stringent rural anti-squatter legislation.

What each of these developments has in common is that they expand the core of potential – and in some cases actual – black commercial farmers.

Effects on the wider economy

For the wider economy, the crisis in agriculture has likewise both negative and positive aspects. On the negative side, agriculture's relative contribution to gross domestic product (GDP) cannot be expected to grow from the present 5 or 6 per cent and, more likely, will continue to wane. This will increase the economy's dependence on urban employment and urban facilities and services – a pattern observable in most industrialising countries. The expected decline in foreign exchange earnings from agricultural exports will also have a negative effect on GDP growth through tightening the balance of payments constraint. If only commodities presently exported at a loss are affected and exports are reduced to a very low level, total foreign exchange earnings might not fall by more than 2 or 3 per cent. Of course, if trade sanctions were effectively imposed, the impact would be far greater. Trade and industrial classifications make it difficult to calculate the contribution of agriculture to exports accurately, but in the early 1980s it was still of the order of 20 per cent.

There are the makings of a vicious circle in this relationship. In the past, agriculture has relied heavily on foreign demand for output growth. In the foreseeable future, domestic demand will play an increasingly important role. So not only will a slow-down in agricultural production have a negative effect on GDP growth, but the latter will also have a negative effect on the former. On the other hand, these developments are not without some positive aspects. Diminishing export markets and greater emphasis on the need to balance domestic demand and supply growth will help reduce the rate of increase of food prices. This should benefit the urban population and, by easing upward wage pressures, assist the growth of urban employment.

Moreover, directly or indirectly, capital is being released from relatively unproductive uses (in agriculture) for deployment elsewhere. However, this is a very difficult process to trace and it is almost impossible to tell whether more productive use of such resources is, in fact, being made. At least some agricultural capital is likely to move abroad so as to bypass sanctions, thereby nullifying any benefit to domestic employment.

One further group of effects is on population distribution. It can be safely predicted that the substantial outflow from 'white' farms will continue. Until recently, most black farm families have had to move to the 'homelands', from which workseekers have had to migrate to urban areas. This has added significantly, though possibly unintentionally, to state-engineered population relocation. The resulting increase in population pressure in black rural areas is making the prospects of raising agricultural productivity in these areas ever more remote.

With the relaxation of influx control, an increasing proportion of farm leavers is likely to move directly to urban areas, often initially to rural towns. As already noted, this movement is widespread amongst coloured farmworkers in the Cape and may now be becoming so amongst black work seekers in the other provinces. Small-town facilities will often be inadequate to cope with such an influx. Together with the slow or negative growth of white residents, this will increase the pressure to de-segregate, or to transfer access to 'whites only' facilities to another 'population group', accompanied in many instances by an increase in white conservatism.

CONCLUSION

Commercial agriculture is currently under great financial pressure, probably greater than at any time since the 1930s. This pressure is manifested in low rates of return on investment (or low profitability), low levels of liquidity and a steady build-up of debt. The most significant impact of these trends has been on the agricultural capital stock: despite the many forms of state assistance to agriculture, the real capital stock in commercial farming appears to have risen little, if at all, over the past two decades, and the proportion of this stock owned by farmers has tended to decline. The average farmer has become less willing or less able to increase, or in many cases even to maintain, the real level of his investment in agriculture. Capital accumulation in agriculture has clearly entered a critical phase.

The roots of the crisis can be traced, on the one hand, to factors which have induced uneconomically high levels of investment in the past, namely, state agricultural policy, negative real interest rates over extended periods, and bank lending policies. In varying degrees, all three seem now to be changing, constituting less of an inducement to invest than before. On the other hand, a distinct but not unrelated group of factors – drought, high nominal interest rates, and the prolonged adverse trend in agriculture's terms of trade – has made further investment more difficult by distending the burden of debt carried by farmers. But what is most significant about this group of factors is that, with the exception of drought, there is reason to believe that they are less cyclical than structural. Interest rates in the foreseeable future are likely to remain higher, both because of the withdrawal of most forms of state interest subsidy and because of shifts in macro monetary policy. And the terms of trade will probably continue to deteriorate gradually – short-term improvements notwithstanding – because, on an international scale, one can expect supply in the medium term to continue to grow rather faster than demand.

The prognosis is, therefore, that commercial agriculture will remain a comparatively unrewarding area of investment. In net terms, capital is unlikely to accumulate in, or flow into, agriculture in significant quantities for some years, though there will always be sub-sectoral exceptions.

For the rural economy, the most important projected trends are a slower rate of growth of farm output, the further consolidation of large farming units, and the transfer of marginal arable land to pastoral production – all of which will tend to reduce employment and keep real wages from rising on farms and in small towns. On smaller farms, the increase in the number of part-time farm operators can be expected to continue.

For the economy as a whole, there is likely to be a continued decline in the relative contribution of agriculture to the gross domestic product, lower foreign exchange earnings on agricultural exports, and an unstemmed flow of rural workseekers and their families into urban areas. More positively, the prices of foodstuffs ought to rise less rapidly, reducing the pressure on urban incomes and assisting urban employment growth.

However, all of this overlooks one other set of consequences, no more than nascent at present, but with the potential to bring about far-reaching changes in production structures and the composition of the farming population. The same market forces that are making agricultural production less attractive to many existing white farmers are starting to generate a core of actual or potential African, coloured and Asian commercial or semi-commercial farmers. This is occurring through at least three distinct channels. First, in the sugar industry large, vertically integrated producers are encouraging small black producers to take over a share of the relatively high-risk, low-return operation of cane-farming. Second, in other sub-sectors part-time farming is not only spreading risks and generating additional cash flows for white farmers but is also transferring responsibility for day-to-day production activities to what are effectively black farm foremen/managers. And third – though this is hard to trace on the ground – the logic of current circumstances dictates that some marginal and sub-marginal commercial farmland is probably being rented covertly to black tenants.

It is also possible that a fourth channel is opening up: scraps of evidence suggest that some – probably isolated – operating commercial farms are being taken over by 'racially disqualified' farmers. Mechanisms for such transfers do exist in terms of the Group Areas Act, but it is not clear whether these are being used, or whether there is under-the-counter circumvention of the law, such as is widely practised in urban areas.

What is perhaps most significant about these processes is that they are mostly being driven by market rather than political forces. And, as was argued above, the essential direction of these forces appears likely to remain the same in the foreseeable future. The number of actual and potential

African, coloured and Asian commercial farmers can therefore be expected to continue to grow. In the absence of the repeal of the Land Acts and the Group Areas Act, this process will constitute the cutting edge of deracialisation in agriculture.

At the level of speculation, one might suggest that, realising the need for the emergence of a black commercial farming class but mindful of the conservative attitude of most white commercial farmers, the present government will – as in urban areas – probably retain the main body of legislation which presently defines rights of access to land in racial terms, but seek ways of accommodating the market forces that are carrying forward the process of deracialisation – what one might call a policy of 'managing the shift of the black–white frontier'.

One element of such a policy would be for the state to purchase parcels of land which it would demarcate for use by groups of individual black farmers or farming communities. Depending on the locality and potential of the land, a variety of schemes could be tried. But it is not unreasonable to assume that commercial agriculture will continue to be dominated by a relatively small group of large farmers, who together will produce the overwhelming bulk of output. Entry to this group will in the future depend less on race classification and more on access to capital.

This identifies a crucial constraint: what will initially most hinder the emergence of a class of smaller black producers is access to capital. For the many black potential farmers in this position, tenancy, in its many forms, offers a way forward. One of the most important tasks to which the government will have to attend is the design and operationalisation of a system or systems of tenancy which will be attractive both to landowners and to tenants. This will, in effect, mobilise 'white capital' for use by black entrepreneurs – a key component in any strategy to preserve the market basis of the economy.

Clearly none of this is imminent. Nor is the widespread collapse of commercial farming as it presently exists. If the term 'crisis' is to be correctly understood, for those individual farmers and farmworkers who have already been or who are in immediate danger of being displaced, it should indeed convey a sense of catastrophe. But for the agricultural sector as a whole and for the wider economy, it describes a combination of developments which are starting to provide the impetus for the restructuring, in part, of agricultural marketing, but, more fundamentally, of agricultural production in South Africa. The winds of change are blowing on the platteland.

APPENDIX A

Valuing the agricultural capital stock

There are few variables in economic analysis as important but as problematic to measure as the capital stock. The importance, in this instance, lies in the dependence of many of the indicators in Table 2 on the value attached to the agricultural capital stock. Assessment of the extent, nature and implications of the 'crisis' in agriculture hinges, in turn, on the values calculated for these indicators. Some discussion of the problems of valuing the capital stock is therefore required. The difficulties arise from two sources: techniques of valuation, and data.

The Department of Agricultural Economics and Marketing divides the stock of capital on farms into four categories: land, fixed improvements, machinery and implements, and livestock. In respect of the 'fixed capital' items (fixed improvements and machinery and implements), the technique of valuation adopted here – despite some arguable shortcomings – is the 'perpetual inventory method' used by both the Department and the South African Reserve Bank. A full description is to be found in De Jager (1973). The details relevant at this point are as follows: (a) the rate of depreciation allowed for machinery and equipment was 10 per cent p.a.; for fixed improvements the rate was 1 per cent p.a. until 1982 and 2 per cent p.a. thereafter – this 'inflates' the estimates of the earlier period marginally (by less than 0.1 per cent of the total value of capital stock); (b) the real fixed capital component of the total real capital stock (Table 2, line 9) was calculated at constant 1980 prices, using the price indices for the respective components calculated by the Department.

This is the least controversial aspect of the calculation.

The current value of livestock is calculated by the Department by multiplying the August quarterly head-count of the various categories of livestock by the respective indices, compiled from auction prices. (The Reserve Bank accepts this valuation.) In the absence of details for each category, the real value of livestock (at 1980 prices) included in the real capital stock (Table 2, line 9) was estimated by adjusting the total current value of livestock by the combined weighted index of the producer prices of livestock products, published by the Department. Any bias inherent in this method is probably small.

More serious is the bias generated by excluding the value of livestock in the 'independent black states' as from 1978. The 'real value of livestock' series estimated as just described shows a sudden fall of 22 per cent – or nearly R1 billion – between 1977 and 1978. At most, only a fraction of this is likely to have been caused by destocking in 'white rural areas'. The resultant relative 'inflation' of the pre-1978 total real capital stock from this source is probably of the order of 3 per cent.

Most difficult of the four asset categories is 'land' – not strictly part of 'capital' in the neo-classical sense, but much the largest single asset for most farmers and therefore essential to include in any meaningful estimate of the capital stock. ('Land' makes up about two-thirds of the total asset value of the 'average farm'.) Until the late 1970s, when much of the information for the annual agricultural census was collected by individual interview and the results were thought to be relatively reliable, the Department simply summed farmers' estimates of the market value of their land to obtain an estimate of the total value of land. Since then, however, census data have been collected by mail. Predictably, the data-collection rate has been too low to allow reasonable estimates of the total market value of land to be made from this source.

In the absence of this input, the Department has fallen back on adjusting the market value of land by the combined index of rural land prices compiled by the Central Statistical Service on the basis of transfers of rural immovable property. Consequently, the 'real value of land' – as calculated by deflating the market value by the relevant price index – has remained more or less constant in the 1980s, with a single downward adjustment of exactly R500m (at 1980 prices) occurring between 1982 and 1983.

However, what makes estimation of the real capital stock still more difficult is the fall of no less than 35 per cent in the real value of land – calculated on the basis just described – that took place between 1979 and 1981. This decline of about R8 billion (at 1980 prices) shows up as a 25 per cent reduction in the value of the total real capital stock (from R31 359 million to R24 805 million) between 1979 and 1981. A decrease of this magnitude is too great to be credible, especially given that these were exceptionally good agricultural years in most parts of South Africa. A small part of this can be explained by the 15 per cent fall in the area of land under tree and field crops between 1976 and 1981, and the transfer of this land to pastoral production. This can be assumed to have led to the depreciation of land values in the areas concerned. But, this apart, one cannot avoid the conclusion that there is a major inconsistency in the measurement of the value of land.

In respect of all the categories of asset, other than machinery and implements (which is the smallest of the four by value), it must therefore be concluded that there has been a relative undervaluation in recent years. Which part of the series is the more reliable, it is hard to say.

Allowing for the disjunctions around 1980, the most that can reasonably be said about the real capital stock (as recorded in Table 2, line 9) is that it has probably changed little over the two decades or so of the study, fluctuating narrowly around the R25 billion level (at 1980 prices). The apparent rise that took place between 1974 and 1979 and the subsequent sharp fall should be regarded with caution.

In respect of the indicators in Table 2, the relative undervaluation of capital in the 1980s implies: (a) a relative overstatement of all of the nominal and real rates of return on assets and owners' equity (lines 1–4) in the 1980s; (b) a relative understatement of the nominal and real net worth of the farmers (lines 5, 8) in the 1980s; (c) a relative overstatement of the burden of debt of agriculture (line 10) in the 1980s; and (d) a relative understatement of the capital–output ratio (line 15) in the 1980s. The implications of these inconsistencies are explored in the text.

Finally, Fiske (1988, personal communication) draws attention to a number of factors that lead to a persistent undervaluation of agricultural capital by the Department of Agricultural Economics and Marketing. Although relatively little agricultural production is undertaken by large corporations in South Africa, a more-than-negligible percentage of capital assets in the farming sector is owned by such corporations and by state and semi-state bodies. While deliveries of produce from these sources are included in 'gross farming income', the greatest part of their farming assets escapes measurement. Secondly, as members of cooperatives, farmers own the reserves of cooperatives, but this is not counted as an asset. Similarly, the reserves held by Control Boards on behalf of farmers are not included in farming assets. Finally, crops on the land, stored produce, stocks of intermediate inputs, and various other inventory items are ignored in the valuation of assets.

The indicator for which these particular shortcomings are most important is the burden of debt. If capital is undervalued, insolvency, for heavily indebted farmers, will appear more imminent than it actually is. This qualification should be borne in mind in the discussion of solvency in the text. There is no immediate reason to suppose that the downward bias was more serious in one period rather than another. Trends, as opposed to levels, are probably not significantly affected by this last group of errors.

10
The restructuring of labour markets in South Africa: 1970s and 1980s[1]

DOUG HINDSON

This chapter describes in broad outline the evolution of labour markets in the 1970s and 1980s, and briefly relates changes in the labour market to current labour, urban and regional policy reforms. What major changes are occurring in patterns of labour demand and supply on a national scale? How has the occupational and racial division of labour changed? What has happened to the geographical distribution and composition of labour demand and supply? How does state strategy seek to respond to these changes?

EMPLOYMENT: SECTORAL SHIFTS

Economic recession and structural reorganisation in the workforce has substantially altered the pattern of labour demand in South Africa during the last three decades. Two broad periods stand out from the data: the 1960s to early 1970s, and the early 1970s to the mid 1980s. The first was a period of relatively rapid growth in employment, while the period since has been one of economic decline and rapid restructuring of the employment process.

As can be seen from Table 1, the annual rate of growth of employment in the whole formal economy (agriculture included) was 2.8 per cent between 1960 and 1970 and then declined to 2 per cent in the next decade. Between 1980 and 1985 it fell to 0.6 per cent.

The demand for labour was restructured in part through changes in the relative importance of economic sectors and in part through changes in the occupational division of labour. As will be shown, one important result of this was the redistribution of workers of different races within the occupational structure.

The share of capitalist agriculture in total employment fell from 22 to 11 per cent, and of mining from 13 to 10 per cent. The share of the secondary and tertiary sectors increased from 17 to 25 per cent and 38 to 48 per cent respectively. Manufacturing, a sub-category of the secondary sector in the table, increased its share from 13 per cent in 1960 to 17 per cent in 1970 to 20 per cent in 1980 and then fell to 19 per cent in 1988. If the South African economy is in a phase of de-industrialisation, this trend has not as yet forcefully expressed itself. The major shifts arise rather out of the relative decline of the primary sectors and the relative growth of the secondary and tertiary sectors.

Table 1. The sectoral composition of employment, 1960–1988[2]

Sector	1960	% total	1970	% total	1980	% total	1985	% total
Agric	1 033	22	1 076	18	1 026	14	868	11
Mines	602	13	658	11	794	11	755	10
Secondary	807	17	1 460	24	1 904	26	1 935	25
Tertiary	2 191	47	2 906	48	3 720	50	4 221	54
Total	4 633		6 100	2,8*	7 444	2,0*	7 779	0,6*

Source: *South African Labour Statistics*, 1989, 250–2.
* Average % increase
Note: Agriculture includes forestry and fishing; mining includes quarrying; secondary includes manufacturing, electricity, gas and water, and construction; tertiary includes commerce, finance, transport, communication and services.

The racial composition of the demand for labour changed in important ways. Table 2 shows that between 1960 and 1988 the share of whites in total employment remained constant, that of coloured and Asian workers increased, and that of African workers declined. Part of the reason for this was that Africans predominated in sectors which shrank or grew slowly while white, coloured and Asian workers had a substantial presence in expanding economic sectors. Another factor, considered in the next section, was the changing occupational and racial composition of the workforce.

Table 2. The racial composition of employment by sector, 1960–1988

Sector percentage	1960				1970				1980				1985			
	W	C	A	B	W	C	A	B	W	C	A	B	W	C	A	B
Agric	11	12	1	76	9	12	1	78	8	13	0	78	7	12	0	80
Mines	11	1	0	88	10	1	0	89	10	1	0	88	10	1	0	89
Secondary	26	14	4	56	23	16	5	56	26	16	5	59	26	15	5	60
Tertiary	31	10	2	56	32	9	3	56	31	10	3	55	29	11	3	57
Total	23	10	2	65	23	11	3	64	23	11	3	63	23	11	3	64

Sources: See Table 1.

THE OCCUPATIONAL–RACIAL DIVISION OF LABOUR

Simkins and Hindson (1979) have shown that between 1969 and 1977 there was an increase in the proportion of professional, semi-professional, clerical, technical and non-manual jobs in the non-primary sectors of the economy. Semi-skilled work remained constant and unskilled jobs declined as a proportion of the total. This alteration in the structure of employment was accompanied by important modifications in the racial division of labour. While whites continued to dominate the skilled, professional, technical and managerial places, coloured, Asian and, in growing numbers, African workers penetrated the semi-skilled, technical and non-manual levels of the occupational hierarchy. Crankshaw (1987) and Crankshaw and Hindson (1990) have produced evidence to show that these trends accelerated during the 1980s.

The sectors in which the trend towards shrinking unskilled and expanding semi-skilled, technical, white-collar and professional work was most marked were manufacturing and construction, the two sectors most severely affected by economic decline in the 1980s. However, a similar pattern of declining unskilled and growing semi-skilled work is evident in the government and service sectors.

The restructuring of occupations was accompanied by major shifts in the racial division of labour; some sections of the black working class suffered severely from declining employment and the restructuring of occupations, while others greatly improved their position in terms of growth in numbers of jobs, the share of employment and, as will be shown in the next section, earnings.

In numerical terms, the hardest-hit sections of the working class were African (and from 1981 coloured) unskilled workers, especially in manufacturing and construction. Job loss in the government and the service sectors was also great for these two groups of workers. Another sector adversely affected by the deracialisation of labour markets was the white working class. The number of whites in semi-skilled and skilled blue-collar and in semi-skilled white-collar jobs increased only slightly from the early 1970s to the early 1980s, and thereafter fell. Although many whites moved up the occupational hierarchy over the period, workers in some of the traditionally protected semi-skilled areas of work lost jobs as a result of competition from black workers.

While some sections of the working class clearly suffered from the declining growth of employment and from the changing racial distribution of jobs, others made substantial numerical and proportional gains. Large numbers of African, coloured and Asian workers took up positions in the expanding semi-professional and technical categories of work. Whites too – a group that has historically dominated this sector – gained from the rapid expansion of such jobs. These occupational categories, which were already large at the end of the 1960s, nearly doubled over the next fifteen years.

One of the most striking changes was the movement of African workers into semi-skilled and skilled jobs during the 1970s and 1980s. Between 1969 and 1985 the numbers of Africans in these positions increased from just over one million to just under two million. It would seem that this new semi-skilled proletariat has superseded the unskilled African proletariat as the numerically dominant stratum of the African working class.

One important outcome of the changing division of labour was that occupational differentiation became increasingly incongruent with racial differentiation in the workplace. The most striking example of this was the African workforce itself, which became more and more differentiated in terms of both skill and the division between the employed and the unemployed.

One of the most significant outcomes of the crosscutting of racial and occupational strata is racial mixing – a form of deracialisation – within the middle strata of the employment hierarchy. This process has gone furthest within the semi-skilled, non-manual occupations and non-degreed semi-professional work. Because jobs in these sectors have grown rapidly both as a share of total employment and in absolute terms, employment for all racial groups has increased within them. Coloured, Asian and African workers have joined rather than displaced whites at these levels.

Different processes appear to be at work at the highest and lowest levels of the occupational hierarchy. Management and the degreed professions remain essentially white preserves, though the numbers of Africans, coloureds and Asians in these jobs have grown very fast from small bases. Semi-skilled operative work, a large but only slowly growing segment of the market, has been characterised by the replacement of whites by African and coloured workers. Africans have always been numerically dominant in the unskilled manual and menial non-manual labour markets, and this pattern has continued over the period.

The growth of the new semi-skilled African workforce and the emergence of the new racially heterogeneous middle strata of the occupational hierarchy represent major shifts in the racial division of labour in South Africa, and have profound implications both within and outside the workplace. Clearly it is not possible on the basis of statistical aggregates to extract the full economic and political significance of these shifts. Further research is needed on their causes and the impact they are having on the experiences and consciousness of different sections of the working class.

From its re-emergence in the early 1970s the democratic union movement has concentrated most of its energy on the organisation of black unskilled and semi-skilled workers. Relatively little interest was shown in the emerging new middle layers of the working population, partly because the size and potential for growth of these layers were underestimated. The opposite is, of course, true of the state. Since the late 1970s state strategy towards the African population focused energies and resources on the creation and cooption of the new middle strata within the black population. The growth of the new multiracial middle strata, their strategic importance within the workplace and as an object of state reform strategy, call for serious consideration of their relationship to the working class and their future role within the national democratic movement.

The overall impact of recession and employment restructuring on the different racial groups can be seen in Table 3, which presents average annual rates of growth of employment for each racial group over the period.

It can be seen that coloureds and Asians experienced above-average rates of growth until 1980 while Africans experienced below-average growth. Between 1980 and 1988, the rate of growth of employment of whites fell to nil, while that of coloureds, Asians and Africans fell to 1 per cent annually.

Table 3. Growth and decline in employment by race, 1960–1985 (percentage per annum)

	1960–70	1970–80	1980–8
Whites	2.8	2.0	0.0
Coloureds	3.1	2.5	1.0
Asians	4.9	3.4	1.0
Africans	2.6	1.9	1.0

Source: See Table 1.

Whites, coloureds and Asians benefited most from economic expansion in the period of classic apartheid, while Africans benefited least. During the period of economic contraction and political reform since the 1970s, all groups experienced declining employment growth, but it is clear that Africans had to shoulder the burden of the economic crisis.

THE CHANGING OCCUPATIONAL WAGE STRUCTURE

Despite the very high and rising unemployment over the 1960s, 1970s and 1980s, average real monthly earnings of African, Asian and coloured workers increased almost every year in all the major sectors, according to the various censuses carried out by the government's Central Statistical Service. Exceptions were the central government sector where real wages of coloured workers declined from the early 1970s, and laundries and drycleaning services, a sector in which there was a dramatic fall in the real wages of all race groups during the 1970s and 1980s (Hendrie, 1986: ch 5).

In contrast to Africans, Asian and coloureds, the relative position of whites as a whole appears to have worsened from the 1970s. Their real earnings began to fall in many of the major economic sectors, in some cases recovering briefly during the boom of the late 1970s–early 1980s, but declining again sharply from the early 1980s. Two important exceptions were manufacturing and local authorities, where the real earnings of whites rose almost continuously over the whole of the 1970s and 1980s. But even in these two sectors the increases in the earnings of Asian and African workers were greater than those of whites, albeit by a small margin. A general feature, then, of the period from the 1970s to the mid-1980s was the narrowing of the gap between white earnings and those of other groups. This happened despite the fact that unemployment was far higher amongst black workers, especially Africans, than amongst whites.

Part of the explanation for the increasing real wages of black workers in the face of growing unemployment was obviously the growth and increasing power of the union movement, but the way management responded was power of the union movement, but the way management responded was equally important. Management's main strategy to counter union militancy since the 1970s has been to shed black unskilled workers and simultaneously upgrade, promote and pay higher wages to the reduced workforce. The effect of this strategy was to alter in very substantial ways the occupational structure, as we have seen above.

But what has been the effect on wages? Wage data produced by the Central Statistical Service do not provide detailed information on occupations and corresponding earnings. The analysis here uses instead the privately sponsored Peromnes survey, which presents annual data from 1981. The sample for this survey is sufficiently comprehensive for one to assume with reasonable confidence that the occupational wage trends which emerge from the surveys reflect the general direction of change in the leading, especially corporate, sectors of the economy.[3]

The Peromnes system grades jobs into 19 ranked groups in terms of their rating on eight criteria: problem solving, consequences of error judgement, pressure of work, knowledge, job impact, comprehension, educational qualifications and training or experience. The 19 grades are aggregated into the following broad occupational categories:

1–3: top executive and the most senior specialists;
4–6: senior management and high-level specialists;
7–9: middle and lower management;
10–12: superintendents and lower-level specialists;
13–16: supervisors and high-level skilled and clerical staff;
17–19: lower-level skilled and clerical staff, and very low-skilled and unskilled workers.

Table 4 shows the ratio of wages of Africans, Asians and coloureds to those of whites for each of these six occupational levels in 1981 and 1988.

Table 4. Earnings gap 1981 and 1988

Grade	1981			1988		
	Af/W	As/W	Col/W	Af/W	As/W	Col/W
1–3	–	–	–	–	–	–
4–6	85	97	85	79	87	77
7–9	68	78	76	90	96	87
10–12	73	82	72	86	93	81
13–16	103	180	112	178	198	126
17–19	86	104	56	–	–	–

Source: FSA Management Consultants, Peromnes Remuneration Surveys, 1981–1988.

At the apex of the occupational pyramid – top executives and the most senior specialists (grades 1–3) – there are so few blacks in the survey sample that no meaningful comparison of earnings levels is possible. At the next level down – senior management and high-level specialist positions (grades 4–6) – there were also relatively few blacks, but here earnings comparisons are meaningful. As can be seen from the table, the wage gap at these levels of the occupational structure was relatively small in 1981 and widened somewhat over the period 1981-8.

The proportion of blacks in middle and lower management positions (grades 7-9), the third level down in the table, was 7 per cent in 1981 and 13 per cent in 1988. The wage gap was fairly large, but narrowed considerably over the period in favour of blacks.

Amongst superintendents and lower-level specialists (grades 10-12), the fourth level down, the proportion of blacks increased from 36 to 40 per cent over the seven years. Here the earnings gap was also substantial, but narrowed somewhat over the period.

In terms of total employment, the most important level was that comprising supervisors and higher-level skilled and clerical workers (grades 13-16). These grades made up 36 per cent of the sample in 1981 and 42 per cent in 1988. The share of blacks in these jobs increased from 60 per cent of the total in 1981 to 77 per cent in 1988. It is here that the most dramatic alterations appear to have taken place in the employment and wage structure. Average earnings of African, Asian and coloured workers were marginally *higher* than those of whites in 1981, and the gap widened rapidly over the period. By far the most rapid advances were made by Africans. Their share of employment in grades 12-16 increased from 40 per cent in 1981 to 56 per cent in 1988. They, too, experienced the most rapid increases in average wages, far outstripping any other group within this occupational level, or, indeed, any other group in any other grade.

A similar pattern occurred at the lowest levels of the occupational hierarchy - lower-level skilled and clerical staff and very low-skilled and unskilled workers (grades 17-19). The vast majority of these jobs were held by Africans, followed by coloureds. The share of Asians and whites fell, in the case of the latter to nil in 1984. In 1981 the highest earners in these grades were Asian workers, followed by whites and Africans. By 1988 the order had changed: on average, Africans earned the highest wages, followed by Asians and coloureds, while whites disappeared from the picture.[4]

What the Peromnes data suggest, then, is that the income gap between whites and blacks closed fastest at the lowest levels of the corporate employment structure. Indeed, a new gap opened up in favour of black workers. In the middle to upper strata of the hierarchy there also occurred a narrowing of earnings differentials but at a more gradual pace. The most dramatic changes took place in the movement of African wages at the semi-skilled and unskilled levels of the structure. It was here that the pressure for change from the democratic union movement was greatest, and here too that management moved fastest to expel unskilled black workers and upgrade and reward those who were kept on.

THE GROWTH AND RESTRUCTURING OF LABOUR SUPPLIES

For some years now there has been wide agreement amongst South African economists that structural unemployment is an endemic and growing feature of the economy. From the early 1960s to the early 1970s, the level of unemployment grew from about 1.2 to 1.8 million, despite the rapid growth of employment experienced during that period. From the mid-1970s to the late 1970s the situation deteriorated as employment growth slowed down and unemployment increased sharply. Charles Simkins, who is considered authoritative in this area, found that the rate of unemployment was over 20 per cent by the late 1970s (Simkins, 1981, 1983).

Simkins's estimates have not been extended into the 1980s. For this period the most readily accessible source is the Current Population Census (CPS), which is seriously flawed. The two main problems are underestimation of the level of unemployment as a result of the narrow definition used; and a creeping downward bias in the estimates over time due to the tendency of the sample to lose unemployed workers. In recognition of this, the Central Statistical Service drew a new sample in 1986 and widened its definition of unemployment. The result was that its estimate for African unemployment in that year was revised from 519 000 to 1.2 million.

Bearing in mind these problems, we may use the CPS data to gain an impression of trends in unemployment since 1980. After rising steeply in the late 1970s, the level and rate of unemployment slowed in growth during the 1978–82 economic upturn, but thereafter rose more or less continuously until the first quarter of 1986 when it again began to fall (from a very high absolute level). It can safely be concluded that while cyclical upturns in the economy slightly alleviated unemployment for short periods, the underlying problem became much worse over the first half of the 1980s. All indications are that the situation will continue to deteriorate for the foreseeable future.

There were also major changes in the patterns of supply and reproduction of labour. I have argued elsewhere that capitalist expansion in the period of orthodox apartheid (from the late 1940s to the late 1960s) was based on the maintenance of a racial hierarchy in industry and on the differentiation of the African urban workforce into permanent and temporary migrant labour (Hindson, 1987). The racial division of labour in production was complemented by residential segregation of the races and the separation of hostel-dwelling migrants from permanent residents in the black townships. As a central feature of this system, residential and territorial space was manipulated not only to separate whites from blacks but also to enforce the disaggregation of African urban labour supplies into permanent and temporary components.

The most important mechanisms used to achieve this end were the township administrative structures, the labour bureaux and the bantustan authority system. Township administration separated migrants and locals residentially, while the labour bureaux divided the African labour force by means of influx control by allocating workers to jobs on the basis of ethnicity and residence qualification (Hindson, 1987, ch 4). The tribal authority system was used to prop up an ailing redistributive economy and to control and discipline the reserve army of African labour (Wolpe and Legassick, 1976).

This combination of measures secured a workable and (in terms of profitability and economic growth) highly successful system of economic reproduction during the 1960s. However, rapid capital accumulation was achieved at this time by a dual labour strategy which nevertheless worked to corrode the basis of future labour supplies. This strategy ensured that urban proletarian labour was reproduced at a bare minimum level of subsistence, yet future supplies from this source were seriously jeopardised by insufficient expenditure on such social overheads as housing, health, education and training. At the same time costs of reproducing migrant labour were displaced onto an increasingly fragile African rural economy, thereby eroding the foundations of this very supply.

Two decades of rapid economic expansion under apartheid left a legacy of rural impoverishment, massive structural unemployment, shortages of trained and skilled labour, and labour market rigidities. By the late 1960s the racial occupational hierarchy within industry presented a growing obstacle to the restructuring of the labour process. As industry modernised a growing demand was generated for skilled, semi-skilled, professional and technical labour and for a residentially stable workforce. The economic advantages of migrant labour, and hence also of the bantustan system, to industrial and mining became increasingly narrowed. As state policy during the 1970s became more and more dysfunctional, new forms of labour supply and reproduction began to develop. These new forces developed at first within the framework of the social and political structures of apartheid, and were constrained by them, but later were able to bypass, override and eventually supersede state controls.

In the 1960s and 1970s the African population in South Africa was redistributed as a result of both the state's resettlement programme and market-impelled migrations. The outcome would rearrange the African labour market, replacing the division between settled migrant labour with new lines of differentiation corresponding to new patterns of regional settlement and labour movement. The interaction of forced resettlement and spontaneous migration can be assessed by examining statistics on removals and African urbanisation in the 1960s and 1970s.

The Surplus People Project has estimated that some 3.5 million people of all races were relocated by government between 1960 and 1982. By far the largest category – 1.7 million – were Africans living on white farms and on African-owned land in white farming districts (areas officially termed black spots). Most of these people were removed to settlements in the rural areas of the bantustans. Another important category was removals from the urban areas. Urban relocation resulted in the shifting of 730 000 people between 1960 and 1980 (Surplus People Project Report, 1, 1983: 6, Table 1). This entailed the deproclamation of established black townships and the removal of their permanent inhabitants to new townships just inside the borders of the bantustans, often within commuting range of their previous home towns and cities.

A further category of removals occurred under the influx control regulations. The numbers of people involved are not accurately known, but were also very great. One estimate for the major cities and towns between 1956 and 1963 is approximately 200 000 (Hindson, 1983: 341, Table 27). Thus, despite the growing impoverishment of parts of the bantustans, their population increased dramatically over the two decades: from 4.4 million in 1960 to 11.1 million in 1980 (Simkins, 1983: 54, 56, Table 1).

The official aim of resettlement within the bantustans was to segregate Africans territorially as a prelude to constructing politically independent and economically self-contained independent or self-governing states. The unintended consequences, however, were to restructure the reserve army of labour within the bantustans and to reintegrate their working populations into the core metropolitan economies in new ways, thereby subverting the long-term aims of territorial segregation. This effect can only be studied once the bantustan population statistics have been disaggregated, and population movement and settlement patterns within the bantustans identified.

In a study of African urbanisation, Graaff has shown how official population censuses grossly underestimate the level of urbanisation within the bantustans, and thus perpetuate the false impression that their populations remain supported by rural production (Graaff, 1985). He has argued that the category 'urban' should be widened to include not only people living in proclaimed towns (the official definition in the 1980 census), but also peri-urban populations (those dependent on commuting to proclaimed towns for employment and shopping) and semi-urban concentrations (settlements of 5 000 and more people). The category 'rural', in Graaff's definition, becomes residual. It includes a very small percentage of people who make a living solely out of agriculture (perhaps 10 per cent of the bantustan population), and many who, while living in a rural setting, derive only a minimal income from agriculture. Combining Graaff's estimates with census material on the non-homeland areas, we obtain the distribution of the African population set out in Table 5.

From the table it emerges that over half of South Africa's African population must be regarded as fully integrated into and dependent upon its urban economy, albeit many of these would fall into the reserve army of unemployed rather than its employed labour force. Only some 24 per cent of this population, Africans living in settlements of less than 5 000 within the rural districts of the bantustans, exist in circumstances that could conceivably continue to support the traditional migrant labour system through non-capitalist rural means of production.

Of the total bantustan population, no less than 15 per cent lived in proclaimed townships. Many of these are the people relocated after townships were deproclaimed in the white urban areas, and can be considered fully a part of the established urban proletariat, despite the distances between their homes and workplaces. Another 15 per cent were located in peri-urban concentrations, strung out on the fringes of the major metropolitan areas. Peri-urban population growth within the bantustans is a highly significant development because it arises out of spontaneous market- (as against state-) induced movement within the bantustans from rural and semi-rural areas to the metropolitan peripheries. Though influx controls held these populations at some distance from the city centres, they too must be regarded as fully integrated into the metropolitan economy, forming part of its active and reserve armies of labour.

Table 5. The distribution of Africans by region, 1980

	Number	Per cent RSA	Per cent bantustans
Non-bantustans			
Urban	5 606 700	26	
Rural	4 310 000	20	
Bantustans			
Urban	1 809 151	8	15
Peri-urban	1 747 934	8	15
Semi-urban	3 011 602	14	26
Rural	5 249 623	24	44
Total RSA	21 735 010	100	
Urban	12 175 387	56	
Rural	9 559 623	44	

Sources: J. Graaff, 1985: 22, Table 1; C. Simkins, 1983: 56, Table 1.
Note: Foreign migrant are excluded from these figures.

The changing interconnections between African settlements of various kinds in the bantustans and urban labour markets have not been widely researched. However, fragmentary material suggests that a hierarchy of access to markets which cross cuts and is beginning to supersede the local–migrant division, has begun to emerge. On the basis of a survey of labour bureaux in 1982, Greenberg and Giliomee identified a labour market hierarchy corresponding closely to the legal distinctions then still operative between Africans who qualified under section 10 of the Urban Areas Act, border commuters and other sections of the African population in the bantustans (Greenberg and Giliomee, 1983).

In a study published in 1984, Graaff also identified a process of hierarchical structuring of the labour market in parts of Bophuthatswana (Graaff, 1984). At the top of the hierarchy were commuters with employment in the metropolitan centres. They were followed by commuters to decentralised and deconcentrated areas on the borders of Bophuthatswana. The poorest wage-earning group comprised commuters to white farms, but the poorest of all were the unemployed who depended on informal activities, agriculture and transferred welfare payments. Migrant labour to the mines would constitute a relatively privileged and distinct sector, but Graaff does not separate this group out for specific consideration.

There is growing evidence to suggest that even in that bastion of the traditional migrant labour system – the mines – a process of stabilisation of African labour is under way. From the early 1970s the mines increased the proportion of domestic labour and reduced their dependence on foreign workers. Greater occupational stabilisation was achieved by lengthening contracts and extending the application of the Valid Re-engagement Guarantee (VRG) system whereby workers agreed in advance to fixed periods of leave. Jonathan Crush has identified a further stage: beginning in the early 1980s, the mines began to stabilise a growing proportion of the workforce residentially by converting compounds into family units and by employing African township residents (Crush, 1987).

As yet this process has not gone far, and the terms and extent of stabilisation are a matter of dispute and conflict between mine-owners and organised mine labour. While the pace and pattern of change on the mines differ in certain respects from the non-primary sectors, the industry shares in the general process whereby employment growth has declined and the workforce has become increasingly skilled and residentially stable. This has meant that a growing number of workers are being permanently excluded from mine employment, an occupation which traditionally provided what Crush calls a 'safety net' for unemployment and the most extreme forms of poverty.

All these findings indicate that labour markets have been substantially transformed during the 1970s and 1980s. In these decades occupational and residential stabilisation, and the rising skill and wage composition of the employed workforce, have been associated with increased differentiation within the labour market, growing unemployment and the permanent exclusion of increasing numbers from the wage economy.

Moreover, labour supplies within the bantustans are being reintegrated into the metropolitan mining, industrial and commercial economy through new patterns of migration, commuting and resettlement. The corollary of structured access to employment and the stabilisation of employment at the centres may be the exclusion of sections of the African workforce from the core of the economy. An increasingly marginalised reserve army of African labour has grown in deconcentration areas on the peripheries of the major metropolitan centres. The bantustans no longer fill the role of providers of 'cheap' labour – in the form of subsidised long-distance temporary migration. They now play an important part in the evolving process of hierarchical restructuring of the labour market and in containing and controlling the growing reserve army of labour.

THE REGIONAL DISTRIBUTION OF INDUSTRY AND EMPLOYMENT

One important mechanism for the restructuring of the occupational racial division of labour in the 1970s was the regional decentralisation of employment. It is thus important to examine the extent and nature of the regional dispersal of economic activity over the period.

Until the early 1980s the prevailing liberal orthodoxy about the South

African government's policy of industrial decentralisation was that it was ineffectual in the face of powerful forces of industrial concentration in the major metropolitan areas, and also costly as a means of creating employment (Bell, 1973; Rogerson, 1982; Tomlinson and Hyslop, 1984). Radical scholars presented it as an adjunct to influx control: they argued that decentralisation served the twin aims of securing territorial apartheid and providing cheap migrant labour in the cities and cheap settled labour for decentralised industry (Lacey, 1982).

The orthodox liberal view has been undermined by findings on the changing pattern of industrial dispersal over the last three decades. In a paper published in 1983, Trevor Bell showed that industrialisation has moved through two distinct phases in the post-Second World War period: up until the late 1960s the forces of spatial concentration predominated and the metropolitan centres of the country increased their share of industrial employment; but from the late 1960s until the early 1980s the forces of dispersal strengthened and the non-metropolitan areas (towns outside the major centres, mainly on the borders of the bantustans or within them) gained an increasing share of manufacturing employment (Bell, 1983: 8).

In support of his periodisation Bell showed that industrial employment in the major metropolitan areas grew as a share of total industrial employment in the country from 1956/7 to 1965/6 (industrial census years). From 1955/6 to 1979 it declined from 82 to 77 per cent of the total (Bell, 1983: 8–9). Conversely, the share of the rest of South Africa, which comprises towns outside the metropolitan complexes, declined until the mid-1960s and then increased steadily until the end of the 1970s.

Bell's more detailed examination of sub-sectoral shifts in manufacturing industry indicates that a process of dispersal in some key sectors – for example, textiles – began to occur well before the government's decentralisation policy was seriously implemented. The typical pattern for these sectors was to move first from the PWV area to the Durban–Pinetown area (and to a lesser extent to Cape Town) and then later to intermediate and smaller towns on the borders of the bantustans and within them.

A debate has developed over Bell's explanation of the causes of industrial decentralisation. In his 1983 paper he argued that the observed industrial dispersal was largely the outcome of spontaneous capital movements and that the state played a relatively minor role in bringing about the changed spatial distribution of industry. Some of his critics have sought to reassert the orthodox liberal position (Tomlinson and Hyslop, 1984). The observed industrial dispersal, they argue, is the outcome of the accelerated decentralisation programme implemented by government since the late 1970s, and has been achieved despite the continued operation of the forces of concentration over the period. Industrial decentralisation has thus occurred at the expense of sharply increased opportunity costs; the loss of jobs, output and welfare in the metropolitan centres has greatly outweighed the gains in African employment in the decentralised areas.

This debate is relevant in the present context because it has a direct bearing on the assessment of the state's current urban and regional strategy. If industrial decentralisation is indeed the outcome primarily of state inducements and incurs steeply rising private and social opportunity costs, this would suggest that there is little more scope for effectively implementing the current urban and regional strategy than the traditional policy pursued in the period of orthodox apartheid. If, however, it is found that an autonomous process of capital dispersal is in operation, the possibility arises that state urban and regional policy may complement or even promote that process, in which case the aims of policy are more likely to be achieved. These questions clearly have a bearing not only on the explanation of past movements of capital, but also on the scope for urban and regional policies in the future.

Bell's argument that the fundamental causes of industrial movement lie in declining profitability and increased competition is well substantiated and convincing. His emphasis on the need to periodise economic processes, and on the spontaneous character of industrial dispersal during the later period, provides an important corrective to the naive liberal orthodoxy that apartheid policies are and have always been fundamentally opposed to market rationality. However, his representation of the state and its role and influence in the economic system is oversimplified.

The positions in the debate over the causes of industrial decentralisation have become too polarised on this issue to be fruitful as a basis for explanation. The increasing share of industrial employment in decentralised areas in the 1970s is clearly the outcome of a combination of both the increased spatial mobility of capital and the vigorous efforts of the state to promote industrial decentralisation. The precise relative influence of each cause has yet to be established.[5]

The mid-to-late 1960s mark a decisive break in the pattern of geographical location of industry in South Africa, after which the tendency towards locational concentration weakened and the tendency towards dispersal strengthened. The mobility of capital increased and the state encountered greater cooperation with its programme of industrial decentralisation. As a result the pattern of industrial location was substantially modified through processes of deconcentration and decentralisation of industry.

In the deconcentrated and decentralised areas of the country, the occupational wage structure could be more easily reorganised to take advantage of the greater supplies of unorganised and legislatively unprotected black labour. While the impact on aggregate employment may have been slight or even negative, industrial dispersal facilitated the deracialisation of the industrial division of labour.

The changing patterns of location of capital in the 1970s represent one dimension of a fundamental reordering of the industrial structure in South Africa, with major implications for employment and labour supply. The effect of these changes has been not to reinforce the bantustan system and territorial apartheid, but to speed up the erosion of that system. Industrial dispersal has effectively integrated parts of the bantustans into the wider metropolitan-centred market economy, allocating to sections of their populations new places in a changing regional, racial and occupational division of labour, while excluding others through a process of expanding unemployment and marginalisation.

CONCLUSION: CHANGES IN URBAN AND REGIONAL POLICY

State policy was slow to respond to these underlying shifts in the character of labour markets. Only from 1979, after well over a decade of population and industrial relocation that ran counter to territorial apartheid, were there substantial responses in the sphere of labour, urban and regional policy.

On the recommendation of the Riekert Report on Manpower Utilisation (RP32/1979) which appeared in 1979, the government abandoned its policy of reversing African urbanisation and accepted the right of the settled African population in white urban areas to remain there. The recommendations of the Riekert Report were not successfully implemented, and after six years of deepening economic crisis and township revolt, and a growing awareness in official, academic and other circles of the scale of *de facto*, illegal and displaced urbanisation, further changes were considered. This re-think resulted in the President's Council report on urbanisation, which appeared in 1985 (PC3/1985).

The PC report went much further than Riekert. It argued that African urbanisation was both inevitable and economically desirable. It recommended that a policy of orderly urbanisation be adopted that should aim to control, but not prevent, this process. The bantustans, it argued, should be reintegrated economically into a number of development planning regions which would supersede the old provincial and homeland boundaries. Within these regions urbanisation should be planned and Africans channelled to new residential sites on the peripheries of the major cities. Industry and commerce would be encouraged to move there too in order to provide the employment to sustain these communities. Within each development region, one or more metropolitan regions would be established to provide the framework for third-tier government, comprising fiscal and administrative structures.

The policy of orderly urbanisation set out by the President's Council report, and accepted by government in a White Paper in 1986, is based on five fundamental principles: racial residential segregation; social differentiation within residential areas; industrial and residential deconcentration within and between development regions; privatisation of house ownership; and devolution of the provision of municipal services. In short, orderly urbanisation as conceived up until the end of the 1980s represents a form of decentralised apartheid, in which market forces are to be given play within a framework of policy based on racial residential segregation.

The new labour, urban and regional strategies seek to provide a spatial framework for an economic strategy that would re-establish the conditions for renewal of economic growth by removing territorial apartheid barriers to geographic mobility and releasing competitive forces in the labour and commodity markets. They also seek to provide an economic and spatial foundation for a reformed political system in which Africans are to be drawn into multiracial local, regional and national political structures with attenuated powers (Cobbett et al., 1986). Questions remain, however, about how this strategy will work.

One of the major functions of a modern capitalist state is to provide a framework within which capitalist development may take place. It is the system of production and distribution, impelled by class forces, which fundamentally determines the patterns of racial and regional inequalities, and the growth of unemployment and poverty. State policy responds to these forces, and in its turn can exercise powerful effects on these processes. Thus, displaced urban development and the concentration of impoverished populations in quasi-urban settlements in the bantustans were made possible in the 1970s by the tendency of capitalist development to generate a surplus population. But state intervention oriented to the aims of territorial apartheid has given the geographical allocation of population a particular shape.

It follows that the abolition of apartheid is not enough to resolve the major social and economic problems arising out of rapid African urbanisation, unemployment and the restructuring of the employment process. The abolition of apartheid will undoubtedly ameliorate racial inequalities, and in this respect reinforce market-directed tendencies already present within the labour market for nearly two decades. But unrestrained market-impelled urbanisation may also reinforce social and regional differentiation and inequality, albeit in forms that partially cut across the specific racial–regional matrix of apartheid in its earlier form. Social, racial and regional inequalities in South Africa will not be resolved simply by removing the state from the market. What is needed is the simultaneous transformation of the South African state and economy and the development of urban and regional planning mechanisms which promote the benign while curbing the malignant effects of the market.

11
Unemployment and the current crisis

DAVID LEWIS

There is a growing consensus that the exceptionally high level of unemployment is the most dramatic evidence of the parlous state of the South African economy. Its effects are, to be sure, experienced differently. For the unemployed themselves, and the workers, as well as for the small traders who service them, unemployment is experienced as grinding poverty. For trade union and popular organisation, high levels of unemployment are debilitating and threatening. And whilst the state and capital have an ambiguous relationship to unemployment, the wastage of resources and the potential for social turmoil implicit in the current level of joblessness counterbalance the more pleasing features – the downward pressure on wages and acute demoralisation and division among the underclasses – that generally follow in its train.

There is no widely accepted measure of unemployment in South Africa. The official statistics are notoriously unreliable, particularly because they exclude the bantustans and rely, as for the most part they do, upon registered workseekers, who clearly constitute a very small proportion of those actually unemployed. In the recent past the official statisticians have revised upwards – by some 100 per cent – their figure for total unemployment, but there is widespread consensus that even the January 1988 official estimate of just over a million unemployed vastly understates the severity of the problem.

In what is widely considered to be a soundly based estimate Charles Simkins (1982: 26) put black unemployment at slightly in excess of 2 million in 1981. Using Simkins's figures as a base – and employing some dubious statistical methods – Sarakinsky and Keenan (1986: 21) estimate upper and lower limits for black unemployment in 1986 as 4 164 510 and 5 258 541. The President's Council in its 1987 report entitled A Strategy for Employment Creation and Labour Intensive Development accepts that in 1982 some 3.6 million – 1.7 per cent of the workforce – were 'without formal employment opportunities'. The Institute for Futures Research at the University of Stellenbosch calculates that in 1980 altogether 3 259 000 people – 30 per cent of the economically active workforce – were unemployed. A more recent study from the Institute for Futures Research estimates that, by the turn of the century, at a growth rate of 3.1 per cent per annum, this figure will rise to 9 800 000 – or 54 per cent of the workforce (*Sunday Times*, 4 December 1988).

However dubious the value of 'by the year 2000' projections, such estimates are useful indicators of unemployment, if only because, in this instance, these sources appear to have no objective interest in overstating the magnitude of the problem.

The extent of unemployment is naturally affected by regional peculiarities. Hence it is generally accepted that the Eastern Cape is very severely affected by unemployment, Port Elizabeth being the hardest-hit metropolitan area in the country. A recent study by Vista University concluded that 56 per cent of the economically active African population of Port Elizabeth was unemployed (cited in Labour Research Service, 1988: 8). The rural areas are particularly blighted: Keenan estimates that 50 per cent of the economically active population of Ciskei, Bophuthatswana, Lebowa, KwaZulu and Gazankulu are unemployed (cited in Labour Research Service, 1988: 8). There are no readily available statistics in respect of major resettlement cities like Botshabelo, but it is generally accepted that unemployment would be high in these areas. A survey of the mid-Karoo area around Cradock, Hanover and Graaff-Reinet calculated that between 50 and 60 per cent of the economically active population was unemployed (cited in Labour Research Service, 1988: 9). But even those areas apparently best placed to provide employment are characterised by massive unemployment. Estimates for the Durban metropolitan area put unemployment there at between 17 and 25 per cent of the economically active population (cited in Labour Research Service, 1988: 8); a random survey of households in Alexandra, Johannesburg found that a total of 11.3 per cent of the economically active population was unemployed.

But possibly more pertinent – and more revealing – than the endless debates around the precise numbers of unemployed is the clear evidence of a decline in the number of jobs available. Official statistics confirm that in mining, construction, manufacturing, electricity, transport, and the Post Office the number of workers declined by some 200 000 in the three-year period from June 1984 to June 1987. Half of this decline is accounted for by the construction industry, a sector characteristically subject to severe cyclical swings, but it is noteworthy that the number of jobs in the manufacturing sector declined by 75 800 – or 5 per cent of total employment in the sector – over the same period. Whilst high rates of population growth undoubtedly impact heavily on the level of unemployment, demographic features should not be used, as they so often are, as a convenient camouflage for the economic factors that underlie joblessness. The underlying problem is clearly economic rather than demographic.

With the growing awareness on the part of the state and capital of the magnitude of unemployment, there is some indication of an apparent shift in the way the factors underlying the massive incidence of joblessness are perceived. A few years ago the notion that the South African economy exhibited the characteristic of a structural crisis and, hence, that the persistent mass unemployment derived from this structural malaise, was the exclusive preserve of the left. However, it now seems generally accepted that the remedies appropriate to the South African economic crisis must address the structural roots of the problem. This is certainly the starting point of some recent government publications and is reflected in contemporary analyses presented by leading representatives of capital. The basis for the dawning realisation that structural unemployment exists locally is its emergence in the advanced industrialised countries, what Goran Therborn refers to as 'permanent mass unemployment'. In South Africa this realisation is underpinned by the emergence of structural unemployment amongst the white, coloured and Indian communities. For the most part, official unemployment statistics were designed to monitor the essentially cyclical fluctuations in employment levels in these three communities. The remedies – in particular the Unemployment Insurance Fund, as we shall outline below – all presupposed brief bouts of unemployment conditioned by the downward phase of the business cycle.

When state policy was based on the geographical marginalisation of the unemployed, it was easier, and indeed consistent with state policy, to ignore the deep-seated structural unemployment which plagued the economy but which affected the African population only. When, however, state policy has been obliged to concede the presence of millions of unemployed African workers in the cities, it becomes necessary to build structural unemployment into policy. When, in addition, the problem begins afflicting the coloured, Indian and particularly the white communities, the problem cannot be ignored. Emphasising the distinction between 'cyclical unemployment' and 'structural unemployment' is indeed a necessary starting point. 'Cyclical unemployment' is consequent upon a downturn in the economic cycle and its causes may reside in a variety of factors. The policy responses are essentially symptomatic and are aimed at addressing the particular problem or at ameliorating the consequences of this temporary bout of joblessness. 'Structural unemployment' on the other hand refers to a deep-seated disjuncture between the growth and structure of the national workforce on the one hand, and the growth and structure of economic activity on the other hand. Critically, structural unemployment is unaffected by different phases of the economic cycle.

This chapter will be concerned with identifying responses – on the part of the state and capital as well as the dominated classes – to the question of unemployment. Underlying the argument of this chapter is the notion that there are more and less appropriate policy responses to unemployment and that one cannot, by merely ascribing joblessness to capitalist relations of production, await the palliative of socialist revolution as the means to end magically this era of mass unemployment. Goran Therborn, in a study of unemployment in the OECD countries, argues for a 'labour comeback scenario' that will put employment on the top of the policy agenda. He supports his quest for a 'labour comeback' under capitalism by arguing that 'as long as a large part of the [potential] working class is unemployed and marginalised, no further advances are likely. People on the dole will not bring about socialism' (1986: 36).

There is, moreover, as Therborn further points out, an alternative scenario to 'labour comeback' and the possibilities that its success embodies for a socialist future. This sinister alternative is the 'brazilianisation' of the workforce, where both policy, and the consequent reality, accept that a high proportion of the workforce is permanently marginalised. It is the barbarous political consequences entailed by this scenario that make urgent the necessity for a policy focus on unemployment.

There is a range of available policy responses to unemployment. For our purposes these policy responses fall into one of two categories: firstly, there are those that are 'welfare'-oriented; and secondly, there are responses designed to create new employment opportunities. This chapter will examine each of these responses in turn.

WELFARE RESPONSES

We are talking here of responses that are essentially aimed at ameliorating the consequences of unemployment. These may range from soup kitchens to vast state unemployment-insurance schemes. They may incorporate short-term work relief programmes; they inevitably — and most far-sightedly — incorporate efforts to maintain employment at levels above that dictated by the 'market'. And these palliatives will be pursued by a variety of institutions: the state, the union movement, pension funds, to name but the most common of the social agencies that participate in the welfare structure, which directs part of its resources to combating the effects of unemployment.

At first sight this set of responses has no place in a chapter concerned with structural unemployment. The measures associated with welfare are essentially designed to partially compensate an individual for the loss of livelihood by ensuring that the state or capital or a collective institution shares the burden of joblessness. They are not, in general, measures that intervene in the structure of the economy or have as their objective the creation of new jobs.

There are, however, at least two good reasons why an examination of welfare policy is important. Firstly, full employment — the resolution of the problem of structural unemployment — is far off, and welfare may constitute the only possible source of livelihood for a growing portion of the population for the foreseeable future. Secondly, forcing the state and capital to devote a larger and larger portion of their resources to ameliorating the consequences of unemployment may concentrate their minds (and resources) on tackling the underlying structural causes of the problem. Whilst there are instances where a substantial social security net coexists with high levels of unemployment (Denmark and Holland, for example), it seems generally true to say that those countries where the state and capital carry a large portion of the burden of unemployment are precisely the countries that have weathered the current crisis and successfully maintained high levels of employment (Japan and Sweden are examples at the opposite end of the capitalist spectrum). Again, those countries where the state and capital bear a relatively small part of the unemployment burden are generally characterised by mass unemployment, Britain being the outstanding case in point. The share of the South African state's resources devoted to the unemployed is minuscule. This would include unemployment benefits directly subsidised by the state and schemes to provide temporary employment. Possibly the largest contribution that the South African state makes to combating unemployment is in its effective subsidisation of 'unnecessarily' high levels of employment in the state and para-state sector. It is instructive that this aspect of state economic policy has been the major whipping boy of capital and is an area where the state has exhibited the greatest willingness to see 'reason' – witness the 60 000 (to just under 200 000) decline in SATS's workforce in the late 1980s, or the paring of some 10 per cent of the Eskom workforce over the same period. What is absolutely clear is that the additional 'efficiencies' consequent upon the privatisation of state enterprises would mean a further decline in employment in these sectors.

The state-financed or -subsidised schemes that address themselves directly to an immediate amelioration of unemployment are, firstly, what the Department of Manpower refers to as 'special employment creation' projects, and, secondly, the Unemployment Insurance Fund.

Direct state subsidisation of employment in the shape of special employment creation projects is a relatively recent phenomenon. It was recommended after a lengthy investigation of unemployment conducted by the National Manpower Commission and the Economic Advisory Council. Whilst the 1984 White Paper, A Strategy for the Creation of Employment Opportunities in South Africa, accepted these recommendations, its acceptance and the ultimate implementation of the state-financed employment creation were heavily hedged by the notion that employment-creation is the responsibility of the 'private sector'.

It is instructive to note that the fiscal allocation to 'special employment creation' is coupled with 'drought relief' – both presumably residing in the natural order of things. The coupling also makes gathering statistics very difficult although there is a healthy basis for assuming that the lion's share of the allocation goes to the farmers in the form of drought relief, rather than to the workers in the form of employment creation. The allocation to 'special employment creation and drought relief' also includes a portion of the allocation to training, which properly falls into the longer-term projects related to restructuring.

With all those qualifications in mind, then, in the 1984/5 financial year R27.5 million was allocated to drought relief and special employment creation. In 1985 this allocation was increased to R100 million, of which R25 million was earmarked for the training of unemployed workers in basic skills.

In September a further allocation of R500 million was allocated for training and job creation. The 1986/7 budget allocated R160 million and this was supplemented by a further R50 million for job creation and training. The 1987/8 budget allocated a further R100 million for job creation. This brought the total amount allocated in the financial years 1985/6–1987/8 to R1 060 million. Whilst it is difficult to isolate the portion spent on employment creation, the Department of Manpower estimates that at any one time during the period June 1985–June 1987 some 250 000 otherwise unemployed persons were working on projects under the special employment creation programme.

At best these figures do, on the face of it, indicate a growing concern with unemployment. However, the sums are still minuscule – the R1 000 million allocated over three years remains an insignificant portion of state expenditure during the period. In addition the figures are budgetary allocations only – they are not amounts actually spent – and a large proportion of the allocation has not been spent.

It must be emphasised, though, that this expenditure on employment creation does not reflect a change in the state's conception of its role in the economy generally or in the creation of employment in particular. And it is this that has ensured that state-financed projects are of no lasting significance. The two underlying principles of the special employment creation projects illustrate this: firstly, projects are designed to be temporary and capable of immediate curtailment; secondly, the state stipulates that these projects must not show an economic return, thus further underlining their temporary nature. It is little wonder, then, that the projects tend to range from the sublime to the downright sinister – from trimming weed on the Durban sidewalks to training 'kitskonstabels'.

The schemes most closely associated with state welfare activity are old-age pensions and unemployment insurance. It is common cause that the benefits available are grossly inadequate. In brief, the beneficiary of unemployment insurance may receive from UIF 45 per cent of his or her income. The period for which this relief is available depends upon the length of time for which the beneficiary has contributed to the fund, but never exceeds six months.

These then are the two key principles underlying UIF. Firstly, it is available to tide over short bouts of unemployment. This should be read against the background of micro-studies that reveal exceptionally long bouts of unemployment. A study undertaken in the Pilanesberg area of Bophuthatswana calculated the mean average time spent between periods of employment was five years and seven months (cited in Sarakinsky and Keenan, 1986: 23). On average, then, even those who received full benefits from the fund would have been without any benefits for at least five years. There are, moreover, penalty clauses that ensure that the full benefits are only available to those who have been retrenched – resignation or dismissal are penalised through the fund.

Secondly, only contributors qualify for benefits from the fund. This excludes all workers in the public service, in agriculture, in domestic service, as well as seasonal workers. Above all it excludes those who have never been employed. Whilst there are few conclusive statistics available, it it clear that the majority of the unemployed population have never been contributors to the fund. The age structure of the population alone – with some 45.5 per cent of the population below the age of 20 – would ensure that the majority of the unemployed population are new, or very recent, entrants to the labour market. The Department of Manpower estimates that over half of unemployed coloureds and Asians and over 40 per cent of unemployed Africans are below the age of 24 (cited in Van der Merwe, 1987: 6–8). In addition, a significant number of the remainder (those currently unemployed but previously employed) would certainly have been employed in one of the excluded sectors, agriculture being the most important case in point.

Table 1. Unemployment Insurance Fund

	1984	1985	1986
No. of applications received	204 982	332 157	345 766
No. approved without penalty	126 783	225 413	221 523
No. approved with penalty	59 342	79 492	87 998
No. refused	18 857	28 567	36 045
Percentage of total contributors receiving benefits	4.38%	8.50%	7.52%
Amount paid in UIF benefits	R105m	R219m	R271m

Source: UIF Reports 1985 & 1986.

With an unemployed population conservatively estimated to exceed 2 000 000, little over 10 per cent are recipients of unemployment benefits. The President's Council report on A Strategy for Employment Creation and Labour Intensive Development clearly spells out the political and ideological constraints within which the South African social security net operates: 'The generation of income through the productive contributions of a limited section of the population coupled with a comprehensive system of welfare payments to a large non-working proportion of the work force is, in the view of the Committee, not in the interests of the long term stability and prosperity of the community.' At most, the UIF is intended to deal with cyclical unemployment, and this only very half-heartedly. It is, above all, not intended to act as a safety net for those in long-term unemployment. The solution for that portion of the unemployed lies, firstly, in longer-term labour market and national economic restructuring; and, secondly, in the informal sector, a concept which functions increasingly as a metaphor for a policy that has accepted the idea of permanently consigning a large proportion of the population to the outer periphery of the economy.

There are certain non-state resources available to ameliorate cyclical unemployment. The greatest potential source of assistance here lies in the mobilisation of the large pension funds. The unions have had limited success in the deployment of pension funds as a short-term cushion against unemployment. Hence, whereas until 1984 members of the Metal Industries Group Pension Fund could not, on ceasing to be contributors to the fund, withdraw their share of the fund until they had left the industry for six months, a union-inspired change in the rules of the fund enables retrenched metalworkers to withdraw their share after six weeks. Another change in the rules enables retrenchees with more than five years' membership to withdraw management's contribution, in addition to the worker's own contribution. There are also several union-administered benefit funds that provide a small cushion against unemployment.

In general, though, assistance from these non-governmental sources is highly limited. Their resources are generally limited and they inevitably address themselves to a very narrowly defined segment of the unemployed population – pension fund members, union members, etc. This is not to say that there are not considerable funds available in the private sector, but the marshalling of these resources and their direction need to be undertaken by the state. Aided by the state there is considerable potential, not only for appropriating part of these resources and, through the fiscus, directing them to the unemployed, but also for directing the investment practices of some of the large institutional investors, most notably the pension funds. The Metal and Allied Workers' Union (now Numsa) provided a glimpse of this potential when it successfully pressurised the Metal Industries Group Pension Fund to invest in housing construction in the Brits area, an investment which certainly produced a low rate of return to the fund, but which not only provides a socially necessary commodity, but is also a labour-intensive economic activity in an area noteworthy for the high level of factory closures.

Unquestionably, though, the most wide-ranging attack upon cyclical unemployment has come as an attempt not to cater for the unemployed, but rather to defend job security against the unilateral decisions of profit-maximising managers. These efforts are, predictably, spearheaded by the unions and take the form of a campaign for 'fair' retrenchment procedures. These campaigns are documented elsewhere and will not be detailed here. Suffice to say, however, that whilst substantial gains have been made in the realm of procedural 'rights' and have effectively challenged a hitherto uncontested managerial prerogative in the area of hiring and firing, it is doubtful whether many jobs have been saved in consequence of these retrenchment campaigns. It is indicative of government attitudes to unemployment that the recent amendments to the Labour Relations Act seek to limit union gains in the area of retrenchment.

Most significantly, however, the advent of large-scale retrenchment has focused attention upon the search for alternatives, as part of the now standard 'fair' retrenchment procedure. The alternatives explored have ranged from the short term and particular – for example, limitations on overtime, unpaid leave – through to general efforts aimed at according employment a higher place on managerial agendas by, for example, reducing the working week. There have been successes in devising short-term alternatives to retrenchment, although for the most part these involve introducing measures that share the burden amongst the workers themselves. The efforts at effecting longer-term solutions of the problem have been markedly less successful. In truth, what has emerged is that, in part, the unions have not constructed particularly creative alternatives. But above all, it is clear that the problem of job loss and, more so, unemployment in general is not soluble at the bargaining table. Rather it lies in the realm of economic policy-making.

EMPLOYMENT CREATION

Goran Therborn's *Why Some Peoples Are More Unemployed than Others* – an examination of unemployment in sixteen advanced capitalist countries – takes as its point of departure the fact that national employment levels have responded to the international economic crisis in highly specific ways. On the one hand, certain of the countries examined in Therborn's study – Sweden, Norway, Japan, Switzerland and Austria – have maintained consistently low levels of unemployment. At the other extreme, certain of the countries – Britain, Holland, Belgium, Canada and Denmark – have become, and are likely to remain, countries of mass unemployment.

The conclusions of Therborn's study are controversial and worth recounting at some length. He concludes that the significant differences in unemployment in these countries cannot be accounted for by differences in overall economic growth and labour supply. Nor is world market dependency and a highly internationalised economic structure – what Therborn refers to as 'crisis exposure and vulnerability' – the culprit. Thirdly, he discerns no significant relationship between unemployment and inflation, or between unemployment and labour costs. Finally, he concludes that wage restraint does not of itself lead to full employment or to successful international competitiveness – that, in short, incomes policies are the creatures of currency policy and exchange-rate fluctuations.

Therborn attributes the differential rates of unemployment in these countries to 'the existence or non-existence of an institutionalised commitment to full employment'. The essential ingredients of this 'institutionalised commitment' are: '(a) an explicit commitment to maintaining/achieving full employment; (b) the existence and use of counter-cyclical mechanisms and policies; (c) the existence and use of specific mechanisms to adjust supply and demand in the labour market to the goal of full employment; (d) a conscious decision not to use high unemployment as a means to secure other policy objectives' (1986: 23).

The conclusions of this study cannot be uncritically extended to South Africa. Above all, Therborn's study examines the economies of the leading capitalist nations, and South Africa occupies a significantly lower rung on the international economic ladder. But many of Therborn's conclusions remain pertinent. In any event, it is possible to incorporate into our analysis

South Africa's particular place in the international economy and this does not presuppose accepting the convenient generalisation so beloved of South African policy-makers, namely the fiction that South Africa is a Third World nation and that unemployment is inextricably associated with a relatively low level stage of development. The converse of this notion is the argument that attaining a higher level of development – 'economic growth' by another name – is the cure-all for unemployment.

Current state strategy

We are examining here current state policy and strategy with respect to unemployment. This policy is located within, and powerfully influenced by, an era of large-scale urbanisation or, what is possibly more pertinent, an era in which the state has accepted the existence of a massive black population in the urban areas. This has a particular impact on state policy towards unemployment, although in one crucial area, this chapter will argue, official policy has remained fairly consistent since the early 1960s. 'Brazilianisation', or the permanent marginalisation of a large proportion of the economically active population, has its clearest expression in the 'separate development' era of apartheid rule. However, it remains at the base of current policy as well. At the risk of overdrawing the analogies – or, indeed, of exaggerating the developments in apartheid strategy – the bantustans of yesteryear are to employment policy of the 1960s what the squatter camps and housing estates are to current strategy. The Dimbazas are to employment in the 1960s, what small business parks in Alexandra may be in the 1990s. At base, in both periods, state policy is predicated upon the continued existence of very high levels of unemployment. The policy measures directed at this 'problem' are designed, firstly, to isolate the unemployed mass politically, ideologically and geographically; and secondly, to articulate the mass of unemployed, on the one hand, with the employed working class on the other hand, so as to lower the reproduction costs of the working class and in this way to address some of capital's general accumulation problems.

A paper presented to the September 1987 Toyota Conference on Unemployment by P. J. van der Merwe, the director general of Manpower, embodies all the stereotypes associated with current government thinking about unemployment. According to this view, the general context of the problem is that 'unemployment is obviously not a problem which is unique to South Africa. It is a world-wide problem facing all countries – particularly the developing ones' (1987: 5). Within this general context certain characteristic constraints operate. In the order presented by Van der Merwe, these are high rates of population growth; an accelerating rate of urbanisation; large numbers of legal and illegal workers from neighbouring states; the youthful age structure of the population; low level of education; and, finally, 'structural imbalances'.

There is also a characteristic set of policy responses. Firstly, the government's view – as contained in a 1984 White Paper on A Strategy for the

Creation of Employment Opportunities and dutifully re-stated by Van der Merwe – is that the state bears no direct responsibility for employment creation. This duty resides 'in an economic system based on private initiative and effective competition', with the state's role effectively limited to an 'elimination of measures that inhibit the satisfactory operation of a market-orientated system'. This 'free-market' strategy has come to be associated with a hodge-podge of state initiatives. Whilst the most tangible of these initiatives is a limited education and training or re-training programme, the most revealing are those that have become the public buzzwords in the employment creation strategy. These are 'the development of the informal sector'; 'small business development'; and 'privatisation and deregulation'.

This is consistent with the second pillar of the state's position, namely that 'there is no instant solution to the unemployment problem ... and that the promotion of economic growth remained the single most important method of employment creation.' Alternative growth strategies are examined in the White Paper and the one best geared to South Africa's high unemployment is identified as 'inward industrialisation', the success of which depends upon 'the elimination of measures that inhibit the satisfactory operation of a market-orientated system'. Whilst export-led growth and import substitution are dismissed as appropriate employment-creation strategies, they clearly remain key elements of economic policy. The overall position is summarised by Van der Merwe: 'The Government's long-term strategy for employment creation revolves around the pursuit of the highest possible, but employment orientated, average economic growth rate that is reconcilable with other national objectives such as equilibrium on the balance of payments and the curbing of inflation.'

Inward industrialisation

'Inward industrialisation' is used here – as indeed it has been used by the state and capital – as a catch-all term for a number of policy initiatives that incorporate both demand and supply side components. It is an approach strongly supported by big capital. In fact, the first real shot in the 'inward industrialisation' campaign was fired by the influential Mercabank *Focus on Key Economic Issues*. In the May 1985 issue entitled 'Industrialisation and Growth', the report questioned the possibility of exports or import substitution providing the engine of growth, largely because of the country's need to generate employment. 'Orderly urbanisation' which will lead to a process of 'inward industrialisation' was identified as the appropriate growth path for South Africa. The Old Mutual has clearly defined inward industrialisation as 'domestically generated growth based upon supplying basic consumer products (e.g. clothing, shoes, furniture, basic foodstuffs) and facilities (e.g. low cost housing) to the rapidly urbanising black population, with the increasing labour force coming from the rural areas simultaneously finding employment in these expanding industries' (1987: 15).

Inward industrialisation received the official seal of approval in the report of the President's Council's Committee for Economic Affairs on a Strategy for Employment Creation and Labour Intensive Development (PC1/1987). Its major policy proposals are contained in the following two observations: 'In a developing country capital-intensive and skill-intensive economic development cannot be reconciled with a rapidly growing work force of unskilled workers.' Secondly, 'structural employment can also originate if the occupational and geographical mobility of labour is restricted'. The latter is dealt with in the President's Council report on urbanisation. The former – the problem of 'capital deepening' in a 'Third World' economy – is the subject matter of the 1987 report. Many of the solutions proposed are premissed upon the 'urbanisation' of South Africa's black population, and the orderly relaxation of controls over geographical mobility. The report identifies the low levels of growth as the most severe contributor to the high levels of unemployment. It then isolates three possible growth paths: firstly, the development of export industry; secondly, import substitution; and finally, 'inward industrialisation'. Whilst the report recognises the importance of developing exports and of import substitution, they are ultimately rejected because of their failure to take sufficient account (or advantage) of the Third World characteristics of the South African economy and society. There are, *inter alia*, a shortage of skilled and a surplus of unskilled, poorly educated labour; secondly, a high rate of population growth; and thirdly, high levels of social tension and 'unrest'. Whilst 'inward industrialisation' and its component parts form part of a general approach to the economy – and critically represent a response to low growth levels – a professedly key reason for favouring it above other potential growth paths is to be found in its employment-creation potential.

Inward industrialisation is then based upon employment of the 'urbanising' black population and the market that this provides. The latter is severely constrained – although the committee does not mention this consideration as one of the structural underpinnings of unemployment – by the low earnings of the black population. There has accordingly to be a 'kickstart', which generates employment and, consequently, disposable income. The solution here is found in the large-scale provision of low-cost housing, and emphatically not the raising of the general wage level.

The supply side component of inward industrialisation has become associated with a wide-ranging – and not always compatible – series of potential policy measures: selling off of state assets; removing regulations that inhibit private sector development across the spectrum; and directly encouraging small business and the informal sector. The Mercabank report warns that 'the success of inward industrialisation depends largely upon the lowering of capital intensity in production'.

In the view of the Committee for Economic Affairs, the two major obstacles in the path of successful inward industrialisation are located on the supply side: firstly, 'labour as a production factor should be made competitive in terms of cost, especially as compared with capital goods' (the

Committee adds that 'the structure of the labour market should be adapted in such a way that it meets the requirements of the economy') and secondly, excessive regulation of the private sector. These two features – 'high wages' and 'excessive regulation' – uncharacteristically ape developed economies, and are inconsistent with the otherwise Third World character of South Africa. Whilst these two features seriously inhibit employment-generating growth, they present particularly severe constraints on the development of 'small industry', which is in the Committee's view a major potential benefactor of 'inward industrialisation' and, hence, generator of employment.

Whilst the Committee does recommend several projects aimed at raising the skill level of the workforce, it is emphatically clear when it comes to lay the major blame for capital deepening: 'In the course of the investigation it was repeatedly pointed out to the committee how important a role excessive wage demands and wage increases and strikes by Black workers – especially strikes instigated by unions – played and were still playing in the process of capital deepening.' Accordingly the Committee 'recommends that an investigation be launched into these activities and specifically into the activities of the unregistered trade unions, general trade unions and trade union federations'.

Dealing with these developments presupposes a 'sustained national awareness campaign [which] should communicate the message to all workers and potential workers that "labour ennobles" and that productive employment creation, whether it be at a high or low level, and whether it be of a permanent or temporary nature, is socially desirable and a source of progress. The campaign must also widely publicise the factors that have a bearing on employment creation so that irresponsible action in the labour field can be combated with knowledge.'

Small business

An important component of the 'inward industrialisation' concept is the development of the small business sector, the success of which is based upon deregulation. An earlier report of the President's Council (1985) identified the 'high cost of formal sector operation' as the major obstacle to small business development. Included in the nine general instances cited are workmen's compensation and unemployment insurance; the cost of satisfying the factories, health, and safety and fire protection inspectors; industrial council regulations; and minimum wage stipulations. The last-mentioned are singled out for particular attention because 'small businesses have problems in paying the minimum wages determined'. The report recommends that the National Manpower Commission coordinate an investigation with a view to the immediate amendment of certain legislation. This includes the Machinery and Occupational Safety Act, particular areas of which (such as 'the acquisition of machinery not complying with the necessary safety provisions') are not deemed 'appropriate to the circumstances of the small industrialist'; the Basic Conditions of Employment Act; the Workmen's Compensation Act;

the Unemployment Insurance Act; and the Health Act.

Big capital has in general welcomed deregulation. Early on in the campaign for inward industrialisation, Zach de Beer called, in the name of employment creation, for the abolition of minimum wage regulation. The Old Mutual has also welcomed the 'inward industrialisation' growth path and its deregulation component, and believes that 'acceptance of "Third World" standards in many areas of South African life is implicit in this policy if developed fully'. Assocom (Association of Chambers of Commerce) sees the President's Council report as 'a major watershed in thinking about the unemployment problem'. To some extent big capital has been willing to devote certain of its resources to supporting the small business project. This has largely been done through the Urban Foundation, the Small Business Development Corporation and other such institutions.

However, although capital has publicly supported deregulation and the small business project, it is also clear that its endorsement of this growth path will remain highly conditional. Capital will support the relaxation of regulations purportedly designed to encourage small business in those instances where it is possible to differentiate the small business sector rigidly from the mainstream sectors of capital, and designate a certain area of economic activity as the terrain of small business. In particular, capital will encourage the development of small business in areas directly complementary to big capital. A recent feature of employment creation in the USA – which was substantial in the Reagan years – is the extent to which jobs have been created in small, low-wage concerns frequently performing specialised, ancillary tasks under contract to a larger firm. This type of employment is generally poorly paid, is frequently part-time or casual, and is extremely vulnerable to an economic downturn or change in the fortunes of a particular sector or even a single firm. At the same time, however, it is possible that this type of sub-contracting arrangement, with the relative degree of stability it permits, is a condition for the successful generation of a small business sector. Needless to say, sub-contractors need to provide a highly reliable service and will often be found at the larger, possibly more capital-intensive end of small business. This is particularly true in those instances where the small sub-contractor is responsible for producing a commodity for, as opposed to delivering a service to, a larger client.

Naturally at the point at which encouraging small business poses a threat to larger capitals, the latter will resist. Lifting restrictions in such a way that permit too great an ease of entry into established areas of domination by big capital will be resisted by these champions of small business – witness big capital's steadfast opposition to the tentative deregulation of the trucking industry. The furore surrounding the road transport industry, the only significant instance of deregulation to date, seems to indicate support for deregulating running costs – wages for example – but not for deregulating capital costs. It is the latter that largely determine entry into industry. Accordingly big capital in the trucking industry has not attempted to counter the entry of these new competitors into the industry by raising the wage levels of the drivers but has rather attempted to persuade the state to tighten up regulations governing the 'quality' of the trucks – or capital – employed, thereby raising entry costs.

There is thus an economic basis underlying capital's support for the development of an ordered, small business sector, especially of the sub-contracting variety. However, as capital and, to a large extent, the state, clearly perceive, the encouragement of small business has more to do with class stratification than employment generation. As the Old Mutual points out, drawing blacks into the 'process of wealth creation' 'can make a massive contribution to a return to greater socio-political stability'. Capital's support for small business is generally accompanied by homilies about the need to spread the benefits of free enterprise, the 'spread' referred to being a greater spread of owners rather than workers. As yet another chapter in the 'black middle class' saga, 'inward industrialisation' has possibilities. It is, however, difficult to treat seriously its employment-generating capacity.

Even on its own terms, small business does not offer much hope for significant employment generation. Small business is not necessarily the capital-saving and labour-intensive project that it is so readily assumed to be. More often than not, small business involves supplying a very localised or specialised product or service, with the enterprise being crucially dependent upon one or other skill or attribute of the entrepreneur rather than upon the labour of many poorly paid workers. What is more, small business accounts for a tiny proportion of total employment in South Africa. Whilst that might point to the possibilities for the rapid expansion of the sector, even a doubling of the size of the sector would have little impact upon total employment.

There is in addition little concrete evidence that substantial resources are being allocated to small business. Booth argues that the institutions designed to underpin small business financially are inevitably geared to assisting the upper, and larger, echelons of small business. In the 1984/5 financial year the Small Business Development Corporation allocated R1.4 million to loans of R2 000 or less. R2.7 million was allocated to loans of up to R30 000, and R22 million was allocated to loans of up to R300 000. Nattrass, who championed the employment-creating potential of small business, claimed that the cost of creating a single job in an enterprise of capital size R10 001–R100 000 is five times greater than in an enterprise of a capital size smaller than R10 000 (1984: 4). Clearly, by Nattrass's criteria the SBDC is not really directing itself at the employment-creating end of small business. Moreover, the extremely high proportion of bad debts incurred by the SBDC would dictate a future orientation increasingly closer to the larger end of small business.

The informal sector

Reliance upon the informal sector as a generator of employment is even more difficult to support, despite the *Financial Mail*'s ludicrous claim that 'if it were easy and legal to run small-scale business in SA – brick-making, taxi driving, carpet weaving, flower selling, house building – the swamp of unemployed would dry up' (*FM*, 'Enriching the Masses', 11 December 1987). Booth (n.d.: 17–19) points out that the faith of policy-makers in intervention in the informal sector is based on three myths. The first assumes that participants in the informal sector are 'burgeoning dynamic entrepreneurs' whereas the majority view their participation in the sector as a 'temporary survival strategy' dictated by poverty rather than entrepreneurial ambitions. The second myth identifies the 'typical' informal sector participant as the hero of a success story which the timeous intervention of a state or financial institution can raise to new heights. The norm is rather a subsistence activity, with far fewer 'rags to riches' cases than the proponents of the informal sector would have us believe.

Thirdly – and most pertinently for our purposes – there is the employment-creating myth, the notion that the informal sector represents a particular case of the small business sector characterised by high labour-intensity. The flowering of this myth is represented by arguments that support the 'formalisation' of the most successful of these informal sector activities and the encouragement of the further development of the informal sector, both of which are to be achieved by a minimal allocation of state and private sector resources and principally by deregulation. There is little evidence to support the employment-creating myth. Indeed, Booth refers to a study by Nattrass and Glass which found that 63 per cent of informal sector enterprises in Durban provided employment to the proprietor only (Booth, n.d.: 19).

However, whilst the employment-creation potential of the informal sector may be myth, capital's general support for the sector does have a material basis. It may in fact be more helpful to view the informal sector in terms of the reproduction costs of the working class rather than as a creator of employment. The informal sector may be able to meet the basic needs of an extremely impoverished market – a market that has little access to transport, storage, or electrical appliances – more effectively and less expensively than big capital.

Support for the development of the informal sector translates, above all, into support for deregulation on a fairly wide front – relaxation of trading-licence regulations; relaxation of building regulations; and the elimination of regulations that essentially protect labour. Capital's support for the elimination of major apartheid legislation, especially the Group Areas Act, though essentially political in its trajectory, contains elements of this generalised quest for deregulation. Underlying all this is capital's major concern with the wage level.

As mentioned above, the President's Council report on A Strategy for Employment Creation emphatically lays the blame for high unemployment levels at the door of high wages, which are, in turn, held to be the responsibility of the trade-union movement. This is a view supported by capital and finds its expression in the recently amended Labour Relations Act, which is not primarily an attack upon the political activities of the unions, but attempts, as it were, to beard the unions on the factory floor.

It is tempting to dismiss this concern with 'high wages' as part of the generalised ideology of capital. It may, however, have a more specific reference. One can argue in fact that the wages of black workers in South Africa are indeed high in relation to many of South Africa's actual or potential competitors. Here the pertinent comparisons are with the newly industrialising Asian countries and also with certain Latin American countries. The union movement may, indeed, be held partly 'responsible'. But surely, then, massive unemployment and the rapid influx into the cities of an unemployed mass previously confined to the bantustans should act as a devastating counterweight to the power of the unions? Well they might, but, whilst clipping the wings of the union movement may advantage capital and the state in a variety of important ways, it will not, of itself, sufficiently depress the general level of wages. This is the accumulation problem of South African capital which unemployment throws into such sharp relief.

The wage-level floor is constituted by the very high costs of reproduction. In short, whilst wages may be relatively high, the black working class is no less, and may indeed be more, impoverished than its counterpart in Asia and Latin America. Housing costs, transport costs, and food prices form the basis of impoverishment and of 'high' wages. Attacking the trade-union movement is thus only one part (and, as the fight over the Labour Relations Act well illustrates, may be the most difficult part) of the answer to capital's 'high' wage problem. The other part is lowering the reproduction costs of labour. Here one can find the material basis for 'black taxis', for new initiatives in housing, for relaxation of trading regulations, in short for capital's support for the informal sector and small business.

These economic considerations notwithstanding, it is clear that, to a substantial degree, the encouragement of small business and the development of the informal sector are politically rooted policies designed to encourage faith in 'free enterprise'. However, neither the economic nor broader political considerations that underlie small business and the informal sector have much to do with employment creation, despite the form within which they are frequently presented. Where they contribute to economic growth, they do so by a direct or indirect contribution to big capital – either by providing the latter with a low-cost specialised service or commodity, or by holding down the reproduction costs of those employed in the large corporations that remain the prime engine of economic growth. Moreover, it is clear that, to the extent that development of small business and the informal sector underpins economic growth, this is achieved by lowering working-class living standards and working conditions. Whatever the possible validity of any claims made about employment creation, 'inward industrialisation' will be opposed by organised workers.

It is this limited ability of the state's particular conception of 'inward industrialisation' to underpin growth and employment – as well as the power of big capital – that accounts for the persistent reminder from the state that this path will be subordinated to balance of payments considerations and to the need to fight inflation, that export-led growth remains at the top of the economic policy agenda.

In reality, little has changed from a policy point of view, especially in so far as employment creation is concerned. 'Inward industrialisation' and its various reflexes are not an attempt to address mass unemployment. It has clear economic content but not directly as a large-scale employer of labour, rather as an important complement to big capital.

Current economic policy does, however, carry with it major political implications. Seeking class support within the black population has been a persistent feature of the 'reform' era. It is rooted in the notion of the black middle class, and though reform has entered other realms – working-class organisation, housing and township development, etc. – the state and capital have always been most comfortable with the idea of facilitating this 'middle class'. 'Small business' and the 'informal sector' represent the material – and hence the most significant – face of this quest for class allies, and carry the additional advantage that they may be presented as strategies directed at generating employment.

ALTERNATIVE PATHS

The economic consequences of unemployment speak for themselves. Suffice to say that it is unemployment rather than the wage–price level that principally accounts for the acute poverty which characterises South African society. The political consequences of unemployment are equally severe. Increasingly, what the vigilantes and the young comrades have in common is lack of employment. Whilst unemployment may not, contrary to the conveniently held view of the state and capital, provide the powder for socialist revolution, it creates fertile ground for despair and anarchy.

The implications for union organisation are severe and direct. They are faced with the task of defending existing jobs, and maintaining wage levels, let alone increasing them, in the face of large-scale unemployment. This task is potentially difficult and exacerbated by an economic policy that attributes a central role to union gains in generating unemployment. However, these problems – the difficulties of improving wages and working conditions against the background of mass unemployment – may pale into insignificance when contrasted with the major political problems that massive levels of joblessness imply for the unions and the broader liberation movement.

In the longer term there exists the real possibility of the unions occupying a difficult position in an increasingly trichotomous society – representing the 'privileged' minority in employment, in a society characterised by permanent mass unemployment. Two interrelated developments exacerbate this tendency. Firstly there are the increasing opportunities available to an identifiable grouping of black workers. These opportunities arise for diverse reasons: the changing structure of the economy and the rapid development of a service sector in particular; shortages of skilled labour; and, in a small but growing number of instances, affirmative-action-type programmes that seek specifically to promote black workers in the job hierarchy. Clearly, not all black workers will benefit from this upward mobility. For the likes of construction workers, the pressure on their earnings will be downward, partly in consequence of the drive towards deregulation. Low wages and poor working conditions will increasingly distinguish the latter from their upwardly mobile comrades. Recent research into the upgrading of Mamelodi, a prime site of state intervention, reveals a startling difference in the relative fortunes of the upper social strata of the township and the employed lower strata. There is little doubt, moreover, that detailed research would confirm the impression that worker leadership of the union movement is increasingly drawn from the higher echelons of the black working class. But the mere fact of employment will distinguish those on the lower rungs of employment from the permanently marginalised mass.

Secondly, certain social fringe benefits are more and more tied to employment and contain within them the seeds of social stratification on a previously unthought-of scale, deepening divisions not only between the various echelons of employment, but also between those in employment and those without work. Housing is the key benefit here. With the growing intervention of capital in black housing – both at the financing and at the construction ends – the provision of housing and its relative quality will become increasingly tied to employment.

It is appreciation of what this already existent trichotomy implies that nurtures the hopes of Conor Cruise O'Brien (*The Atlantic Monthly*, March 1986) for an eventual alliance between society's 'ins' – above all, capital and the unions – defending their interests against the economically distant 'outs'. What above all impedes this – the politics of 'brazilianisation' – is, in O'Brien's view, an overly cautious attitude on the part of the state to reform. Repeal of the Group Areas Act, for example, would allow a 'natural' geographical distance to accompany the economic gap already evident between the working class and the unemployed.

Despite the severity of the problem, responses to unemployment by representatives of the dominated classes and progressive intellectuals are sporadic and unsystematic. It has largely fallen to the unions to generate, often inadvertently, alternative positions and to take action aimed at alleviating unemployment.

At the 1985 inaugural congress of Cosatu, the resolution on unemployment was premissed on the assertion of the 'right to work'. Its programme of action against unemployment consists of eight elements, namely to fight retrenchments, factory closures, and for the right to participate in decisions concerning the implementation of new technology; to campaign for a 40-hour week; to fight for free and increased unemployment benefits; to fight for rent, transport and medical concessions for the unemployed; to demand a public works programme that will provide jobs and improve services and facilities in working-class communities; to fight for work-sharing on full pay when threatened by retrenchments; to establish a national unemployed workers' union as a full affiliate of Cosatu; and, finally, to 'struggle for a fair, democratic and rational political and economic system which can guarantee full employment for all people in Southern Africa at a living wage'.

The Secretariat's report to the 1987 second congress noted growing unemployment and the extent to which the unemployed were being used to break strikes and as recruits for 'vigilante and other anti-democratic forces'. These considerations underlay the need 'to cement unity in action between employed and unemployed workers, manifest in the presence at the congress of observers from the National Unemployed Workers' Co-ordinating Committee, which had been set up by Cosatu.

As outlined above, much of Cosatu's activity with respect to unemployment has concentrated upon the organisation of the unemployed, whilst affiliates have focused upon defending existing jobs.

It is difficult to assess the success, or otherwise, of the efforts to organise unemployed workers. It is a path fraught with difficulty, and its major possibilities for success would appear to lie in exerting pressure for improved welfare benefits. It appears, however, that the major thrust behind the organisation of unemployed workers is the attempt to prevent scabbing. Whilst this may be successful in small, tightly knit communities, it will, under most circumstances, be of little significance.

In the process of fighting retrenchments some of the affiliates have raised certain demands – infrequently realised – that are intended to increase the total number of jobs. Chief among these is the demand to curtail overtime. More self-consciously directed to creating, as opposed to saving, jobs is the demand for a reduction in working hours. This demand forms part of the Cosatu 'living wage' campaign and is explicitly rationalised as an attempt to create employment. There are unquestionably many sound reasons for reducing the working day but increasing total employment is not one of them. Therborn has found in his comparative study that it is a fairly common response amongst unions in Europe but that, even when successful, it contributes nothing to total levels of employment. On the contrary, Therborn finds that a relatively short working day may accompany mass unemployment and may be the peculiarly European form of the 'brazilianisation' referred to elsewhere.

This is course not always the case. Sweden is a clear example of high employment and short working hours, although there is no evidence that connects the two phenomena. As with state policy, the only chance of the unions mounting a successful attack on unemployment will come from a direct hit, as it were, rather than an oblique attempt to intervene in unemployment by manipulating a secondary variable.

Direct intervention in the spiral of unemployment will come in three forms: firstly, as an effort to alleviate the devastating poverty that follows in its train; secondly, by direct participation in the creation of jobs in specific communities or sectors; and thirdly, in the attempt to articulate and support organisationally a macro-economic policy that accords priority to employment creation.

There is a remarkable quiescence – from the unions as well as from other organisations of the dominated classes – with respect to the whole question of welfare. Remarkable, firstly, because welfare payments are a means of alleviating gross poverty. Indeed, they are clearly the only means of alleviating poverty in the short-to-medium term. And remarkable because, given South Africa's particularly low levels of welfare and the relatively 'apolitical' nature of a welfare-oriented campaign, substantial gains can be won. Innes refers to a study which demonstrates that whilst 40 per cent and 16 per cent of government expenditure in the industrialised countries and the developing countries respectively go on social security, the equivalent figure in South Africa is only 8 per cent (1987: 12–13).

The second aspect of a general approach to unemployment is to be found in attempts to create employment in blighted areas. This has found expression in efforts to pressurise the large pension funds to invest in the production of socially necessary commodities in specific areas. Numsa's attempt – successful it appears – to force the Metal Industries Group Pension Fund to invest in housing developments in the Brits area is a case in point. A second form in which this direction appears is in the current interest in production co-ops.

The unions are then left with their resolve to 'fight for a fair, democratic and rational political and economic system which can guarantee full employment ... at a living wage'. The Cosatu resolution lays the blame for high levels of unemployment at the door of 'capitalist conditions of exploitation' manifest in the 'introduction of new technologies for profiteering purposes'. Neither is a particularly helpful pointer. The one serves as a vague vision of the future, whereas there are current policy positions at stake. The other repeats a Luddite view of technological progress shared by both sides of the political equation. Whilst there is space for discussion of 'appropriate technologies' (or indeed 'small business'), it does not constitute a response to mass unemployment.

Cosatu's clearest policy response at present to unemployment is probably contained in its 'living wage' campaign. Implicit in the demand for a living wage – especially when contrasted with a view of economic growth and employment creation that relies upon, *inter alia*, the further weakening of minimum wage regulations – is an eminently respectable notion which asserts that an important cause of South Africa's economic stagnation is the limited size of its internal market. Erwin has in fact suggested that the growth and dispersal of greater spending power that would accrue from a rapid increase of wages may be more encouraging of 'small business' than the alternative policy of reducing wages as a cost-cutting device (1986: 33). Kaplan, in arguing for a high wage thrust by the unions, has also pointed to the stimulatory potential implicit in a rapid increase of wages (1986: 42–43). Much the same may be argued in favour of deficit budgeting. In fact the

President's Council, in arguing for a 'kickstart' or injection of income – in the form of mass housing developments – so as to generate the market required for successful 'inward industrialisation', has accepted the Keynesian notions inherent in the 'living wage' campaign. Goran Therborn's study supports this view: 'With regard to policies, our findings show that countries successful in the sphere of employment have all pursued expansive Keynesian-type policies' (1986: 26–27).

But there is more to economic stimulation than increasing wages and budget deficits. Therborn continues: 'But Keynesianism is not enough; a comparative overview of budget policies demonstrates that expansionary, demand-boosting policies were necessary but not sufficient to resist the onslaught of the crisis in the mid-seventies.... Firstly, successful Keynesianism has to be accompanied by consistently complementary monetary policy – in particular a policy of low real interest rates. Second, in all the low unemployment countries, expansive fiscal and monetary policies have been crucially supplemented by nationally specific direct interventions in the market economy.'

An expansionary policy in South Africa would clearly have to be accompanied by additional controls and interventions. The openness of the economy would ensure that much uncontrolled expansion would leak out of the South African economy and, in order to prevent this, a regime of tariffs and outright prohibition of certain imports is required.

Above all, it would have to be accompanied by direct intervention on the supply side, though not in the form that has come to be associated with 'supply side' economics in the 1980s (in particular, the introduction of regressive tax systems and tight monetary policy). More creative interventions range from subsidisation of R and D facilities, through to controls that direct private investment to particular products – the state has itself identified mass housing developments as one such area – and particular regions. None of these controls are foreign to capitalism. Moreover, the forms of capitalist state (and the strength of working-class organisation) in which these strong, employment-creating interventions take place are highly divergent.

What Sweden, Japan and Austria share is a high degree of state intervention and low levels of employment. There is little *a priori* reason why coherent and powerful working-class organisation should not succeed in placing this, albeit reluctantly, on the policy agenda of South African capital and the state. There is indeed sound reason for arguing that the struggle to place a working-class perspective of unemployment on current policy agendas may be an indispensable aspect of the struggles for national liberation and socialism.

The trajectory of state economic policy appears to move in the opposite direction – in the conduct of fiscal policy which emphasises expenditure cuts rather than deficit budgeting; in the conduct of a monetary policy that has as its major objective dampening inflation rather than fanning growth; in the context of a general macro-economic policy which raises the spectre of an incomes policy rather than a policy that adopts a benign and permissive attitude towards wage increases; and in the teeth of an intense ideological campaign that stresses the sanctity of 'free enterprise' (with all the accoutrement of 'deregulation' and 'privatisation').

The state's positions are not unassailable and are probably more vulnerable in the area of economic policy than most others. In the quest to define an economic policy the postures that seem only capable of evoking keywords like 'nationalisation' and 'mixed economy' appear inconsequential and divorced from the everyday reality of mass unemployment.

On the ground the unions have several obvious points of entry: for example, in investment policy through their potential influence over the large pension funds; they certainly have the capacity to resonate loudly in the general area of welfare; above all, in the unions' unique capacity to intervene in the institutions of the labour market. None of these opportunities has been seriously explored. Both, if located within the context of a coherent and sustainable economic policy, embody greater possibilities of employment creation and economic growth than the rhetorical options currently on offer from the state and its various opponents.

Notes and References

1. South Africa's economic crisis: an overview

1. The weight of mining output in South Africa's total production has produced some debate about the actual levels of GNP and GDP in the past two decades, since the world prices of gold and other minerals have fluctuated markedly, affecting the valuation of output (Abedian and Kantor, 1988). But this issue does not negate the broader picture of long-term shifts.

2. The average data used to distinguish the longer period naturally reflect the endpoints, so that with different endpoints, a different picture could emerge. But the identification of turning-points, whether for short-run cyclical fluctuations or longer-run waves, involves an assessment by the analyst, based upon some theoretical considerations, rather than being self-evident from any particular set of empirical facts. No analysis of cyclical instability and fluctuation, of whatever periodicity and for whatever purpose, can rest on empirical data relating to the time-series alone – this must be supplemented by concepts and explanation, since shorter- and longer-term fluctuations can be identified in any random time-series. In this chapter the peaks and troughs of the short-run fluctuations are taken from the work of Smit and Van der Walt (1970, 1973, 1982).

3. Fine and Harris (1979) and Wright (1977) present useful summaries of these debates.

4. This problem is not specific to marxist economics, of course. Neo-classical theory faced the same dilemma, but resolved it by completely separating the theory of business cycles from the theory of long-run growth.

5. Since the approach was first introduced by Aglietta (1979), regulation theory has become widely used. But the broad direction of its adherents, as well as their critics (for example, Clarke, 1988), has been away from a rigorous approach to the analysis of economic crises, towards a greater emphasis on the political dimensions of the theory (see especially the important work of Bob Jessop on these aspects). This has meant that some of Aglietta's valuable pointers for the analysis of economic fluctuations have not really been developed.

6. Lipietz refers in this context to the work of the French sociologist Pierre Bourdieu on the concept of 'habitus'. There is however a large and growing literature within economic theory which develops similar ideas. A valuable introduction is Hodgson (1988).

7. The concept of stability should be understood here in a particular sense, deriving from economic theory. It means that a departure from an equilibrium characterised by a set of initial conditions leads not to infinite divergence, but rather to another equilibrium, characterised by a different set of defining conditions.

It is important to underline also that the concept of equilibrium used by 'regulation' theorists is dynamic, not static, as in neo-classical economics. In other words, equilibria are not characterised by a 'state of rest', but rather by constant changes in the forces acting upon each other. The state of equilibrium thus reflects a balance in the changes in these forces. This is the meaning of the concept of 'metastable' equilibrium, or changing, shifting stability (see Aglietta, 1979: 9–17, 353–6).

8. See also Lipietz (1987: 32). Here he defines a regime of accumulation as mathematically representable in the form of a 'reproduction scheme', a concept used by Marx in *Capital*, to discuss the distribution of overall production between departments. The movement through time of the reproduction scheme is constantly 'disrupted' by technical change. The regime of accumulation describes how these disruptions are contained to allow accumulation to proceed. This underlines the point that the regime of accumulation is a macroeconomic concept concerned with aggregate economic analysis.

9. This ambiguity is endemic, and perhaps inevitable, in marxist theory, or indeed any other systems approach.

10. The concept of mode of regulation is close to that of 'social structure of accumulation', introduced by the US theorists, Bowles, Gordon and Weisskopf (1983, 1986), in the context of their 'challenges to capitalist control' theory of economic crisis. In earlier versions of this chapter, I used the concepts of mode of regulation and social structure of accumulation interchangeably. Despite the common and essential emphasis on institutional considerations, which was previously missing from marxist economics, it no longer seems appropriate to elide the two concepts. The general theories of accumulation and crisis of the two schools are quite different, the 'challenges to capitalist control' view being based heavily on a concept of 'power' that seems too narrow. However, the concept of non-reproductive cycle, introduced by Gordon *et al.* (1983), is retained.

11. The mode of regulation is based upon the more abstract concept of the 'capital circuit', developed by Marx in *Capital*, in that the elements of the mode of regulation reflect the various stages passed through by capital as it moves around the circuit. An important analytical aspect of the concept is its macroeconomic nature, its concern with aggregate economic relations. (Indeed, this is true of regulation theory as a whole.) This is not intended to suggest that the mode of regulation is fully determining: it provides the macro context for accumulation, imposing (positive and negative) constraints. But it does imply that the importance of the concept within the theoretical framework is linked to its total coherence.

Neither should the mode of regulation be seen as being purposively constructed – the theory is functionalist, but in an *a posteriori* sense (Lipietz, 1987: 16). It is not being suggested that the mode of regulation springs into existence fully formed and then remains static – clearly there is constant 'tinkering' with and adjustment of the institutions. But the theory abstracts from these processes, focusing on the economic effects of a mode of regulation seen to be fundamentally stable, in its major features, over a long-run phase.

12. For a thorough explication of the rise and decline of Fordism in the advanced capitalist economies, see the work of Aglietta, Armstrong *et al.*, Boyer, de Vroey, Glyn and Lipietz.

13. The interactions between these variables were regarded by Kaldor as amongst the critical 'stylised facts' of capitalist economic growth. A 'stylised fact' is a 'conceptualised phenomenon ... interpreted to hold in a way, or to a degree, that the researcher regards as significant enough ... that prima facie explanation is called for... [They] often provide a starting point for the analysis of enduring structures and mechanisms' (Lawson, 1989: 65–6).

14. This section is elaborated in my paper 'Economic Crisis and Growth Models for the Future', presented to the Workshop on South Africa: Towards a Post-Apartheid Economy, Harare, April 1990.

Abedian, I. and B.S. Kantor. 1988. Relative Price Changes and Their Effects on Sectoral Contributions to National Income. Unpublished. Cape Town

Aglietta, M. 1979. *A Theory of Capitalist Regulation*. London: New Left Books

Armstrong, P., A. Glyn and J. Harrison. 1984. *Capitalism since World War 2*. London: Fontana

Bell, T. 1975. Productivity and Foreign Trade in South African Development Strategy. *South African Journal of Economics*, 43 (4)

Bell, T. 1984. *Unemployment in South Africa*. Durban: Institute for Social and Economic Research, University of Durban-Westville

Black, A. 1990. Manufacturing Development and the Crisis: A Reversion to Primary Production? This volume

Block, F. 1977. *The Origins of International Economic Disorder*. Los Angeles: University of

California Press

Bowles, S., D. Gordon and T. Weisskopf. 1983. *Beyond the Wasteland: A Democratic Alternative to Economic Decline*. Garden City, New York: Doubleday

Bowles, S., D. Gordon and T. Weisskopf. 1986. Power and Profits: The Social Structure of Accumulation and the Profitability of the Postwar US Economy. *Review of Radical Political Economics*, 18 (1/2)

Boyer, R. (ed.) 1988. *The Search for Labour Market Flexibility: The European Economies in Transition*. Oxford: Clarendon Press

Clark, N. n.d. The State Corporations in the Postwar Period: The Limits of Industrialisation. Unpublished mimeo, Stanford University

Clarke, S. 1988. Overaccumulation, Class Struggle and the Regulation Approach. *Capital and Class*, 36

Cloete, J. J. 1986. Recent Exchange Rate Policy in SA: A Critique. *South African Journal of Economics*, 54 (3)

De Kock, G. 1975. The Business Cycle in South Africa: Recent Tendencies. *South African Journal of Economics*, 43 (1)

De Kock, G. 1980. The New South African Business Cycle and Its Implications for Monetary Policy. *South African Journal of Economics*, 48 (4)

De Kock, G. 1986a. Unpublished speech. Johannesburg: Afrikaanse Sakekamer

De Kock, G. 1986b. Market-oriented Economic Policy Versus Quantitative Controls. *South African Reserve Bank Quarterly Bulletin*, December

De Vroey, M. 1984. A Regulation Approach Interpretation of the Contemporary Crisis. *Capital and Class*, 23

Dickman, A. 1973. The Financing of Industrial Development in South Africa. *South African Journal of Economics*, 41 (4)

Dickman, A. 1982. Corporate Financing and Monetary Policy. *South African Journal of Economics*, 50 (4)

Dreyer, J. P. and S. S. Brand. 1986. 'n Sektorale Beskouing van die Suid-Afrikaanse Ekonomie in 'n Veranderde Omgewing. *South African Journal of Economics*, 54 (2)

Driver, C. 1981. Review Article: 'A Theory of Capitalist Regulation'. *Capital and Class*, 15

Evans, D. and P. Alizadeh. 1984. Trade, Industrialisation and the Visible Hand. *Journal of Development Studies*, 21 (1)

Fine, B. and L. Harris. 1978. *Re-reading Capital*. New York: Columbia University Press

Frieden, J. 1981. Third World Indebted Industrialisation: International Finance and State Capitalism in Mexico, Brazil, Algeria and South Korea. *International Organization*, 35 (3)

Gelb, S. 1987. Making Sense of the Crisis. *Transformation*, 5

Glyn, A. 1982. The Productivity Slowdown: A Marxist View. In *Slower Growth in the Western World*, ed. R. C. O. Matthews. London: Heinemann

Gordon, D., R. Edwards and M. Reich. 1982. *Segmented Work, Divided Workers*. New York: Cambridge University Press

Gordon, D., T. Weisskopf and S. Bowles. 1983. Long Swings and the Nonreproductive Cycle. *American Economic Review*, 73 (2)

Griffith-Jones, S. and O. Sunkel. 1986. *Debt and Development Crises in Latin America: The End of an Illusion*. Oxford: Clarendon Press

Helleiner, G. K. 1972. *International Trade and Economic Development*. Harmondsworth: Penguin

Hindson, D. 1987. *Pass Controls and the Urban African Proletariat*. Johannesburg: Ravan Press

Hodgson, G. M. 1988. *Economics and Institutions*. Cambridge: Polity Press

Holden, M. 1985. Exchange Rate Policy for a Small Open Economy in a World of Floating Rates: The Case of South Africa. Occasional Paper No. 17, Economic Research Unit. Durban: University of Natal

Jessop, B. 1983. Accumulation Strategies, State Forms, and Hegemonic Projects. *Kapitalistate*, 10/11

Jessop, B. 1989. Conservative Regimes and the Transition to Post-Fordism: The Cases of Great Britain and West Germany. In *Capitalist Development and Crisis Theory: Accumulation*,

Regulation and Spatial Restructuring, ed. M. Gottdiener and N. Komninos. New York: St. Martin's Press

Knight, J. B. 1979. Black Wages and Choice of Technique in South Africa. *Oxford University Bulletin of Economics and Statistics*, 41 (2)

Krogh, D. C. 1985. From a Windfall to a Whirlwind in Less than 5 Years. *Transactions of the Actuarial Society of South Africa*, 6 (III)

Lawson, T. 1989. Abstraction, Tendencies and Stylised Facts: A Realist Approach to Economic Analysis. *Cambridge Journal of Economics*, 13

Lipietz, A. 1985. *The Enchanted World: Inflation, Credit and the World Crisis*. London: Verso

Lipietz, A. 1986. Behind the Crisis: The Exhaustion of a Regime of Accumulation. *Review of Radical Political Economics*, 18 (1/2)

Lipietz, A. 1987. *Mirages and Miracles: The Crises of Global Fordism*. London: Verso

Lipietz, A. 1989. The Debt Problem, European Integration and the New Phase of World Crisis. *New Left Review*, 178

Lombard, J. A. 1988. Housing Finance and the National Economic Scenario. Pretoria: Conference on 'Finance, the Pathway to Housing'

Lombard, J. A. and J. P. van den Heever. 1989. The Optimal Allocation of Savings in the South African Economy and the Role of Monetary Policy. Johannesburg: Biennial Meeting of the Economic Society of South Africa

Lombard, J. A. *et al.* 1985. Industrialisation and Growth. *Mercabank Focus*, 36

McCarthy, C. 1988. Structural Development of South African Manufacturing Industry. *South African Journal of Economics*, 56 (1)

McGrath, M. 1977. *Racial Income Distribution in South Africa*. Durban: Department of Economics, University of Natal

Mjoset, L. 1987. Nordic Economic Policies in the 1970s and 1980s. *International Organization*, 41 (3)

National Productivity Institute. 1986. *Productivity Focus*. Pretoria

Nattrass, J. 1981. *The South African Economy*. Cape Town: Oxford University Press

Nieuwenhuysen, J. 1972. South Africa. In *Macroeconomic Policies: A Comparative Study*, ed. J. O. N. Perkins. Toronto: University of Toronto Press

Noel, A. 1987. Accumulation, Regulation and Social Change: An Essay on French Political Economy. *International Organization*, 41 (2)

Parboni, R. 1981. *The Dollar and Its Rivals*. London: Verso

Ranis, G. and L. Orrock. 1985. Latin American and East Asian NICs: Development Strategies Compared. In *Latin America and the World Recession*, ed. E. Duran. Cambridge: Cambridge University Press

Seidman, J. 1980. *Facelift Apartheid: South Africa after Soweto*. London: IDAF

Smit, D. J. and B. E. van der Walt. 1970. Business Cycles in South Africa During the Period 1946-1968. *South African Reserve Bank Quarterly Bulletin*, September

Smit, D. J. and B. E. van der Walt. 1973. Business Cycles in South Africa During the Period 1968-1972. *South African Reserve Bank Quarterly Bulletin*, June

Smit, D. J. and B. E. van der Walt. 1982. Growth Trends and Business Cycles in the South African Economy 1972-1981. *South African Reserve Bank Quarterly Bulletin*, June

Swanepoel, C. J. and J. van Dyk. 1978. The Fixed Capital Stock and Sectoral Capital-Output Ratios of South Africa, 1946-1977. *South African Reserve Bank Quarterly Bulletin*, September

Walsh, V. and H. Gram. 1980. *Classical and Neoclassical Theories of General Equilibrium*. New York and Oxford: Oxford University Press

Williamson, J. 1985. On the System in Bretton Woods. *American Economic Review*, 75 (2)

Wright, E. O. 1977. Alternative Perspectives in the Marxist Theory of Accumulation and Crisis. In *The Subtle Anatomy of Capitalism*, ed. J. Schwartz. Goodyear Publishing

Yudelman, D. 1984. *The Emergence of Modern South Africa*. Cape Town: David Philip

Official publications

Central Statistical Services. *South African Statistics*, various issues. Pretoria: Government

Printer
International Monetary Fund. *International Financial Statistics*, various issues. Washington, DC
RSA Department of Finance. 1990. *Budget Review*. Cape Town: Government Printer
RSA Department of Industries and Commerce. 1983. Report of the Study Group on Industrial Development Strategy. Pretoria: Government Printer
South African Reserve Bank. *Quarterly Bulletin*, various issues. Pretoria

2. State, capital and growth: the political economy of the national question

1. I am indebted to Stephen Gelb, Dave Kaplan, Dave Lewis, Bill Freund and Vishnu Padayachee for their assistance in writing this chapter. I naturally take full responsibility for what appears in print, however.

2. See the oft-cited piece by Saul and Gelb, 1986.

3. See Gelb's discussion in chapter 1 on regulation theory. I have depended on Lipietz, 1986, 1987; Jessop, 1983a, 1983b, 1988, 1989; Haeusler and Hirsch, 1989.

4. See Jessop, 1983a, 1983b, 1989; and Haeusler and Hirsch, 1989, for further discussion of these concepts.

5. The following section draws very heavily from Kaplan, 1977, 1980; Gelb, 1987, 1988, 1989; Morris and Padayachee, 1988, 1989; and Hindson, 1987.

6. See Morris and Padayachee, 1988, and Gelb, 1989. Conceptualising the state–capital relation in these terms comes from initial ideas put forward by Dave Kaplan in the Economic Trends Group.

7. See Jessop, 1989: 263–4 for a comprehensive list of the general features of Fordism in advanced industrial countries.

8. See Dave Kaplan, 1980 for an extended discussion of the concept of racially exclusive limited democracy in South Africa.

9. See Giliomee, 1979 and O'Meara, 1983.

10. 'English' capital was organised through the Federated Chamber of Industries and the Associated Chambers of Commerce whilst 'Afrikaner' capital was organised through the Afrikaanse Handelsinstituut. See Noel, 1987, for a discussion of the importance of organisation in class formation.

11. See Poulantzas, 1976 and Lipietz, 1987.

12. See Le Roux, 1987.

13. Kaplan, 1980 has put this point very succintly.

14. See O' Meara, 1982

15. See Morris and Padayachee, 1988, 1989 for an extended discussion of the elements of reform.

16. Reform involved a process of racial elimination as well as racial addition; of destructuring and of restructuring Verwoerdian apartheid; in short, a process of 'de-racialisation/ re-racialisation' of social and political life.

17. This problem is also discussed from another theoretical perspective in Huntington, 1982, which according to Rhoodie, 1989, was the model P. W. Botha closely followed.

18. See Bernstein and Godsell, 1988.

19. See Morris and Padayachee, 1988 for the details of this story.

20. This is an issue that requires further detailed analysis. The ANC in its Kabwe documents adopted a position of insurrectionism which it had consistently fought against as a political mistake.

21. See Morris and Padayachee, 1988, 1989; Jochelson, 1988.

22. See Morris and Padayachee, 1988, 1989; Swilling and Phillips, 1989, for a discussion of the differences and conflicts between Heunis and Malan.

23. See Hindson, 1987.

24. In examining the evidence available from a search of the mass media Lee and Buntman (1989: 117) conclude: 'Most calls from business people for the "dismantling of apartheid", "power sharing", and "reform" are criticisms of apartheid as discrimination. They do not attempt to tackle the issue of the apartheid state.'

25. There is a wealth of information of the NSMS and the way it related to the state, the

military, and Malan, as well as its social welfare interventions in the townships. See for example, Boraine, 1988, 1989; Swilling and Phillips, 1989; Morris and Padayachee, 1988, 1989.

26. See Boraine, 1988, 1989, for a detailed discussion of the role of the NSMS in the upgrading of townships.

27. *Die Suid-Afrikaan*, Winter 1986. See also Swilling and Phillips, 1989 for statements from Malan justifying placing the provision of social services above political incorporation.

28. See Gavin Relly's review of the activities of the Anglo American Corporation in the financial year ended March 31, 1989. This appeared in August 1989, significantly before the election. In a very sympathetic statement Relly makes clear that the task of the 'new leader of the National Party, Mr de Klerk, supported by the younger generation of Nationalists, is to consummate the process of reform and, even more important, to give content to a post-apartheid vision for our country.... Clearly, if after the election, government takes steps to alter the political landscape by releasing political prisoners and addressing the issues of the state of emergency and the status of exiled movements, opposition groupings will have to move away from protest politics, to the need to recognise the hard realities of the South African situation, and to negotiate and find compromise solutions that do not satisfy abstract ideological positions.'

29. See Cock and Nathan, 1989, particularly the chapter by Swilling and Phillips.

30. This was clearly perceived at the time by Van Zyl Slabbert in his speeches throughout the public debate over the tricameral parliament, as well as by André du Toit in a topical series of excellent newspaper analyses of the new constitution.

31. The need for a powerful executive to manage a process of reform from above in order to ideologically reinterpret accepted policies and pilot through new state interventions which are in fact contrary to the governing party's historically stated policies and principles is aptly described by Huntington, 1982.

32. Confidential parliamentary sources. It is also interesting that the radical shift in policy was sold to the party on the basis of the extremely favourable reaction from Western governments.

33. See Morris and Padayachee, 1989; Gelb, 1989.

34. See Hindson, 1988 on major changes in the occupational division of labour.

Bernstein, A. and B. Godsell. 1988. The Incrementalists. In *A Future South Africa: Visions, Strategies and Realities,* ed. P. Berger and B. Godsell. Cape Town: Human & Rousseau

Boraine, A. 1988. Security Management Upgrading in the Black Townships. In The Economic Crisis: Recent Economic Trends in South Africa, ed. S. Gelb. Unpublished report to COSATU

Boraine, A. 1989. Security Management Upgrading in the Black Townships. *Transformation*, 8

Cock, J. and Nathan, L. 1989. *War and Society. The Militarisation of South Africa.* Cape Town: David Philip

Gelb, S. 1987. The South African Crisis. *Transformation*, 6

Gelb, S. 1988. The Economic Crisis: An Overview. In The Economic Crisis: Recent Economic Trends in South Africa, ed. S. Gelb. Unpublished report to COSATU

Gelb, S. 1989a. Growth, Crisis and Development in South Africa. Paper presented at Colloquium on the South African Economy. Lausanne, July 8–13

Gelb, S. 1990. Economic Crisis and Growth Models for the Future. Presented at Workshop on Economic Policy for a Post-Apartheid South Africa. Harare, May 1990

Giliomee, H. 1979. The Economic Advancement of Afrikanerdom. In *Ethnic Power Mobilised*, ed. H. Adam and H. Giliomee. New Haven: Yale University Press

Heusler, J. and J. Hirsch. 1989. Political Regulation: The Crises of Fordism and the Transformation of the Party System in West Germany. In *Capitalist Development and Crisis Theory: Accumulation, Regulation and Spatial Restructuring,* ed. M. Gottdiener and N. Komninos. New York: St. Martin's Press

Hindson, D. 1988. The Restructuring of Labour Markets in the 1970s and 1980s. In *The Economic Crisis: Recent Economic Trends in South Africa,* ed. S. Gelb. Unpublished report to COSATU

Hindson, D. 1987. *Pass Controls and the Urban African Proletariat.* Johannesburg: Ravan Press

Huntington, S. 1981 Reform and Stability in a Modernising Multi-Ethnic Society. *Politikon*, 8

Jessop, B. 1983a. The Capitalist State and the Rule of Capital: Problems in the Analysis of Business Associations. *West European Politics*, 6 (2)

Jessop, B. 1983b. Accumulation Strategies, State Forms, and Hegemonic Projects. *Kapitalistate*, 10 (11)

Jessop, B. 1988. Regulation Theory, Post-Fordism and the State: More than a Reply to Werner Bonefield. *Capital and Class*, 34

Jessop, B. 1989. Conservative Regimes and the Transition to Post-Fordism: The Cases of Great Britain and West Germany. In *Capitalist Development and Crisis Theory: Accumulation, Regulation and Spatial Restructuring*, ed. M. Gottdiener and N. Komninos. New York: St. Martin's Press

Jochelson, K.J. 1988. Urban Crisis, State Reform and Popular Reaction: A Case Study of Alexandra. Honours thesis, University of Witwatersrand

Kaplan, D.E. 1977. Class Conflict, Capital Accumulation and the State. Ph.D. thesis, Sussex University

Kaplan, D.E. 1980. The South African State: The Origins of a (1980) Racially Exclusive Democracy'. *The Insurgent Sociologist*, X (2)

Kaplinsky, R. 1990. A Policy Agenda for Post-Apartheid South Africa. Presented at Workshop on Economic Policy for a Post-Apartheid South Africa. Harare, May 1990

Lee, R. and F. Buntman 1989. The Limousine Lizard and the Taxi Proletariat. In *South Africa at the End of the Eighties*. Johannesburg: Centre for Policy Studies, University of Witwatersrand

Le Roux. P, 1986. The State as an Economic Actor: A Review of Divergent Perceptions of Economic Issues. Paper Presented at a Conference 'South Africa Beyond Apartheid'. University of York

Lipietz, A. 1986. New Tendencies in the International Division of Labor: Regimes of Accumulation and Modes of Regulation. In *Production, Work, Territory*, ed. A. Scott and M. Stoper. London: Allen and Unwin

Lipietz, A. 1987. *Mirages and Miracles: The Crises of Global Fordism*. London: Verso

Morris, M.L. and V. Padayachee, 1988. State Reform Policy in South Africa. *Transformation*, 7

Morris, M.L. and V. Padayachee 1989. Hegemonic Projects, Accumulation Strategies and State Reform Policies in South Africa. *Labour, Capital and Society*, 45

Noel, A. 1987. Accumulation, Regulation, and Social Change: An Essay on French Political Economy. *International Organization*, 41 (2)

O' Meara, D. 1982. 'Muldergate': The Politics of Afrikaner Nationalism. *Work in Progress*, 22

O' Meara, D. 1983. *Volkskapitalisme*. Johannesburg: Ravan Press

Poulantzas, N. 1978. *State, Power, Socialism*. London: NLB

Poulantzas, N. 1978. *The Crisis of the Dictatorships*. London: NLB

Relly, G. 1989. Extracts from Gavin Relly's Review of the Activities of the Anglo American Corporation in the Financial Year ended March 31, 1989. *Leadership*, 8 (6)

Rhoodie, E. 1989. *P. W. Botha: The Last Betrayal*. Melville: S.A. Politics

Saul, J. and S. Gelb. 1986. *The Crisis in South Africa*. London: Zed Books

Swilling, M. and M. Phillips, 1989. The Powers of the Thunderbird. In *South Africa at the End of the Eighties*. Johannesburg: Centre for Policy Studies, University of Witwatersrand

Van Zyl Slabbert, F. 1989. *The System and the Struggle: Reform, Revolt and Reaction*. Johannesburg: Jonathan Ball

3. The crisis and South Africa's balance of payments

1. See De Vroey (1984).

2. See Aglietta (1982), who provides an analysis of the international monetary system within the regulation framework.

3. This effect only holds if domestic price increases do not equal or exceed the amount of the depreciation. Thus a 10 per cent depreciation confers a competitive advantage as prices are now cheaper in foreign currency terms. If, however, domestic prices increase by 10 per cent the

competitive advantage is offset.

4. See for example Krugman and Taylor (1978).

5. The import penetration ratio is also sensitive to cyclical fluctuations. It should be noted moreover, that a declining i.p.r. could be a sign of increased self-sufficiency of the economy or increased protection. Van Zyl (1984) argues that where i.p.r. have declined, this has been caused by a combination of these factors.

6. Direct investment involves ownership and control over assets in South Africa. Indirect investment does not involve control and includes portfolio investment and foreign loans.

7. Estimates quoted in the Commonwealth Committee Report on South Africa (1988).

8. Real effects are changes in physical variables, such as the volume of output, level of employment or distribution of income. By contrast, a nominal effect is a change in prices only.

9. See also McCarthy (1988).

10. By 1980 South African investment abroad was in excess of one-third of foreign investment into South Africa (Kaplan, 1983).

Aglietta, M. 1982. World Capitalism in the Eighties. *New Left Review*, 136

Commonwealth Committee. 1988. *South Africa's Relationship with the International Financial System*. Report of the Intergovernmental Group of Officials

De Vroey, M. 1984. A Regulation Approach Interpretation of Contemporary Crisis. *Capital and Class*, 23

Diaz-Alejandro, C.F. 1985. *Exchange Rate Devaluation in a Semi-Industrialized Country*. Cambridge, Mass.: M.I.T. Press

Gidlow, R.M. 1988. Workings of a Managed Floating Exchange Rate System for the Rand. *South African Journal of Economics*, 56

Grant, C. 1985. The Banks Abandon South Africa. *Euromoney*, December

Kahn, S.B. 1987a. Exchange Controls and Exchange Rate Policy in the South African Economy. *Social Dynamics*, 13

Kahn, S.B. 1987b. Import Penetration and Import Demands in the South African Economy. *South African Journal of Economics*, 55

Kaplan, D. 1983. South Africa's Changing Place in the World Economy. In *South African Review One*. Johannesburg: Ravan Press

Khan, M. and N. Ul Haque. 1985. Foreign Borrowing and Capital Flight: A Formal Analysis. *International Monetary Fund Staff Papers*, 32

Krugman, P. and L. Taylor. 1978. Contractionary Effects of Devaluation. *Journal of International Economics*, 8

McCarthy, C.L. 1988. Structural Development of South African Manufacturing Industry. *South African Journal of Economics*, 56

Seidman, A. 1986. *Money, Banking and Public Finance in Africa*. London: Zed Press

Strydom, P.D.F. 1987. South Africa in World Trade. *South African Journal of Economics*, 55

Van Zyl, J.C. 1984. South Africa in World Trade. *South African Journal of Economics*, 52

Official publications

Republic of South Africa. 1978, 1984. Reports of the Commission of Inquiry into the Monetary System and Monetary Policy in South Africa. Pretoria: Government Printers (De Kock Commission): (i) Exchange Rates in South Africa. Interim Report (RP 112/1978); (ii) The Monetary System and Monetary Policy in South Africa. Final Report (RP 70/1984)

Republic of South Africa. 1985. Report of the Study Group on Industrial Development Strategy. Pretoria: Government Printer, for the Department of Industries and Commerce (Kleu Report)

Central Statistical Services. *Quarterly Bulletin of Statistics*, various issues

Central Statistical Services. *South African Statistics*, various issues

South African Reserve Bank. *Quarterly Bulletin*, various issues

4. The politics of South Africa's international financial relations, 1970–1990

Bell, R.T. 1975. Productivity and Foreign Trade in South African Development Strategy. *South*

African Journal of Economics, 43 (4)

Baughn, W.M. and D.R. Mandich (eds). 1983. *The International Banking Handbook*. Homewood, Illinois: Irwin

Center for International Policy (CIP). 1978. *How the IMF Slipped $464 Million to South Africa*. Washington, DC: CIP Publication

Center for International Policy. 1983. *A Billion Dollars for South Africa*. Washington, DC: CIP Publication

Center for International Policy. 1984. *A Victory over Apartheid*. Washington, DC: CIP Publication

De Kock, G.P.C. 1987. The Gold Price and the South African Economy. Paper presented to the Cape Town Sakekamer

De Vroey, M. 1984. A Regulation Approach to the Interpretation of the Contemporary Crisis. *Capital and Class*, 23

Financial Analysts Report (FAR). 1985. *The South African Rand. The Reasons for the Rand's Degradation*. New York: CitiBank

Flight, H. and B. Lee-Swan. 1988. *All You Need to Know About Exchange Rates*. Sedgwick and Jackson. London

Gelb, S. 1987. Making Sense of the Crisis. *Transformation*, 5

Gelb, S. 1988. Moving Forward on Sanctions. A Reply to Fleshman and Cason. *Transformation*, 7

Hirsch, A. 1988. The Causes and Consequences of the 1985 Credit Crisis. Paper subsequently published in *Transformation*, 7, 1989

Holden, M. 1989. A Comparative Analysis of Structural Imbalance in the Face of a Debt Crisis. *South African Journal of Economics*, 57 (1)

Jenkins, C. 1989. Trade Finance and South Africa's Foreign Debt: Another Pressure Point? Paper presented to the Biennial Conference of the Economic Society of South Africa, 6 September

Johnson, R.W. 1977. *How Long Will South Africa Survive?* London: Macmillan

Kaplan, D. 1987. The Limited Development of the South African Machine Tools Industry: Causes and Consequences. *Social Dynamics*, 13 (1)

Ladd-Hollist, W and F. LaMond Tullis (eds). 1985. *An International Political Economy*. Boulder, Colorado: Westview Press

Lipietz, A. 1987. *Mirages and Miracles: The Crisis of Global Fordism*. London: Verso

Lind, J. and D. Espaldon. 1986. *South Africa's Debt at the Time of Crisis*. San Francisco: CN–ICCR Publication

Militz, E. 1985. *Bank Loans to South Africa mid-1982–end 1984*. World Council of Churches

Ovenden, K. and Cole T. 1989. *Apartheid and International Finance: A Program for Change*. Penguin, Australia

Padayachee, V. 1986. Apartheid South Africa and the International Monetary Fund. *Transformation*, 3

Padayachee, V. 1988. Private International Banks, the Debt Crisis and the Apartheid State. *African Affairs*, 87 (348)

Petras, J.F. and M.H. Morley. 1981. Development and Revolution: Contradictions in the Advanced Third World Countries – Brazil, South Africa and Iran. *Studies in Comparative International Development*, XVI (1)

Piore, M. and C. Sabel. 1979. *The Second Industrial Divide*. New York: Basic Books

Stalling, B. 1985. International Lending and the Relative Autonomy of the State: A Case Study in Twentieth Century Peru. *Politics and Society*, 14, 3

Official publications
International Monetary Fund. 1945. *Articles of Agreement*. Washington DC
International Monetary Fund. 1976. *IMF Survey*. Washington DC
South African Reserve Bank. March 1985. *Quarterly Bulletin*
United Nations. 1982a. Reports of the Secretary-General, 4 October 1982 and 28 October 1982
United Nations. 1982b. UN General Assembly Report A37/552.1904
United Nations. 1982c. UN Press Releases GA/6695, 21 October 1982 and 4 November 1982

United States State Department. 1982. Confidential telegram. (Made available to the author in April 1986)

5. South African gold mining in transformation

1. In becoming more critical of management views on technology and their interpretation of their own history, I am grateful to Jeff Guy for valuable comments. See his 'Technology, Ethnicity and Ideology: Basotho Miners and Shaft-Sinking on the South African Gold Mines', *Journal of Southern African Studies*, 14, 2, 1988, written with Motlatsi Thabane. Jean Leger has taught me a lot about gold mining also.

2. Gold, little of which is lost or destroyed and much of which is hoarded, operates in part outside the normal rules that govern exchange transactions. Its relationship to currency values has become more and more fluctuating and unpredictable since 1972. See Figure 2 for the gold price in rands and US dollars.

Aglietta, M. 1979. *A Theory of Capitalist Regulation*. London: New Left Books
Burawoy, M. 1985. *The Politics of Production*. London: New Left Books
Crush, J. 1987. Restructuring Migrant Labour on the Gold Mines. *South African Review*, 4
Elliot, H. 1986. Down Towards the Styx. *Optima*, XXXIV (2)
Gordon, R. 1977. *Mines, Masters and Migrants*. Johannesburg: Ravan
Hermanus, M. 1987. The Gold Mining Industry: Changing Technology and Its Implications. Labour Studies Workshop, University of the Witwatersrand. Johannesburg
James, W. 1987. A Political Economy of Migrancy. Labour Studies Workshop, University of the Witwatersrand. Johannesburg
Johnstone, F. 1976. *Class, Race and Gold*. London: Routledge Kegan Paul
Leger, J. 1986. Safety and the Organisation of Work in South African Gold Mines: A Crisis of Control. *International Labour Review*, CXXV (5)
Leger, J. and M. Mothibeli. 1987. South African Gold Miners' Perceptions of Safety, 1984 to 1987. Mine Safety and Health Congress, Johannesburg
Lever, J. and W. G. James. 1987. Towards a Deracialised Labour Force: Industrial Relations and the Abolition of the Job Colour Bar on the South African Gold Mines. University of Stellenbosch, Department of Sociology, Occasional Paper 12. Stellenbosch
Lipton, M. 1986. *Capitalism and Apartheid, South Africa 1910–86*. Aldershot: Wildwood House
Lipton, M. 1980. Men of Two Worlds: Migrant Labour in South Africa. *Optima*, special issue
Markham, C. and M. Mothibeli. 1987. The 1987 Mineworkers Strike. Unpublished
Moodie, T. D. 1980. The Formal and Informal Social Structure of a South African Gold Mine. *Human Relations*, XXXIII (8)
Rafel, R. 1987. Job Reservation on the Mines. *South African Review*, 4
Ruiters, G. 1987. NUM and the Miners' Strike. *Work in Progress*, 50/51
Salamon, M. 1986. Research in South Africa's Gold Mining Industry. *Optima*, XXXIV (2)
Wilson, F. 1972. *Labour in the South African Gold Mines*. Cambridge: Cambridge University Press
Yudelman, D. and A. Jeeves. 1986. New Labour Frontiers for Old: Black Migrants to the South African Gold Mines 1920–85. *Journal of Southern African Studies*, XIII (1)

Official publications
Republic of South Africa. Report of the Commission of Inquiry into the Tax Structure of the Republic of South Africa. RP 34/1987. Pretoria: Government Printer. (Margo Commission)

6. Coal mining: past profits, current crisis?

The generous assistance of Stephen Gelb and Martin Nicol is gratefully acknowledged.

1. In 1950 the Government Mining Engineer put the average cost of coal at the pit head in South Africa as 7s 4d a short ton, compared with 45s in Britain, France and Germany, 65s in Belgium, 34s in the US and 17s in Australia.

2. This is an important consideration. Producers who were not part of the initial investment

involved in each phase are only able to make use of unutilised capacity. Since the costs of transporting and handling through Richards Bay are considerably lower than any other port, the large corporations enjoy considerable price advantages. They can thus squeeze out or at least delay the entry into the export market of new, smaller producers.

3. At the end of 1987 BP withdrew from the venture.

4. Department of Mineral and Energy Affairs, *Annual Statistics*. Unfortunately figures published by the Department of Mineral and Energy Affairs for coal employment are somewhat misleading. They appear to include contractors involved in mine construction and possibly even synfuel plant construction. The total employment figure climbs from 80 000 in 1976 to a peak of 130 000 in 1981, dropping precipitously to 95 000 in 1984. This is probably due to the inclusion in the statistics of workers constructing the Secunda operations and other mines as coal workers. By contrast, the number of employees underground (including opencast mine workers) increases steadily but only slightly throughout this period. Hence the apparent decline in total coal mining employment between 1981 and 1984 appears to be an artifact of the statistics.

5. Estimated from Graph 4 of Centro de Estudos Africanos, 1986. Unfortunately territorial breakdowns for this period for the Natal mines have not been obtained.

6. Colliery management was more easily able to dispense with Mozambican workers because there were no labour shortfalls as had happened in gold with the withdrawal of Malawian nationals.

7. Cited in SAIRR, 1979: 212.

8. The following comments draw heavily on the analysis of Hill (1987).

Anon. 1970. Export Profitability Must be Increased. *CGB*, August
Anon. 1974a. Historic Contract Signed. *CGB*, January
Anon. 1974b. Mining Chiefs Fire First Salvoes. *CGB*, March
Anon. 1975. Richards Bay – the 'super-port'. *CGB*, January
Anon. 1976. RBCT. *CGB*, March
Anon. 1977. Economics Apply Brakes to Coal Mechanisation. CGB, May
Berning, F. S. 1972. Chairman's Review. *CGB*, April
Boers, R., D. Kennedy and J. Walker, 1989. Coal: A South African Success Story. *Mining Survey*, 1
Centro de Estudos Africanos. 1986. *Changing Labour Demand Trends on the South African Mines, with Particular Reference to Mozambique*. Maputo
Coetzee, A. C. 1975. Interview. *CGB*, December
Dutkiewicz, R. F. and K. F. Bennett, 1976. Coal Critique. *CGB*, May
Eskom. 1988. *1987 Statistical Yearbook*. Johannesburg
Gilbertson, B. P. 1988. Chairman's Statement. *TransNatal Coal, Annual Report*. Johannesburg
Hill, A. 1987. *Coal Review*, Johannesburg: Max Pollak & Freemantle
Hoddinott, P. J. 1987. *Coal*. In *Mining Annual Review*
Joint Coal Board. 1989a. *Black Coal in Australia, 1987–88*, Sydney
Joint Coal Board. 1989b. *Forty-first Annual Report, 1987/88*, Sydney
Malan, T. 1985. Migrant Labour in Southern Africa. *Africa Insight*, 15 (2)
McQuaid, J. 1988. Conference Review: International Conference of Safety in Mines Research Institutes, 2–5 November 1987, Beijing, China. *Mineral Resources Engineering*, 1 (2)
Mehliss, A. T. 1986. Energy minerals – Coal. In *South Africa's Mineral Industry 1985*, Department of Mineral and Energy Affairs. Johannesburg
Mine Safety and Health Administration, 1988. Table 2. Injury experience in coal mining, 1987, IR 1164
Newman, S. C. 1972. Interview. *CGB*, April
Oliver, R. B. 1982. *Steam Coal in Southern Africa*. London: Economist Intelligence Unit Report No. 122
Sealey, A. A. 1974. Interview. *CGB*, March
South African Institute of Race Relation. 1979. *Survey of Race Relations in South Africa 1978*. Johannesburg
Tew, A. D. 1971. Consistent sales pressure can prise open the doors. *CGB*, December

Wassenaar, A. D. 1977. Coal industry is economy's sick man. *CGB*, May
Wilson, F. 1972. *Labour in the South African Gold Mines, 1911–69*, Cambridge: Cambridge University Press

Official publications
RSA. Department of Mines 1975. Report of the Commission of Inquiry into the Coal Resources of the Republic of South Africa. (Petrick Report)

7. Manufacturing development and the crisis: a reversion to primary production?

1. In 1987 manufacturing contributed 23.2 per cent of GDP and 27 per cent (1.3m) of total non-agricultural employment.

2. See the major study by the World Bank (1987) which classifies developing countries into four groups according to their degree of outward- or inward-orientation. The strongly outward-oriented group (South Korea, Singapore and Hong Kong) had the best growth record during the period 1963–85 while the strongly inward-oriented group had the weakest growth performance. A number of studies have questioned the correlation between outward-orientation and high growth in the case of very low income countries (Helleiner, 1986) or argued that it only holds when external demand is strong (Singer and Gray, 1988).

3. This figure is based on the 1976 manufacturing census and has probably declined since then, given the large number of foreign firms which have recently sold out to domestic interests. The foreign-controlled share of assets would be greater than that of employment because foreign firms are generally larger and more capital-intensive than domestically controlled firms.

4. 'Fully manufactured commodities' are defined as those in which raw materials constitute less than 25 per cent of the input–output ratio (Van Zyl, 1984: 51).

5. See Schatz (1981) for a discussion of 'assertive pragmatism'.

6. Average labour output in manufacturing grew by only 1.6 per cent per annum from 1970 to 1986.

7. The advent of microelectronics in the clothing industry may reduce the competitive advantage of low-wage producers and hence slow or even reverse the tendency to relocation (Hoffman, 1985).

8. For example, the recent Board of Trade and Industries' proposals regarding the textile and clothing industries.

9. In terms of narrowly defined 'fully manufactured' goods, only 3 per cent of production is exported. If a broader definition is used, the proportion of output exported is roughly 10 per cent.

Bell, R.T. and V. Padayachee. 1984. Unemployment in South Africa: Trends, Causes and Cures. *Development Southern Africa*, 1 (3 & 4)

Black, A. 1985. Government Policy and Employment Creation in the South African Manufacturing Sector. Durban: University of Natal, Development Studies Unit, M.Soc.Sci. thesis

Black, A. and J. Stanwix. 1987. Manufacturing Development and the Economic Crisis: Restructuring in the Eighties. *Social Dynamics*, 13 (1)

Dreyer, J.P. and S.S. Brand. 1986. 'n Sektorale Beskouing van die Suid-Afrikaanse Ekonomie in 'n Veranderende Omgewing. *South African Journal of Economics*, 54 (2)

Fitzgerald, E. 1983. The State and the Management of Accumulation in the Periphery. In *Latin America in the World Economy*, ed. D. Tussie. Aldershot: Gower

Freund, W. 1986. South African Business Ideology, the Crisis and the Problem of Redistribution. Inaugural Lecture. Durban: University of Natal

Fröbel, F., Heinrichs, J. and O. Kreye. 1980. *The New International Divison of Labour*. London: Cambridge University Press

Helleiner, G. 1986. Outward Orientation, Import Instability, and African Economic Growth: An Empirical Investigation. In *Theory and Reality in Development*, eds. S. Lal and F. Stewart. London: Macmillan

Hoffman, K. 1985. Clothing, Chips and Competitive Advantage: The Impact of Microelectronics on Trade and Production in the Garment Industry. *World Development*, 13 (3)

NOTES AND REFERENCES

Jenkins, R. 1984. Divisions over the International Division of Labour. *Capital and Class*, 22

Jenkins, R. 1987. *Transnational Corporations and the Latin American Automobile Industry*. London: Macmillan

Jones, D. and J. Womack. 1985. Developing Countries and the Future of the Automobile Industry. *World Development*, 13 (3)

Kahn, B. 1987. Import Penetration and Import Demands in the South African Economy. *South African Journal of Economics*, 55 (3)

Kaplan, D. 1983. The Internationalisation of South African Capital: South African Direct Foreign Investment in the Contemporary Period. *African Affairs*, 82

Kaplan, D. 1987. Machinery and Industry: The Causes and Consequences of Constrained Development of the South African Machine Tool Industry. *Social Dynamics*, 13 (1)

Kaplinsky, R. 1984. The International Context for Industrialisation in the Coming Decade. *Journal of Development Studies*, 21 (1)

Kaplinsky, R. 1988. Restructuring the Capitalist Labour Process: Some Lessons from the Car Industry. *Cambridge Journal of Economics*, 12 (4)

Krueger, A. 1985. Import Substitution versus Export Promotion. *Finance and Development*, 20 (2)

Lal, D. 1984. *The Poverty of 'Development Economics'*. London: Hobart

Levy, B. 1981. *Industrialisation and Inequality in South Africa*. University of Cape Town, SALDRU Working Paper 36. Cape Town

Lombard, J.A. (ed.) 1985. Industrialisation and Growth. *Mercabank Focus on Key Economic Issues*, 36

Luedde-Neurath, R. 1984. State Intervention and Foreign Direct Investment in South Korea. *IDS Bulletin*, 15 (2)

Marais, G. 1981. Structural Changes in Manufacturing Industry, 1916 to 1975. *South African Journal of Economics*, 49 (1)

McCarthy, C. 1987. Structural Development of South African Manufacturing Industry. A Policy Perspective. Conference of the Economic Society of South Africa, Pretoria

McCarthy, C. 1988. Structural Development of South African Manufacturing Industry. *South African Journal of Economics*, 56 (1)

McGrath, M. and C. Jenkins. 1985. The Economic Implications of Disinvestment for the South African Economy. Conference of the Economic Society of South Africa, Durban

Morley, S. and G. Smith. 1973. The Effect of Changes in the Distribution of Income on Labour, Foreign Investment and Growth in Brazil. In *Authoritarian Brazil*, ed. A. Stepan. New Haven: Yale University Press

Reynders, H.J.J. 1975. Export Status and Strategy. *South African Journal of Economics*, 43 (1)

Rogerson, C. 1982. Multinational Corporations in Southern Africa: A Spatial Perspective. In *The Geography of Multinationals*, eds. M. Taylor and N. Thrift. London: Croom Helm

Schatz, S.P. 1981. Assertive Pragmatism and the Multinational Enterprise. *World Development*, 9 (2)

Scheepers, C.F. 1982. The International Trade Strategy of South Africa. *South African Journal of Economics*, 50 (1)

Singer, H. and P. Gray. 1988. Trade Policy and Growth of Developing Countries: Some New Data. *World Development*, 16 (3)

Stewart, F. 1983. Macro-Policies for Appropriate Technology: An Introductory Classification. *International Labour Review*, 122 (3)

Thompson, A.M. 1987. Technological Choice as a Determinant of Inequality, Poverty and Unemployment in South Africa. University of Cape Town, Centre for African Studies. Cape Town

UNIDO. 1986. Industrial Policies and Strategies in Developing Countries: An Analysis of Local Content Regulations. *Industry and Development*, 18

Van Zyl, J.C. 1984. South Africa in World Trade. *South African Journal of Economics*, 52 (1)

Viljoen, S.P. 1983. The Industrial Achievement of South Africa. *South African Journal of Economics*, 51 (1)

Wade, R. 1984. Dirigisme Taiwan-Style. *IDS Bulletin*, 15 (2)

Wellings, P. and A. Black. 1986. Industrial Decentralisation under Apartheid: The Relocation of Industry to the South African Periphery. *World Development*, 14 (1)

Official publications
Board of Trade and Industries. 1988. Interim Report of the Investigation into the Industry Manufacturing Passenger Cars and Light Commercial Vehicles. Pretoria: Government Printer
Republic of South Africa. 1983. Summary of Report of the Study Group on Industrial Development Strategy. Pretoria: Government Printer
Republic of South Africa. 1985. White Paper on Industrial Development Strategy in the Republic of South Africa. Pretoria: Government Printer
World Bank. 1987. *World Development Report*. Oxford: Oxford University Press

8. The South African capital goods sector and the economic crisis

1. Technological capability includes the capacity to manufacture, but even more crucially, the capacity to undertake product design and development.

2. The capital goods sector produces a wide range of products with different technological requirements. For an attempt at categorising these products in terms of some measure of technological complexity and especially distinguishing between simple and complex capital goods, see Chudnovsky and Nagao (1983), Chapter 1. For South Africa and the NICs, where the production of simple capital goods is widespread, the key issue is the extent to which these countries are able to satisfy their capital accumulation requirements both through indigenous production and through the indigenous design and development of complex capital goods.

3. Gross output is the total value of all goods manufactured. Since the products of an industry may be the inputs of another industry, the measure of gross output contains an element of duplication. Net output is gross output minus the cost of all material inputs. Net output is equivalent to value added.

4. The constant share of capital goods in the import bill over a lengthy period has been noted elsewhere, e.g. 'the percentage of imports related to machinery has remained steady at about 28% over the past twenty years' (White, 1984: 4).

5. The true extent of this is likely to have been overstated in the data, since the Department of Customs and Excise makes no distinction between exports and the re-export of previous imports. A considerable number of formerly imported machine-tools, for example, have recently been leaving South Africa categorised as exports.

6. It is for this reason that machine-tool production is regarded as having 'a decisive effect on a country's manufacturing strength' (Melman, 1983: 56–7).

7. For more detail on the history and development of the South African machine-tool industry, see Kaplan (1987).

8. For example, 'South Africa is almost completely dependent on foreign technology and capital equipment' (Sisulu, 1985: 20).

9. The material presented here is drawn from my present research project on the South African telecomms industry. Most of the material was obtained through interviews with senior management in SAPO and the telecomms equipment companies. See Kaplan (1990).

10. The problem with making an accurate assessment is that the TVBC 'countries' have their own Post Office administrations and the data for these are not included in the SAPO reports. Including the TVBC 'countries' would add approximately 67 350 telephones to the network size and 6m to the population (see Kaplan, 1988: 33).

11. This information was obtained from International Telecommunications Intelligence, Chichester, UK.

12. Data obtained from correspondence with SAPO, Senior Director, Telecommunications Commercial, 1 December 1987.

13. These two figures are respectively from a market study by Business and Marketing Intelligence (BMI, 1985) and the study of the South African Electronics Industry by the Board of Trade and Industries (BTI, 1987).

14. This was also true in the case of a number of Third World countries such as Brazil, Mexico and India. See Hobday (1986b: 29).

15. This excludes more minor agreements, such as for coin telephones with Telkor (Reunert Group), and the producers of cable.

16. Particularly large and very low cost components industries and highly skilled and productive labour.

17. The *Financial Mail* currently reserves its harshest rhetoric and shrillest pleas for privatisation for SAPO. Under the title of 'Time to Sell the Shop', a recent leader called SAPO 'this hoary old avaricious bureaucracy, this slumbering maternal viper' (*FM*, 4 December 1987).

Amin, S. 1974. *Accumulation on a World Scale: A Critique of the Theory of Underdevelopment.* New York: Monthly Review Press

Business and Marketing Intelligence. 1985. *The South African Electronics Industry. Management Report and Global Scenario.* Pretoria: BMI

Chudnovsky, D. and Nagao, M. 1983. *Capital Goods Production in the Third World. An Economic Study of Technology Acquisition.* London: Frances Pinter

Fransman, M. 1982. Learning and the Capital Goods Sector under Free Trade: The Case of Hong Kong. *World Development*, 10

Fransman, M. 1984a. Promoting Technological Capability in the Capital Goods Sector: The Case of Singapore. *Research Policy*, 13

Fransman, M. 1984b. Technological Capability in the Third World: An Overview and Introduction to Some of the Issues Raised in This Book. *In Technological Capability in the Third World*, ed. Fransman, M. and King, K. London: Macmillan

Fransman, M. 1986. *Technology and Economic Development*. Brighton: Wheatsheaf Books

Hobday, M. 1985. *The Brazilian Telecommunications Industry: Accumulation of Microelectronic Technology in the Manufacturing and Services Sectors.* New York: UNIDO

Hobday, M. 1986a. Telecommunications and the Developing Countries: The Challenge from Brazil. Brighton: University of Sussex, D.Phil. thesis

Hobday, M. 1986b. Telecommunications – A Leading Edge in the Accumulation of Digital Technology? Evidence from the Case of Brazil. *Information Technology for Development*, 1 (1)

Kaplan, D. 1987. Machinery and Industry: The Causes and Consequences of Constrained Development of the South African Machine Tool Industry. *Social Dynamics*, 13 (1)

Kaplan, D. 1988. Technological Change and Telecommunications in South Africa – An Industry at the Crossroads. Cape Town (unpublished)

Kaplan, D. 1989a. *The South African Capital Goods Industry with Special Reference to the Manufacture of Machine Tools and Telecommunications Equipment: A Comparative Study.* Report Prepared for the HSRC 15/1/3/3/643

Kaplan, D. 1989b. State Policy and Technological Change: The Development of the South African Telecommunications Industry. *Journal of Southern African Studies*, 15 (4)

Kaplan, D. 1990. *The Crossed Line: Technological Change and Telecommunications in South Africa.* Johannesburg: WUP

Mansell, R.E. 1988. Telecommunication Network-based Services. Regulation and Market Structure in Transition. *Telecommunications Policy*, September

Melman, S. 1983. How the Yankees Lost Their Know-How. *Technology Review*, October

Rosenberg, N. 1976. *Perspectives on Technology*. London: Cambridge University Press

Sisulu, M.V. 1985. *Transnational Corporations Involvement in South Africa's Electronic Industry.* Govan Mbeki Foundation, Amsterdam: University of Amsterdam

Telecommunications Industry Research (TIR). 1987. *Market File: South Africa*. Barnham UK: TIR

Westphal, L.E., Rhee, Y.W. and Pursell, G. 1984. Sources of Technological Capability in South Korea. In *Technological Capability in the Third World*, ed. M. Fransman and K. King. London: Macmillan

White, M.D. 1984. To Automate or Not to Automate. *GTES Newsletter* (CSIR), 2 (3)

Official publications
Board of Trade and Industries. 1972. *The Machine Tool Manufacturing Industry in the Republic*

of South Africa. Report No. 1425. Pretoria
Board of Trade and Industries. 1987. *Investigation into the South African Electronics Industry. Report No. 2455.* Pretoria
International Bank for Reconstruction and Development (IBRD). 1988. *World Development Report 1988.* Oxford: Oxford University Press
Postmaster General. 1988. *Annual Report of the Postmaster General of the Republic of South Africa 1987–88.* Pretoria: South African Department of Posts and Telegraphs
UNIDO. 1984. *World Non-Electrical Machinery. An Empirical Study of the Machine Tool Industry.* New York: United Nations
UNIDO. 1986. *The Machine Tool Industry in the ASEAN Region: Options and Strategies. Main Issues at Regional Level.* New York: United Nations. Sectoral Working Paper Series, 49 (1 and 2)

9. The accumulation crisis in agriculture

1. I wish to thank the following: for making available statistical data, officials of the Department of Agricultural Economics and Marketing, Pretoria; for comments on the first draft, members of the Economic Trends Research Group; staff of the School of Economics, UCT; Professor W. E. Kassier and colleagues of the Department of Agricultural Economics, University of Stellenbosch; Symond Fiske of AgriAfrica; and Nicoli Nattrass of the Research Unit for the Sociology of Development, University of Stellenbosch; for assistance with production, Brian Kahn, Wendy Foulkes, and my wife Janeen. The opinions expressed, and any errors, are of course my own.

De Jager, B.L. 1973. The Fixed Capital Stock and Capital–Output Ratio of South Africa from 1946 to 1972. *South African Reserve Bank Quarterly Bulletin,* 108
De Klerk, M.J. 1984. *The Incomes of Farm Workers and their Families: A Study of Maize Farms in the Western Transvaal.* Cape Town: Southern Africa Labour and Development Research Unit, University of Cape Town (Carnegie Conference Paper 28)
De Klerk, M.J. 1985. The Labour Process in Agriculture: Changes in Maize Farming during the 1970s. *Social Dynamics,* 11 (1)
Farmer's Weekly. 1986. Be Prepared for a Dry Year. 76047 (21 November)
Farmer's Weekly. 1987. NAMPO Warning on Maize-Price Cut. 77019 (8 May)
Farmer's Weekly. 1987. Maize Growers Agree to Cut Production. 77037 (11 September)
Farmer's Weekly. 1988. State Aid takes Effect – Farming on Way to Recovery. 78006 (5 February)
Farmer's Weekly. 1988. Wheat Warning. 78016 (15 April)
Farmer's Weekly. 1988. State Aid Keeps Many on the Land. 78032 (5 August)
Farmer's Weekly. 1988. Maize Board Plays Safe with Prices Scenario. 78037 (9 September)
Fenyes, T.I., J. van Zyl and N. Vink. 1988. Structural Imbalances in South African Agriculture. *South African Journal of Economics,* 56 (2–3)
Groenewald, J.A. 1987a. South African Food Resources for the Future. *Development Southern Africa,* 4 (2)
Groenewald, J.A. 1987b. Agriculture: A Perspective on Medium-Term Prospects. *Development Southern Africa,* 4 (2)
Helfert, E.A. 1967. *Techniques of Financial Analysis.* Homewood, Illinois: Richard D. Irwin
Janse van Rensburg, B.D.T. 1984. Land Prices in South Africa: 1969 to 1979. *Agrekon,* 23 (2)
Maize News. 1988. Uitverkopings van Boere se Eiendom Wek Groot Kommer. 95 (September)
Nel, M. and J. A. Groenewald. 1987. An Efficiency Comparison between Part-time and Full-Time Farmers on the Transvaal Highveld. *Agrekon,.* 26 *(1)*
Potgieter, J.T. 1987. Boerderyfinansiering. Agricultural Economics Association of South Africa (unpublished conference paper)
S.A. Agricultural Union. 1984. *Die Finansiële Posisie van Boere in die R.S.A.* Pretoria
Schuh, G.E. 1986. Strategic Issues in International Agriculture. *Economic Impact,* 53
Seleoane, M. 1984. *Conditions in Eight Farms in Middelburg, Eastern Transvaal.* Cape Town: Southern Africa Labour and Development Research Unit, University of Cape Town (Carnegie

Conference Paper 29)

Stadler, J.J., J.P. van den Heever and J.A. Lombard. 1983. *Die Koste, Voordele en Finansiering van Beskerming in Suid-Afrika: Die Invloed van Beskerming op die Land*. Pretoria: Buro vir Ekonomiese Politiek en Analise, Universiteit van Pretoria (Navorsingsverslag 17)

Tyson, P.D. and T.G.J. Dyer. 1978. The Predicted Above-Normal Rainfall of the Seventies and the Likelihood of Droughts in the Eighties in South Africa. *S. A. Journal of Science*, 74 (10)

Van der Vyver, A. 1988. World Grain Outlook and its Effect on South Africa. *Effective Farming*, 3 (8)

Van Zyl, J. 1986. Duality and Elasticities of Substitution: An Empirical Application. *Agrekon*, 25 (3)

Wilson, F. and M. Ramphele. 1989. *Uprooting Poverty: The South African Challenge*. Cape Town: David Philip

Official publications

International Monetary Fund. 1977, 1978, 1981, 1984, 1988. *International Financial Statistics*. Washington, DC. 30(7), 31(7), 34(7), 37(1), 41(7)

International Monetary Fund. 1986. *Supplement on Price Statistics*. Washington, DC. Supplementary Series 12

R.S.A. Central Statistical Services. 1986. *South African Statistics*. Pretoria: Government Printer

R.S.A. Central Statistical Services. 1988a, 1989. *Statistical News Release P0141. 1: Consumer Price Index*. Pretoria: Government Printer

R.S.A. Central Statistical Services. 1988b. *Bulletin of Statistics*. Pretoria: Government Printer

R.S.A. Central Statistical Services. 1988c. *Statistical News Release P1101: Agriculture Survey 1987*. Pretoria: Government Printer

R.S.A. 1970. Commission of Enquiry into Agriculture. Second Report. RP84/1970. Pretoria: Government Printer (Marais Report)

R.S.A. Department of Agricultural Economics and Marketing, Division of Agricultural Economic Trends. 1988, 1989. *Abstract of Agricultural Statistics*. Pretoria: Government Printer

R.S.A. Economic Advisory Council of the State President. 1986. Report of an Investigation into the Restructuring of Agriculture. Pretoria: Government Printer

S.A. Reserve Bank. 1973, 1980, 1981, 1988, 1989. *Quarterly Bulletin*. Pretoria

S.A. Wool Board. 1986–7, 1987–8. *Annual Report*. Pretoria

10. The restructuring of labour markets in South Africa: 1970s and 1980s

1. The financial assistance of the Human Sciences Research Council towards this research is hereby acknowledged. Opinions expressed in this publication and conclusions arrived at are those of the author and do not necessarily represent the views of the Human Sciences Research Council.

2. The data in this Table, as well as Tables 2 and 3, is based on population censuses, the most recent of which was carried out in 1985.

3. The Peromnes Survey is conducted by FSA Remuneration (Pty) Limited. These surveys have produced comparable material on wages and grades of work for every year from 1981 onwards. The number of employees covered by the survey was 96 000 in 1981, and increased to just under 180 000 in 1988. The companies sampled are mainly large, concentrated in the Witwatersrand and in the manufacturing and financial sectors. Furthermore, the higher grades of work (mostly occupied by whites) were more comprehensively covered than the lower grades (mainly occupied by blacks), although increasing attention was paid to lower grades as time passed.

4. It is unfortunate that the Peromnes system does not make it possible to distinguish clearly between semi-skilled and unskilled work. We have seen above that the major casualties of the economic crisis and restructuring since the 1970s were unskilled workers and that semi-skilled jobs continued to grow, albeit at a slow rate.

5. There is some debate about whether the trends identified by Bell have continued into the 1980s.

Bell, R.T. 1973. *Industrial Decentralisation in South Africa*. Cape Town: OUP
Bell, R.T. 1983. *The Growth and Structure of Manufacturing Employment in Natal*. University of Durban-Westville, ISER, Occasional Paper 7. Durban
Bell, R.T. circa 1985. Is Industrial Decentralisation a Thing of the Past? Unpublished mimeo
Cobbett, W., Glaser, D., Hindson, D. and Swilling, M. 1986. South Africa's Regional Political Economy: A Critical Analysis of Reform Strategy in the 1980s. In *South African Review, 3*. Johannesburg: Ravan Press
Crankshaw, O. 1987. The Racial and Occupational Division of Labour in South Africa, 1969–85. Labour Studies Workshop, University of the Witwatersrand. Johannesburg
Crankshaw, O. and Hudson, D. 1990. South Africa's Changing Class Structure. Paper presented to ASSA conference, University of Stellenbosch, 1–4 July
Crush, J. 1987. Restructuring Migrant Labour on the Gold Mines. In *South African Review*. Johannesburg: Ravan Press
De Villiers Graaff, J. 1985. Function, Dependency and Reformist Potential: The Case of Bophuthatswana. Conference of Development Society of Southern Africa, Bloemfontein
De Villiers Graaff, J. 1985. The Present State of Urbanisation in the South African Homelands and Some Future Scenarios. Conference of Development Society of Southern Africa, Cape Town
Glaser, D. 1988. The State, Capital and Industrial Decentralisation Policy in South Africa, 1932–1985. Johannesburg: University of the Witwatersrand, MA Thesis
Greenberg, S. and Giliomee, H. 1983. Labour Bureaucracies and the African Reserves. *South African Labour Bulletin*, 8
Hindson, D. 1983. The Pass System and the Formation of an Urban African Proletariat: A Critique of the Cheap Labour Power Thesis. Brighton: University of Sussex, D.Phil. thesis
Hindson, D. 1987. *Pass Controls and the Urban African Proletariat*. Johannesburg: Ravan Press
Lacey, M. 1982. *Locating Employment, Relocating Unemployment. Homeland Tragedy: Function and Farce*. Johannesburg: Southern Africa Research Service
Legassick, M. and Wolpe, H. 1976. The Bantustans and Capital Accumulation in South Africa. *Review of African Political Economy*, 7
Mastoroudes, C. 1982. *Die Geraamde Getal Grenspendelaars per Nasionale Staat en Tipe Vervoer, 1976–1981*. Pretoria: Buro vir Ekonomiese Navorsing, Samewerking en Ontwikkeling (BENSO)
Rogerson, C. 1975a. Some Aspects of Industrial Movement from Johannesburg, 1960–1972. *South African Geographical Journal*, 57 (1)
Rogerson, C. 1975b. Industrial Movement in an Industrializing Economy. *South African Geographical Journal*, 57 (2)
Rogerson, C. 1982. Apartheid, Decentralization and Spatial Industrial Change. In *Living Under Apartheid*, ed. D.M. Smith. London: George Allen and Unwin
Simkins, C. 1981. *The Demographic Demand for Labour and Institutional Context of African Unemployment in South Africa: 1960–1980*. University of Cape Town. SALDRU Working Paper. Cape Town
Simkins, C. 1983. *Four Essays on the Past, Present and Possible Future Distribution of the Black Population of South Africa*. University of Cape Town, SALDRU. Cape Town
Simkins, C. and Hindson, D. 1979. The Division of Labour in South Africa, 1969–1977. *Social Dynamics*, 5
Surplus People Project. 1983. *Forced Removals in South Africa, Vol 1*. Cape Town: Surplus People Project
Tomlinson, R. and Hyslop, J. 1984. Industrial Decentralisation, Bantustan Policy, and the Control of Labour in South Africa. University of the Witwatersrand, African Studies Institute, Johannesburg
Wellings, P. and Black, A. 1986. Industrial Decentralisation in South Africa: Tool of Apartheid or Spontaneous Restructuring? *Geo-Journal*, 12 (2)

Official publications
Republic of South Africa. 1979. Report of the Commission of Enquiry into Legislation Affecting

the Utilisation of Manpower, RP 32. Pretoria: Government Printer (Riekert Report).
President's Council. 1985. Report on an Urbanisation Strategy for the Republic of South Africa. PC 3 1985. Pretoria: Government Printer
Central Statistical Service. 1986. *South African Statistics 1986*. Pretoria: Government Printer

11. Unemployment and the current crisis

Booth, D. n.d. *Measuring the 'Success' of Employment Creation Strategies in the Apartheid State*. Unpublished

Erwin, A. 1986. Trade Unions, Unemployment and Urbanisation. In *The Crisis: Speeches by COSATU Office Bearers*. Johannesburg, COSATU

Innes, D. 1987. *Unemployment: Perspectives, Problems, Solutions – The Position of Organised Labour*. Paper presented to Toyota Conference for Concerned Leadership, Durban

Kaplan, D. 1986. Unions and the Current Crisis. *South African Labour Bulletin*, 12 (1)

Nattrass, J. 1984. Approaches to Problems of Unemployment in South Africa. *Indicator South Africa*, 2 (1)

Labour Research Service. 1988. *Review of Unemployment Statistics*. Cape Town: Labour Research Service

Old Mutual. 1987. *Economic Monitor*. January

Sarakinsky, M. and Keenan, J. 1986. Unemployment in South Africa. *South African Labour Bulletin*, 12 (1)

Simkins, C. 1982. *Structural Unemployment Revisited*. University of Cape Town, SALDRU Working Paper. Cape Town

Therborn, G. 1986. *Why Some Peoples Are More Unemployed Than Others*. London: Verso

Van der Merwe, P.J. 1987. Paper presented to Toyota Conference for Concerned Leadership, Durban

Official publications

President's Council. 1985. A Strategy for Small Business Development and Deregulation, PC4/1985. Pretoria: Government Printer

President's Council. 1987. A Strategy for Employment Creation and Labour Intensive Development, PC1/1987. Cape Town: Government Printer

Republic of South Africa. 1984. White Paper on a Strategy for the Creation of Employment Opportunities in the Republic of South Africa, WPC/1984

Index

accumulation strategies, 35–58 *passim*
African National Congress (ANC), 3, 31, 48, 49, 54, 56, 57, 108, 109, 127
Aglietta, M., 60, 111
agricultural sector, 17, 69–70, 72, 82, 172, 198–227 *passim*, 228, 237; symptoms and nature of crisis in, 198–207; causes of crisis in, 208–17; implications of crisis for economy, 217–22
Anglo American Corporation, 41, 48, 54, 110, 112, 114, 115, 118, 122, 123, 125, 126, 170, 190
anti-apartheid movements, 99, 106
apartheid, 1, 3, 13, 16, 25, 28, 36–58 *passim*, 91, 107, 154, 232, 235, 236, 241, 243, 253
Argentina, 15, 85, 100, 157, 158, 161, 184
armaments: imports, 69, 101; exports, 73–4
Armscor, 73–4
Asian NICs, see South-east Asia
Australia, 123, 124, 125, 147, 148, 152, 153, 184
balance of payments, 14, 18–19, 20, 26–7, 88–93 *passim*, 98, 100, 106, 108, 173–4, 176, 183, 209, 221, 261; adjustment process, 62, 64–7, 103; effects of international mode of regulation on, 59–62; structure of, 62–7; components of, 67–80; exchange-rate and exchange-control policies, 80–6

banks, private international, 93–104, 106, 107
bantustans, 17, 36, 43, 45, 143, 156, 171, 222, 235–6, 237, 238, 239, 240, 241, 242, 243, 253
Barclays National Bank, 98
Barlow Rand, 41, 190
big business, 3, 29, 30 35–58 *passim*
Board of Trade and Industries, 166, 193
Botha, P.W., 43, 54, 55
Botswana, 120
Brazil, 15, 85, 92, 94, 100, 158, 161, 165, 169, 175, 180, 182, 186, 189, 192, 194, 196
Bretton Woods, 14, 21, 59–60, 80, 113
Burawoy, M., 112
business cycles, 4, 6, 17–18, 62, 64, 70, 74, 235; reproductive, 12; non-reproductive, 12, 13, 22–3, 90
Canada, 123, 124, 125, 153
capital account, 64–5, 79, 80, 84, 85, 108; shocks, 65–7
capital goods, 62, 67, 70, 100, 175–97, 255; capital goods sector since 1970, 177–81; machine-tool industry, 181–8; telecomms, 188–96; future of sector, 196–7
capital–labour ratio, 16, 17, 18, 19, 21, 26, 67
capital–output ratio, 17, 18, 19, 21, 169, 203, 204, 226
Central Statistical Service, 226, 233, 235
Chamber of Mines, 116, 117, 118, 121, 127, 132, 138, 140, 143, 152

INDEX

Chase Manhattan Bank, 100
Chile, 15, 85, 100, 157, 158
Citibank, 102, 104
coal mining, 129–55; importance of to economy, 129–30; growth of, 131–8; and oil crisis, 133–4; mechanisation of, 134–6; transportation, 136–7; new links with international capital, 137–8; changes in labour relations, 138–45; wages in, 140–2; labour recruitment for, 142–4; labour organisation in, 144–5; markets and profitability, 145–9; current crisis and future prospects, 149–55
colour bar, 18, 121
Congress of South African Trade Unions (Cosatu), 108, 262–4
Consultative Business Movement, 52
crisis: of accumulation, 2; as 'turning-point', 2; as period of social conflict, 10; origins of, 2, 19–23; its development, 23–8; its effects, 3; 'organic crisis', 33–6
current account, 60, 62–4, 86, 91, 95; shocks, 65–7; imports, 67–72; exports, 72–5; factor service receipts and payments, 75–7
Current Population Census, 235
De Beer, Zach, 54, 127, 257
debt crisis, 28, 54, 66, 75, 78, 86, 96, 100–4, 107–9, 172
De Klerk, F. W., 29, 54, 55, 56, 127
De Kock, Gerhard, 26, 84
De Kock Commission, 81, 84–5
depreciation, 65–7, 85
deregulation, 257, 262, 266
Deutsche Bank, 102
disinvestment, 28, 78, 84, 99, 162, 170, 171
division of labour, 229–32, 241
Du Plessis, F., 42
Economic Advisory Council, State President's, 208, 212, 248
employment creation, 6, 31–2, 252–61, 264
Eskom, 40, 78, 96, 99, 102, 130, 131, 136, 145–6, 150, 159, 248
Eurobond market, 93

Eurocurrency market, 78, 86, 93–4
European Economic Community (EEC), 129, 215
exchange-control policy, 84–6, 97–8
exchange-rate policy, 80–4
Federated Chambers of Industry (FCI), 49
fiscal policy, 1, 105, 114, 126–8, 265–6
Fordism, 13, 14, 15, 17, 19, 20, 36, 38
Franszen Commission, 84
Gencor, 112, 115, 135, 137, 144, 150
gold mining, 110–28, 159, 173; importance of in economy, 110; history to 1960s, 111–13; divide of early 1970s, 113–14; exploration, 114–15; technological change, 115–18, 122; changing labour market, 118–23; international context, 123–6; and the state, 126–7
gold price, 2, 18 21, 22, 23, 26, 27, 28, 61, 62, 65, 66, 74, 81, 82, 83, 91, 94, 95, 111, 120, 123, 125, 173
Gramsci, A., 33
Great Depression, 13
gross domestic product, 4, 75, 86, 157, 160, 221, 223
gross national product, 4, 6, 26
Group Areas Act, 223, 224, 259, 262
growth model, 2, 10, 11, 12, 160; of sub-Fordism, 15–16; of 'racial Fordism', 2–3, 13–19, 24, 36; emergence of new, 3, 13, 28, 29–32, 57–8
hegemonic projects, 35–58 *passim*
homelands, see bantustans
Hong Kong, 75, 104
import penetration ratios, 68, 180–1
import-substitution, 14, 15, 20, 24, 36, 41, 62, 159, 160–1, 165–9, 171, 214, 216, 255
India, 196, 215
industrial decentralisation policy, 239–42
Industrial Development Corporation, 159
industrialisation in SA, 13, 156–74 *passim*
inflation, 6, 21, 65–6, 81, 95, 124, 209, 213, 214

INDEX

influx control, 37, 48, 50, 92, 172, 222, 235, 237, 238, 240
informal sector, 172, 250, 254, 258–61
interest rates, 208–12, 216, 222
international division of labour, 11, 15, 59, 62, 72, 170
International Monetary Fund, 75, 88–93, 94, 97, 103, 105, 106, 108, 158
international monetary system, 59–62
inward industrialisation strategy, 31, 172, 197, 254–6, 257, 260, 265
Iran, 92, 107
Iscor, 36, 40, 78, 136, 145, 150, 159
Japan, 72, 75, 99, 100, 104, 129, 133, 136, 186, 215, 248, 265
Keynesian economics, 8, 24, 265
Klen Report (Report of the Study Group on Industrial Development Strategy), 67, 82, 171
labour market, 228–43 *passim*, 250, 256, 266; sectoral shifts in employment, 228–9; occupational–racial division of labour, 229–32; changing occupational wage structure, 232–4; growth and restructuring of labour supplies, 234–9; regional distribution of employment, 239–42
Labour Relations Act, 251, 259–60
Land Bank, 205, 207, 209, 212, 219
Lesotho, 120, 143
Leutwiler, F., 101–2
machine-tool industry, 181–8
Malan, General M., 50, 52, 53
Malawi, 113, 118, 119, 120, 143
Mandela, N., 108
manufacturing sector, 2, 15, 17, 20, 26, 27, 67–75, 79, 82–3, 110, 156–74 *passim*, 228, 230, 245; history of development, 159–60; roots of stagnation, 160–3; case-study of motor industry, 163–9; reshaping of investment patterns, 169–72
Marais Commission (Commission of Enquiry into Agriculture), 210
Margo Commission, 126, 127, 212
marxist economics, 8, 9, 10
Mexico, 90, 94, 100, 158, 161, 165

microelectronics, 175, 188–96 *passim*
middle class, black, 30, 43, 258
migrant labour, 26, 37, 45, 113–14, 118–23, 130, 142–3, 238, 239
mining sector, 15, 17, 27, 69, 72, 82, 172–4, 228, 239, 245
mode of regulation, 10, 11–12, 14
monetarist policies, 27
monetary policy, 24, 26, 65–7, 81, 105, 208–12, 222, 265–6
motor vehicles, 68–9, 161, 163–9 *passim*
motor vehicle industry, 162, 163–9
Mozambique, 113, 118, 143
nationalisation, 109, 112, 266
'national question', 33–58 *passim*
National Party, 42– 58 *passim*
National Security Management System, 53, 54, 55
National Unemployed Workers' Co-ordinating Committee, 263
National Union of Mineworkers (NUM), 114, 118, 121, 127, 129, 144
Nedbank, 98
neo-classical economic analysis, 7, 158
oil price rise (1973), 6, 19, 22, 24, 69, 90, 130, 132, 133, 137, 212; (1979), 6, 25, 26, 27, 94, 153
Oppenheimer, H., 19–20, 26, 143
'orderly urbanisation', 51, 52, 53, 172, 242, 254
pension funds, 251, 264
Petrick Commission, 131, 132
Poulantzas, N., 34, 46, 48
President Council's Report on a Strategy for Employment Creation, 244, 250, 255, 259–60
President Council's Report on An Urbanisation Strategy for South Africa, 50, 242, 255
privatisation, 29, 46, 52, 195, 266
racial Fordism, 2–3, 13–23, 25, 28, 44
redistribution, 29–32, 57–8
reform discourse, 43–58 *passim*, 127, 231, 261
regime of accumulation, 10–11, 13, 14, 19; Fordism as, 13–15
'regulation' theory, 8–13, 14, 88, 111; its emergence, 8; meaning of term, 9

INDEX

research and development (R & D), 183, 185, 186, 187, 193, 194, 195, 196, 197, 265
resettlement, forced, 236–7
Reynders Commission, 20, 162
Riekert Commission, 44, 45, 46, 48, 51
Riekert Report on Manpower Utilisation, 242
sanctions, 54, 92, 107, 125, 130, 137, 150, 151–3, 162, 218, 221
Sasol, 24, 36, 40, 69, 134, 135, 145, 150, 159
savings: personal, 6, 28, 95; corporate, 28
Sharpeville (1960), 66, 84, 86
Simkins, C., 244
Singapore, 158
small business development, 254–8, 259, 260, 261
Small Business Development Corporation, 257, 258
South African Agricultural Union, 203, 204, 205, 206, 210
South African Communist Party, 54
South African Post Office (SAPO), 188–96 *passim*, 245
South African Reserve Bank, 64, 79, 80, 81, 83, 84, 85, 96, 97, 100, 105, 173, 225
South African Transport Services (SATS), 97, 136, 248
South-east Asia, 15, 17, 24, 37, 83, 151, 158, 162, 171, 175, 182, 184, 185, 188, 189, 194, 195, 197, 260
South Korea, 15, 104, 151, 158, 169, 175, 182, 183, 185, 186, 187, 189, 194
Soweto uprising (1976), 24, 25, 42, 66, 90, 94
Stals, C., 101, 102
Standard Bank, 98
Standstill Co-ordinating Committee, 101–2
state, SA, 17, 29–31, 34–58 *passim*, 126–8, 195, 196, 216–17, 231, 236, 240–1, 242, 252–61 *passim*
State of emergency (1985), 99
stock market crash (October 1987), 27
structural unemployment, 245–6

sub-Fordism, 15–16, 25
Surplus People Project, 236
Swaziland, 120
Sweden, 248, 263, 265
Switzerland, 99
Taiwan, 15, 75, 104, 151, 158, 175, 182, 183, 185, 186, 187, 189, 194
telecommunications, 188–96
trade account, 75
trade-union movement, black, 25, 28, 44, 98, 122–3, 144–5, 163, 231, 232, 234, 244, 251, 252, 256, 260, 261–5
Transvaal Coal Owners' Association, 132, 133, 136, 151
Trustbank, 97
unemployment, 3, 19, 23, 30, 67, 83, 220, 232, 234–5, 239, 244–66 *passim*; welfare responses to, 247–52; employment creation, 252–61; alternative responses, 261–6
Unemployment Insurance Fund, 246, 248, 249
United Democratic Front (UDF), 49, 98
United Kingdom, 72, 75, 90, 99, 148, 154, 248
United Nations, 91, 97
United States of America, 72, 75, 89, 91, 99, 107, 123, 136, 148, 151, 153, 257
Urban Areas Act, 238
Urban Foundation, 41, 43, 48, 50, 51, 257
urbanisation, African, 51, 237, 242–3, 255
Vaal Triangle uprising (1985), 48, 66, 97
Volkskas, 97
wages, 83, 156, 163, 232–4; on gold mines, 112–14, 118–23; on coal mines, 129–30, 138–42; in agriculture, 218–21; in manufacturing, 232
welfare policy, 38, 52, 247–52, 264–5
West Germany, 72, 75, 99, 133, 136, 154, 188
Wiehahn Commission, 44, 45
World Bank, 108, 158
Zambia, 75
Zimbabwe, 75